Bioclimatic Housing

BIOCLIMATIC HOUSING
INNOVATIVE DESIGNS FOR WARM CLIMATES

Edited by Richard Hyde

publishing for a sustainable future

London • New York

First published by Earthscan in the UK and USA in 2008

ISBN: 978-1-84407-284-2

Typeset by Domex e-Data, India

Cover design by Susanne Harris

For a full list of publications please contact:

Earthscan
2 Park Square, Milton Park, Abingdon, Oxon OX14 4RN
Simultaneously published in the USA and Canada by Earthscan
711 Third Avenue, New York, NY 10017
Earthscan is an imprint of the Taylor &Francis Group, an informa business

Earthscan publishes in association with the International Institute for Environment
and Development

A catalogue record for this book is available from the British Library

Library of Congress Cataloging-in-Publication Data:

Bioclimatic housing : innovative designs for warm climates / edited by Richard Hyde.
p. cm.
Includes bibliographical references.
ISBN-13: 978-1-84407-284-2 (paperback)
ISBN-10: 1-84407-284-3 (paperback)
1. Ecological houses. 2. Solar houses. 3. Architecture and climate. 4. Vernacular
architecture. 5. Temperate climate. 6. Tropics--Climate. 7. Arid regions. 8. Construction
industry--Appropriate technology. 9. Global warming. I. Hyde, Richard, 1949-
TH4860.B57 2006
720'.47--dc22
2006006997

Contents

PART II – LOCATION, CLIMATE TYPES AND BUILDING RESPONSE

List of Figures and Tables

TABLES

List of Boxes and Case Studies

Foreword

This is an extremely timely book as we grapple with the growing challenge of staying cool in our buildings in a rapidly warming world. Alarm bells are now ringing as the pace of climate change escalates and temperatures create records on a daily basis around the world. 2005 was the hottest year in Australia since records began in 1910. Record temperatures swept across parts of South Asia, Southern Europe, North Africa and the south-western US, and long-term droughts are devastating the Horn of Africa, north-western US and many parts of Australia. In 2005, these droughts also led to severe food shortages in a number of regions around the world.

Reports of such devastation seldom touch us deeply as designers. What does make an impact is when these facts are translated into design criteria for our buildings. When we hear of extreme temperatures experienced in cities around the world – 55°C recorded in Kuwait city in 2005; 54°C in Karachi and Basra; 52°C in Islamabad – we start to think about how we would keep people cool indoors in such climates.

In the much cooler climates of Europe, in two weeks during July 2003, temperatures soared to over a mere 40°C and over 35,000 people died there of heat-related causes. These were largely the vulnerable elderly, in traditional buildings that were no longer able to provide adequate cooling for their occupants. Mankind can adapt to an extraordinary range of temperatures as a species if we are given the time to do so; but the speed of these rises in temperature is the real killing factor.

Global warming is now known to be speeding up. In 2005, global temperatures were 0.75°C above the 1950 to 1980 average, and some sources are now predicting that by 2026 this may increase to 2°C hotter than this average. The only way in which we can cope with this level of climate change is to adapt rapidly to living in a warmer world, and fundamental to this adaptation in the built environment is the adoption of more effective, and widely used, methods for passively cooling buildings.

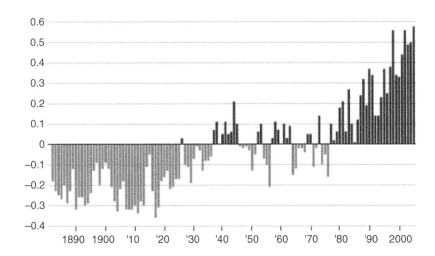

Source: Goddard Space Institute

Differences in global land-sea temperature, in degrees Celsius. This chart shows how much warmer or colder a year was compared with the average temperature between 1951 and 1980

Air-conditioning systems are increasingly seen as a part of the climate change problem, as well as its solution, as the yawning gap grows between the amount of fossil fuels used in the world and the falling amounts of fossil fuels that are available to run these machines. Not only is the rising cost of energy a problem, perhaps least to those who could not afford air conditioning anyway, but the fact that the energy used to run these systems is a major contributor to greenhouse gas emissions. We now have no option but to adapt the actual fabric of our buildings to withstand higher temperatures – hence, the importance of this book for all involved in the built environment. This book sets out the design challenges we face and a range of effective solutions to those challenges, and it provides excellent case studies to show how the cooling strategies proposed have been successfully applied in different climates. The chapters are clearly set out and relate the design process to a broader 'sustainability' agenda. This International Energy Agency (IEA)-inspired publication is a valuable sourcebook for all designers as we try to build a new 'cool vernacular' for the 21st century in which we can all stay cool, comfortable and alive in the buildings in which we spend so much of our lives, in these rapidly changing times.

Professor Sue Roaf
Oxford Brookes University
May 2006

Acknowledgements

This book is the outcome of a five-year International Energy Agency (IEA) Solar Heating and Cooling Programme Task 28 on Solar Sustainable Housing. Within this task are a number of experts, who comprise a 'Cooling Group' that represents countries where the climate requires cooling strategies in buildings. The group comprises many building design professionals who have given their time to the project. The group is self-funded and a debt of gratitude should be given to the organization that has supported the experts in this task. In particular, special thanks should be given to Robert Hastings, task leader for Task 28.

Special thanks should also be given to Professor Michael Keniger, Deputy Vice-Chancellor (Academic), University of Queensland, Australia, who supported this task and facilitated the publication of this book.

Funding for the publication of this book has been provided, in part, by the Australian Research Council.

Thanks should be given to Geoff Foster for editing the book, and to Catherine Watts and Angela Hair, who worked on the final proofing and illustrations. Also special thanks to the many contributors of images and illustrations mentioned as sources in the text.

The images for the Laverack Barracks project were supplied by David Sandison and Jon Florence, and drawings are courtesy of Bligh Voller Nield. Thanks to Shane Thompson for assisting with supplying material for this case study.

Edward Halawa also provided input for the review of solar energy systems in Australia in Chapter 2.

List of Contributors

Márcia Agostini Ribeiro, Lab Green Solar, PUC-Minas Gerais, Brazil

Giovanni Bianchi, Italy

Floriberta Binarti, Indonesia

Valerio Calderaro, University La Sapienza of Rome, Italy

Shailja Chandra, University of New South Wales, Sydney, Australia

Tamaki Fukazawa, Tokyo Metropolitan University, Tokyo, Japan

Vahid Ghobadian, Azad Islamic University, Tehran, Iran

Mehrnoush Ghodsi, Iran

Nathan Groenhout, Queensland manager, Bassetts Applied Research, Brisbane, Australia

Ken-ichi Hasegawa, Akita Prefectural Unverisity, Akita, Japan

Robert Hastings, AEU Ltd, Wallisellen, Zurich, Switzerland

Motoya Hayashi, Miyagigakuin Women's University, Sendai, Japan

Richard Hyde, University of Sydney, Australia

Sanae Ichikawa, Air Cycle Sangyo Co Ltd, Japan

Lars Junghans, AEU Ltd, Wallisellen, Zurich, Switzerland

Joel Kelder, University of Queensland, Brisbane, Australia

Katherine Khoo, University of Queensland, Brisbane, Australia

Nardine Lester, University of Queensland, Brisbane, Australia

Lim Chin Haw, SIRM Bhd, Shah Alam, Selangor, Malaysia

Luca Pietro Gattoni, Politecnico di Milano, Milan, Italy

Deo Prasad, University of New South Wales, Sydney, Australia

Indrika Rajapaksha, University of Moratuwa, Katubedda, Sri Lanka

Upendra Rajapaksha, University of Moratuwa, Katubedda, Sri Lanka

Harald N. Røstvik, Sunlab/ABB, Stavanger, Norway

Yoshinori Saeki, Daiwa House Industry Co Ltd, Japan

Francesca Sartogo, PRAU Architects, Rome, Italy

Massimo Serafini, Italy

Sabarinah Sh. Ahmad, University of Technology, MARA, Kuala Lumpur, Malaysia

Veronica Soebarto, University of Adelaide, Adelaide, Australia

Nobuyuki Sunaga, Tokyo Metropolitan University, Tokyo, Japan

Neda Taghi, Iran

Luke Watson, University of Queensland, Brisbane, Australia

Peter Woods, Multimedia University, Cyberjaya, Malaysia

Preface

Focusing on countries in which houses require cooling for a significant part of the year, this book covers creative, vernacular architecture backed up by practical and applied good science. Having set out new definitions of bioclimatic housing, the book interweaves the themes of social progress, technological fixes and industry transformation within a discussion of global and country trends, climate types, solution sets and relevant low-resource utilization technologies.

With concepts, principles and case studies from Indonesia, Iran, Malaysia, Brazil, Australia, Japan, Sri Lanka and Italy, this is a truly international and authoritative work, prepared under the auspices of a five-year International Energy Agency (IEA) SHC Task 28/BCS Annex 38 project.

Bioclimatic Housing provides a primer for building designers, builders, developers and advanced students in architecture, environmental management and engineering, mapping out the factors that are at work in reshaping housing for a more sustainable future.

Richard Hyde
Professor, Architectural Science
Faculty of Architecture, Design and Planning
University of Sydney
Australia
May 2007

List of Acronyms and Abbreviations

°	degree
3D	three dimensional
ABNT	Brazilian Association of Technical Standards
AC	alternating current
ACH	air changes per hour
AGO	Australian Greenhouse Office
AIVC	Air Infiltration and Ventilation Centre
ANEEL	National Electric Energy Agency (Brazil)
ASHRAE	American Society of Heating, Refrigerating and Air-Conditioning Engineers
BCA	Building Code of Australia
BEA	building environmental assessment
BFC	body-fitted coordinate
BIPV	building integrated photovoltaics
BMS	building management systems
BP	British Petroleum
C	Celsius
CBD	central business district
CEFET/MG	Federal Centre of Technological Education of Minas Gerais (Brazil)
CFD	computational fluid dynamics(s)
CEMIG	Energy Company of Minas Gerais (Brazil)
clo	average clothing
cm	centimetre
CO_2	carbon dioxide
COP	coefficient of performance
CT	current transducer

D	degree day is a quantitative index used to reflect demand for energy to heat or cool houses, e.g. D_{18-18} where the two figures of 18 are the limit of outdoor temperature and room temperature
DBT	dry bulb temperature
DC	direct current
DNA	deoxyribonucleic acid
ECBCS	Implementing Agreement on Energy Conservation in Buildings and Community Systems
ELETROBAS	state-controlled electricity company of Minas Gerais (Brazil)
EPA	US Environmental Protection Agency
ESD	ecological sustainable development
EU	European Union
GHG	greenhouse gas
GJ	gigajoules
GW	gigawatt
ha	hectare
HVAC	heating, ventilating and air conditioning
Hz	hertz
IEA	International Energy Agency
IFCO	Iranian Fuel Conservation Organization
INMETRO	Brazilian national certification board
IPT-SP	Technological Research Institute of São Paulo
ISIRI	Institute of Standard and Industrial Research of Iran
ISO	International Organization for Standardization
IT	information technology
J	joule
K	kelvin
kg	kilogram
kJ	kilojoule
km	kilometre
kW	kilowatt
kWh	kilowatt hours
kWh/m^2	kilowatt hours per square metre
kWp	kilowatt peak
LCA	life-cycle assessment
LEO	low energy office
m	metre
m^2	square metre
MBE	bio-energetic building efficiency
met	metabolic rate
MJ	megajoule

mm	millimetre
m/s	metres per second
MW	megawatts
MWh	megawatt hour
NORAD	Norwegian Agency of Development Aid
NPV	net present value
Pa	Pascal
PAC	photocatalytic air cleaning
PBE	Brazilian Labelling Programme
PBQP-H	Brazilian Programme of Quality and Productivity in Housing Construction
PCM	phase change material
PLEA	passive low energy architecture
PPD	percentage of people dissatisfied
PPDS	Precinct Planning and Design Standard
PROCEL	National Programme for Electricity Conservation (Brazil)
PT	power transducer
PUC-MG	Pontificate University of Minas Gerais (Brazil)
PV	photovoltaic
PV/T	photovoltaic and thermal system
R-value	a measure of the thermal resistance of a material, calculated from the thermal conductivity (k) of the material and its thickness. SI units are $K.m^2/W$. Some countries use non-SI units, the conversion between the two is $1 ft^2. F.h/Btu \approx 0.1761 K.m^2/W$
SET	standardized environmental temperature
SHC	Solar Heating and Cooling programme (*of the* IEA)
TBL	triple bottom line
T_n and T_m	the thermal neutrality temperature (T_n) is the statistical midpoint of the comfort zone and is the function of the prevailing climatic conditions. It can be related to the monthly mean temperature (T_m): $T_n = 17.6 + 0.31 T_m$
TSS	typical solution sets
U-value	a more user-friendly measure of the thermal conductance of a material; the reciprocal of the R-value
UEA	Urban Ecology Australia Inc
UFMG	Federal University of Minas Gerais (Brazil)
UK	United Kingdom
UPS	uninterruptible power supply
US	United States

USDOE	US Department of Energy
UV	ultraviolet
V	volt
VAC	variable air control
VOC	volatile organic compound
W	watt
WBT	wet bulb temperature
W/m^2	watts per square metre
WSUD	water-sensitive urban design
WWF	World Wide Fund for Nature

Overview

Richard Hyde and Nobuyuki Sunaga

Source: Richard Hyde

0.1 Healthy Home Project, Surfers Paradise, Queensland Australia: use of passive design provides an energy efficient solution through a bioclimatically defensive and interactive building, as well as a healthy environment through natural ventilation and selection of sustainable materials (architects: Richard Hyde and Upendra Rajapaksha)

Housing designed according to bioclimatic principles is becoming an important part of the journey towards achieving sustainable ecological development. The term 'bioclimatic' has traditionally related to the relationship between climate and living organisms, or to the study of bioclimatology. In the context of buildings, in general, and housing, in particular, it is concerned with a third factor in the relationship between the living organisms and climate – that is, the form and fabric of the building. Attempts to redefine bioclimatic housing are best examined not in theory, but in practice, and a key part of the investigation can be seen in built work (see Figure 0.1).

In our current age of the environment, where sustainability is emerging as a key issue in society, it is argued that it is necessary to achieve a balance between the needs of living organisms, housing and climate (see Figure 0.2). Yet, material and life systems (energy, water and waste) and the lifestyles of building users form an integral part this relationship. This complex relationship is leading to a need to redefine bioclimatic housing in terms of its form and fabric. Early definitions of bioclimatic housing emphasized the overlapping fields of biology, climatology and architecture (Olgyay, 1963). During recent years, this development of Passive Low Energy Architecture (PLEA) has reframed bioclimatic design in the context of sustainability. This can be seen in terms of the aims of PLEA:

- a commitment to the development, documentation and diffusion of the principles of bioclimatic design and the application of natural and innovative techniques for heating, cooling and lighting;
- serving as an international interdisciplinary forum to promote discourse on environmental sustainability in architecture and planning;
- a global invitation to the undertaking of ecological and environmental responsibility in architecture and planning; and
- the establishment of the highest standard of research and professionalism in building science and architecture in the interests of sustainable human settlements (see www.plea-arch.org/whatisplea.htm).

Definitions of a conceptual model of bioclimatic housing show sustainability as a triangle that outlines the boundary of the relationship between biophysical and climatic elements (see Figure 0.2).

NEW CONCEPTS: REDEFINING BIOCLIMATIC HOUSING

Part I of this book examines how new priorities in the area of the environment have redefined the direction of housing in order to achieve sustainability. Definitions, concepts and principles are identified, and it can be seen that bioclimatic principles are a cornerstone of sustainable housing. Trends in sustainable housing in a number of warm climate countries are discussed.

One trend that has begun to redefine concepts for housing is the process of identifying the predicted or actual environmental performance of buildings. Quantification of performance is seen as being necessary in assessing our ability to meet environmental targets generated by international policy organizations, such as the Kyoto Protocol. Methods of monitoring and computer simulation are emerging as valuable tools in this area.

A measure for sustainable housing

Research using tools such as ecological foot-printing shows a current imbalance in relationships where a range of factors concerning the biophysical environment are being measured (Rees and Wackernagel, 1996).

Overbay (1999) argues that the methodology can be used to describe the present unsustainable exuberant consumption of our cities.

The methodology was developed because of the difficulty of measuring human consumption choices and their direct effect on the biophysical environment – hence, a new instrument, 'ecological footprint analysis', was developed to determine the ecological impacts that various human populations have on the environment.

This basic concept measured the activities of humans, such as energy and material consumption and waste minimization, and related them to the continuing support of a measurable area of land and water. Ecological foot-printing yields an area-based estimate of the natural resource requirements of any given human population, from an individual to an entire country. Factors such as physical occupancy of the land, types of construction materials, the amount of land for city streets and intercity highways, and carbon dioxide production determine the size of a household's eco-footprint.

A common use for this tool has been to examine a household's eco-footprint for educational purposes. The larger the household's footprint, the more resources (in spatial units of land and water) required to sustain its current level of consumption. According to Overbay (1999), it has been estimated that:

> ... the housing and commuting component of the footprint of a person living in a high-rise or walk-up apartment would be 0.7

(a)

(b)

(c)

Source: Richard Hyde

0.2 (a) Extensive housing development on the Gold Coast, Australia, with little consideration of its ecological footprint; (b) bioclimatic housing can set important benchmarks for energy, water and waste recycling onsite – this house can consume 45 per cent less energy and 80 per cent less water, with a low eco-footprint lifestyle by its occupants; (c) design dimensions to bioclimatic housing – sustainability is represented by the preferred balance in the triangle on the left, contrasted with the current imbalance on the right. The ecological footprint created by the construction and operation of the housing sector dominates both the needs of living organisms and the influences of climate. Bioclimatic housing is a corner stone to achieving sustainability

acres; a person living in a modest co-housing community would use about 1.1 acres, and a person living in a large suburban home would use about 5.6 acres. Each estimate assumes the person is living as part of a family of four. These housing estimates do not include other impacts on the environment, including furnishings, recreation, food and clothing. The average American has a total ecological footprint of about 25 acres, meaning that if everyone consumed like an average American, we would need several additional Earths to live on. There are about 5.5 acres of biologically productive land per capita in the world.

This type of 'holistic' assessment of the environmental impacts of housing includes not only the form and fabric of the building, but also a wider set of parameters, such as the lifestyle of the occupant. A study by the Australian Greenhouse Office involving a number of houses also alludes to this problem:

The case studies illustrate a range of real solutions to specific challenges faced by people wanting to design, build or buy a more sustainable home. It is important to note that none of them 'get everything right'. There are few major challenges that humans are unable to overcome; but building a totally sustainable home is one important goal that still eludes us. (Australian Greenhouse Office, www.greenhouse.gov.au/ yourhome/)

Building a sustainable house may necessitate moving back to pre-industrial times in order to draw more strongly on the lessons of the past to achieve a balance; yet, the aim of redefining bioclimatic housing is to move a step further to achieve this goal. Sustainability is therefore not quantified but symbolized – the metaphor of the triangle. Sustainability is seen as an outcome, rather than as a goal or a process. Hence, one aim for designer professionals is to consider a redefinition of housing as a means of not only reducing ecological footprints, but also of improving quality of life.

This methodology is challenging, both in terms of the broad parameters to be considered and the reflective analysis needed. The value of redefining concepts, principles, strategies and resolutions has been demonstrated in order to communicate effectively with building professionals, users and other stakeholders.

Bioclimatic design as a means of footprint reduction
Chapter 1 examines some of the new concepts concerning the design of housing from a bioclimatic perspective.

Components of bioclimatic design

Current themes centre on a range of issues concerning the relationship between the biological and physical domains, such as:

- climate types and requirements;
- adaptive thermal comfort;
- vernacular and contextual solutions;
- tools and assessment methods;
- microclimate: sun path, wind and rain;
- working with the elements, such as passive and active systems; and
- development of a responsive form (Price & Myers, 2005).

This can be conceptualized as building design that utilizes a range of biophysical elements. These biophysical elements are primarily drawn from the ecosphere, rather than the lithosphere – that is, heat, light, landscape, air, rain and materials (see Figure 0.3).

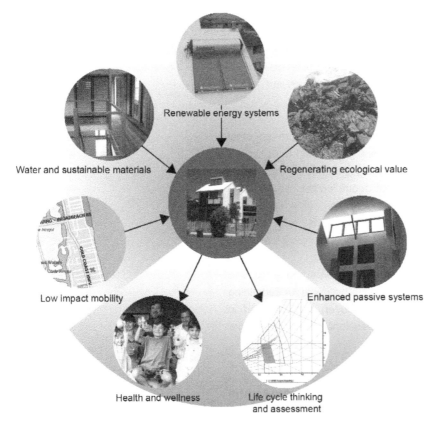

Source: adapted from Price & Meyers (2005)

0.3   New parameters for bioclimatic housing

Box 0.1 Policy on bioclimatic design for residential buildings for Turin, Italy

Focus: policy and standards
Scope: residential sector

Objective

The energy use in buildings contributes to a large amount of harmful emissions; thus, it is a useful starting point for implementing an overall strategy for improving the natural environment through the utilization of clean energy and renewable energy systems (RES).

Description

- Introduction to the concepts of bioclimatic design for buildings: this policy has as a technical objective: the reduction of the need for energy for winter heating, summer air conditioning and lighting. In reality, introducing concepts of bioclimatic design for buildings within the traditional planning process incorporates objectives that involve changes in regulation, knowledge of the professional culture, improvement of housing comfort and, more importantly, the improvement of urban quality.
- Promotion of solar systems in residential building: this policy refers to a dissemination activity of the innovative building culture, aiming at the integration of thermal solar systems and photovoltaic systems in both the building envelope and the building installation.
- Energy retrofit for the existing building stock: actions foreseen here strengthen the intervention policy, which has already been initiated by Turin. These are aimed at reducing the energy consumption for heating and air conditioning of public and tertiary buildings, as well as improving the quality of buildings by introducing energy efficiency criteria in maintenance interventions and reducing the maintenance costs of existing installations.
- District heating: a policy on district heating has already been established by Turin – namely, the extension of district heating and the diversification of thermal energy, aiming to heat around 50 per cent of the built area in Turin by the year 2010.

Conclusion/recommendations

The strategies identified here are an excellent approach for implementation at city level as they address several sectors that contribute substantially to carbon dioxide (CO_2) emissions. By focusing on residential buildings, these strategies target a significant source crucial in the campaign to promote clean heating and energy technologies.

Source: Solar Cities EHT (2005)

Bioclimatic housing as a path to sustainable urban settlement

As human settlements have become more urbanized during recent years, it has been necessary to redefine bioclimatic design in terms of this context. In simple terms, it can be regarded as input and output measures (see Figure 0.3).

Input measures to design

These can be considered as input measures, such as regenerating ecological value through water-sensitive urban design and habitat conservation; use of enhanced passive systems rather than active mechanical systems; and low impact mobility, such transit-oriented housing development, which takes us from a car-based system to a more multi-mobility approach to transport. The linkage of transport to housing is now an urgent priority; therefore, development of housing should consider options for transportation in terms of siting and city infrastructure provision. The following measures should be taken into account:

- low impact mobility;
- sustainable materials and water;
- renewable energy systems;
- regenerating ecological value; and
- enhanced passive systems through integration of microclimate and active systems.

Bioclimatic housing is therefore a multidisciplinary problem requiring planning and the involvement of a range of other building design professionals. The final parameter to this approach is to select sustainable materials that are less energy intensive in their production and that are not harmful to the environment.

Output measures from design

The output from the design of bioclimatic housing using these input measures includes improvement in the performance of two main parameters:

1   The comfort and well-being of the occupants: the biophysical definition of comfort has been expanded to include a range of issues to do with social and economic factors.
2   The life cycle of the building and its infrastructure: this includes reduced environmental impacts over the life cycle and reduced whole-of-life costs.

A step forward in the approach to bioclimatic housing is to see it as an overall objective within the city-planning and building regulatory framework. It is important to start at the policy level prior to institutionalization within planning and local government codes. Policy changes are beginning to occur to make this possible through initiatives such as the Solar Cities Programme in Europe; here, the bioclimatic approach is adopted primarily to conserve energy and to support the use of renewable energy sources (see Box 0.1).

Bioclimatic housing: Improving life-cycle performance

Chapter 2 examines the trends towards improving the sustainability of housing through a variety of measures. This includes government initiatives regarding planning and energy use that achieve a level of sustainability through bioclimatic principles and promotion. These issues are discussed with examples from a range of countries.

Increasingly, with the design of buildings focusing on a performance model with regard to environmental impacts, a concern for life-cycle performance is at the centre of bioclimatic housing. Life-cycle assessment as framed in ISO 14040-43 provides a basis for this kind of investigation. Measuring the performance of housing is increasingly used as a way of assessing environmental impact. Chapter 2 describes the use of some of these design and monitoring tools, while Case study 2.1 demonstrates how these tools can be employed to assist with design.

LOCATION, CLIMATE AND SOLUTIONS

Part II provides studies of housing design from a range of locations in warm climates.

(a)   (b)   (c)   (d)

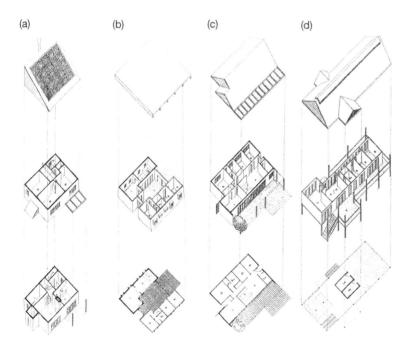

Source: Greenland (1991)

0.4 Form matching: matching house form and fabric specifically to a climate type

a: Cold – very high thermal mass, very well insulated, controlled ventilation

b: Hot arid – high thermal mass, ground coupling, light colour and shaded

c: Temperate – moderate equator-facing windows to collect winter sun, shading of summer sun and thermal mass, well ventilated in summer

d: Warm humid – maritime meso climate, lightweight, elevated and well ventilated

Form matching

Conventional work on housing has identified a particular detached house form for a specific climate. Hence, for cold climates, a square compact plan is recommended; for temperate climates, a rectangular plan is recommended; and for a hot arid climate, a courtyard form is suggested with thermal mass. Meanwhile, a house in a hot humid climate should be based on a linear form, one room deep (see Figure 0.4).

The aim of this kind of typology is to improve the functional efficiency of the house by making a building responsive to the climate in which it is located.

Evolving solution sets

During recent years, the evolution of house forms has created a typology that now responds less to the climate and more to the needs of social and economic criteria. Two major determinants can be identified: first, the densification of cities means the evolution of row houses and apartments; second, the abundance of cheap energy means that form need not be climate dependent but can be energy dependent for comfort. A typology of compact detached houses, row houses and apartments has developed for warm climates. The question arises as to how to make these new typologies climate interactive and less energy dependent for comfort.

Part II provides a description of different climates and the optimum building response. This is described as a 'solution set', which can be used as a general set of principles, strategies and technologies for a particular climate. Case studies of designs in these climates are provided to demonstrate some of the aspects of the solution sets. More details of these parameters of bioclimatic housing design are provided in Chapters 3 to 8. It is worth noting the importance of passive and active systems in these parameters.

STRATEGIES, ELEMENTS AND TECHNOLOGIES

Part III invites further work on developing new solution sets for warm climate from the available range of strategies, systems and technologies that are emerging.

Climate change and the need for bioclimatic design

Roaf (2003) has identified four main arguments for the need to utilize these recommendations:

1   The rate of change in the level of climate variability and modification is increasing, requiring human adaptation to a rapidly warming world.
2   The fundamental means to this adaptation in the built environment is the adoption of more effective, and widely used, methods for passively cooling buildings.
3   Air-conditioning systems are increasingly seen as a part of the climate change problem, as well as its solution, as the yawning gap grows

between the amount of fossil fuels used in the world and the falling amounts of fossil fuels that are available to run these machines. Not only is the rising cost of energy a problem, perhaps least to those who could not afford air conditioning anyway, but the energy used to run these systems is a major contributor to greenhouse gas emissions.

4   It is imperative to create a new 'cool vernacular' building approach, which matches human and environmental needs.

Some general recommendations have arisen from Parts I and II of this book that address these issues; these recommendations provide a checklist of considerations for designing more climate-friendly housing:

1   *Better interpretation of location and climate parameters*. Climate interpretation involves examining the macro-conditions, particularly insolation, humidity, airflow and rainfall on the building typology. The layout of the building, if it is to be bioclimatic, is highly influenced by the macroclimatic conditions.

2   *Greater synthesis between building elements and local climate conditions*. The influence of location involves an interpretation of meso- and microclimatic conditions. This has a profound influence on the extent to which bioclimatic strategies can be achieved.

3   *Harmonizing passive and active cooling/heating strategies*. This may include sun shading, thermal insulation and cross-ventilation in combination with active systems such as air conditioning. Harmonization through mixed-mode design allows better use of energy. Higher thermal performance of the envelope can significantly reduce energy use; the current practice of low performance for the building envelope leads to high energy use and greenhouse gas emissions. The use of high efficiency equipment does not necessarily address this problem.

Chapter 9 examines issues concerning the benefits of bioclimatic design. Areas that are discussed deal with the relationships between climate parameters, building elements and the integration of passive and active systems. Chapter 10 examines a range of opportunities to improve the effectiveness of design outcomes. These are:

•   Verify predicted and operational performance. Working from a performance model is crucial to tuning the building over its life cycle. Bioclimatic buildings require verification performance through commissioning and post-occupancy study to assist with enhancing performance.

•   Integrate appropriate green technologies, which utilize sustainable materials and life systems.

• Engage users in design and operation. In order to achieve the maximum performance, it is crucial that users understand the mechanism of the building and control it based on the concept of the design. Involving the users in the design is central to making this concept workable.

Advanced strategies for passive and active systems

Bioclimatic design focuses on the development of both passive and active systems using the following aspects of the biophysical context. It is important to investigate the use of passive and active systems for the site conditions (see Chapter 9 and 10). The development of more advanced strategies for passive and active systems is a key part of the move to redefining bioclimatic design. Some opportunities are as follows.

Microclimate and ecological enhancement of the site layout

Conventionally, the microclimate of the building has been largely ignored as an opportunity for advanced passive and active strategies. Cities are dominated by hard and reflective surfaces (Givoni, 1998). Yet, a key aspect of the designer's knowledge base is to know the site's biophysical conditions prior to development. This includes climate conditions and other influences, such as topography, soils, flora and fauna. *Site analysis* is carried out to determine the biophysical conditions. At the briefing and feasibility stage of the design, the microclimate should be taken into account. At the sketch design stage, it is possible to plan the layout of the building and development in order to achieve *microclimate enhancement*. For example, if the summer sun is problematic from the west, landscaping can be used to protect the building.

Shading of the sun at the site is necessary in order to enhance the effects of cross-ventilation. Lack of shading has been found to create a localized heat island on the site, increasing the temperature of the air used to cool the buildings.

Selection of appropriate plant species is a major factor in providing shade on the site. In hot climates, evergreen trees provide permanent shading. In temperate and middle latitudes, deciduous plants are suitable for the equator-facing façade. When heat is needed, the plants lose their leaves, and when shade is required, they grow their leaves.

Research work to enhance microclimates has been identified by Givoni (1998). The main aim is to reduce the effect of 'heat islands' (a heat island is the phenomenon of higher temperatures in urban areas compared to outlying rural areas) through the use of plant material and reflective light-coloured building materials. This reduces energy use due to their ability to moderate the environment immediately adjacent to buildings. The direct benefits (for example, lower building energy costs) are complemented by the indirect benefits of plant installation and the use of light-coloured surfacing. Larger indirect effects are felt

(a)

(b)

Source: Nobuyuki Sunaga

0.5 (a) Trees provide shade on the west side of Higa residence in Fukugi, Okinawa, Japan; (b) passive house in Malaysia (architect: William Lim) – mature evergreen trees are used to provide permanent shading in the tropical climate

throughout the community in the form of lower average air temperatures and reduced air pollution.

Questions have arisen with regard to whether enhanced microclimates create indirect benefits for the health and well-being of users and occupants.

Linking building planning and envelope design to microclimate enhancement

The fundamentals of passive design are commonly seen as the creative use of building planning to link buildings with their environments and their microclimates. Yet, a range of social and economic considerations, rather than climatic or environmental considerations, often drives current housing typology. Traditional house form is more a cultural response than a climatic one (Rapoport, 1969). In fact, during recent years, climatic and environmental factors have largely been forgotten.

How best can current types of housing, such as detached housing, row housing and apartments, be adapted to become more climate responsive? What solution sets should be used to create advanced passive and active strategies for this typology?

For example, with regard to detached housing, advantages can be found in breaking up the building form into pavilions in order to allow breezeways across the site. Increased permeability creates places of refuge and external rooms for the occupants to use during periods of high temperatures. Linked to water features, these create a passive alternative to air conditioning.

Selecting mechanical systems that work with building planning and the building's envelope

Plant and equipment are defined as the mechanical systems in a building. The types of systems favoured in housing are those that are driven by renewable energy. Renewable energy, often called 'natural energy', is sourced from renewable sources, unlike non-renewable energy, which is derived from fossil fuel sources, which are not rapidly replaceable. Unfortunately, many of the active systems such as air conditioning are driven by fossil fuel sources through the grid.

A solution would be to use electricity generated by photovoltaic (PV) systems to power air-conditioning systems; unfortunately, the efficiency and cost make this impractical. At present, a photovoltaic system in Australia can cost Aus$15,000 for a 1.5kW system – theoretically, all of this power would be needed to drive the system, with no power for other services.

Hence, to make the system practicable, it is important to reduce the demand for energy within the capacity of the system, rather than to develop systems that can service every housing demand.

(a)

(b)

(c)

Source: Nobuyuki Sunaga

0.6 (a) Oshiro residence in Okinawa, Japan, has operable windows at the top and at the ground level to improve single-sided ventilation; (b) Hunt residence, Davis, California, US: water wall system with no windows on the east and west sides; (c) Urambi Solar House, Canberra, ACT, Australia: direct gain system with insulation shutters

It is argued in Chapter 9 that bioclimatic passive cooling can offer a solution to reducing energy demand.

Passive cooling systems

Examples of cooling systems are as follows:

- Basement rooms and cool tubes utilize geothermal heat for passive cooling. The temperature over 5m from the ground surface is steady and equals the annual average air temperature, and is considerably lower than the air temperature in summer. The cool tube is the air duct that is laid underground and is used to cool fresh air. The under-floor space and the basement passage are used instead of the tube and they are called a cool pit and a cool tunnel. The cooled air is able to lead directly to the inside of the house; but consideration must be given to the fact that the relative humidity rises when the air is cooled.
- Thermal mass and night ventilation is an air-based system, which draws in cool night-time air and purges the thermal mass in the building of heat gained during the day. This is called a flywheel system as it stores and releases energy. The aim is to keep internal air temperature stable in daytime by using the thermal mass in the building as a heat sink. At night the thermal mass is cooled through ventilation. A good range of temperature between night and day is necessary for this system to operate effectively. Measures for ensuring crime prevention, protection against harmful insects or animals and privacy protection are also indispensable.
- A water wall is aimed at using water for insulation and evaporative cooling. An issue in water-scarce climates is the amount of water used for this purpose.

Renewable energy heat sources are mainly solar energy and geothermal energy for pre-heat; those for cooling are evaporative cooling, night radiation, geothermal energy and outdoor temperature during night time. It is important to note in warm and hot climates that the available energy for cooling is not as high as that for heating. For example, solar energy can create heat loads of $1000W/m^2$, but the available cooling is much less. Therefore, preventing the infiltration of solar heat is very important for passive cooling. Housing by virtue of its programme has small occupancy loads and high environmental loads. With an occupancy rate of four persons in a house, the heating loads from people in terms of sensible heat is around 400W. This is a fraction of the environmental loads. As a result, with regard to defensive strategies, a priority is to reduce heat gains from environmental factors, driving the need for higher efficiencies in the envelope and more advanced passive designs. Mapping heating and cooling demands is a good way of designing and managing the demand of active systems (see Table 0.1).

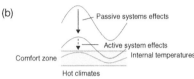

Source: Nobuyuki Sunaga

0.7 (a) In moderate climates passive systems are all that are needed to achieve comfort zone conditions; mechanical system design should involve sizing demand to suit the passive system effects, rather sizing the system to meet possible demand; (b) passive systems can be used to reduce external temperature to provide less work for active systems; this is crucial if energy efficiency is to be achieved

Mapping energy flows

Table 0.1 Mapping energy flows

Energy flows	Heat gain	
Environmental loads		
Solar radiation	1000	W/m²
Building envelope		
Roof: skillion, tile, cavity, plasterboard without insulation	65	W/m²
Roof: skillion, tile, cavity, plasterboard with insulation	25	W/m²
External walls: unshaded, solid brick	105	W/m²
External walls: unshaded, brick veneer	55	W/m²
Windows: unshaded, single glazed, clear glass, aluminium frames	150	W/m²
Windows: unshaded, single glazed, heat-reflecting glass, aluminium frames	130	W/m²
Roof lights: unshaded clear glass	142.5	W/m²
Roof lights: plastic dome, double glazed	80	W/m²
People		
Depends upon activity level	150	W
Appliances		
Lights: 200W per room	200	W
Cooking: rings	1500	W
Cooking: oven	3000	W
Shower	19,000	W

Source: Richard Hyde

Source: Nobuyuki Sunaga

0.8 (a) Akamine residence, Japan: the storage wall is made of coral stones; (b) TMU house I, Japan: water flows onto the roof creating a thin film which evaporates enhancing cooling of the roof; (c) 'Echos I', Japan, has a flat plate collector for hot water supply (left roof) and a crystal-type photovoltaic panel (right roof)

Many warm climates have a need for winter heating, which can be achieved by connecting *passive* systems for heating and cooling to the house. These are called passive plant and equipment because they use natural forces, such as temperature differences, radiation, convection and conduction, with no mechanical systems such as electric fans or pumps. The advantage of these systems is that they can be self-regulating and use no electricity; but they are often complicated to design since the dynamics of the systems are difficult to predict. A level of experimentation and empirical work is needed to optimize these systems. *Active* systems involve the use of mechanical devices such as electric motors and fans to drive heating and cooling systems. These can be used in combination with passive systems, creating *hybrids*. Some examples of these types of systems are outlined in the following section and are discussed in more detail in Chapter 10.

Life systems

Green technologies for housing have made advances in two main directions. First, they have developed life-supporting systems that use natural resources available onsite – in particular, solar energy and water. From the case studies presented in this book, it can be seen that water use can be reduced to 60 to

Source: Nobuyuki Sunaga

0.10 TMU house II, Japan, has the air-type solar floor heating system; the roof is also the amorphous-type photovoltaic panel

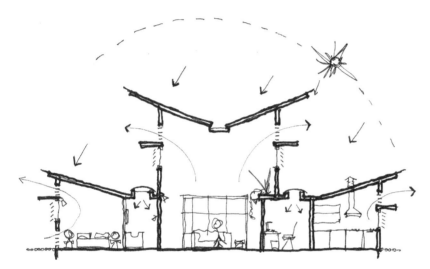

Source: Richard Hyde

0.9 Heat loads mapping can be used to assist with managing the heat demand for mechanical systems. Mechanical systems, such as air conditioning, pump heat from inside the building to the outside; therefore, managing the heat gains into the building is essential to make these systems more energy efficient

80 per cent of normal consumption. In houses where occupants' behaviour reduces demand for energy, renewable systems can provide sufficient power through PV and solar hot water systems alone. Reducing demand can thus result in the creation of zero energy houses – that is, houses where no energy comes from fossil fuel sources.

Active and hybrid systems

Examples of active and hybrid systems, which use a combination of passive and active systems, include the following:

• Water walls and roofs use water sprayed onto the surface of the wall or roof for evaporative effects.

- Cool tubes (cool pit, cool tunnel) are used for pre-cooling the air-conditioning system. This system is used for heating, as above, and therefore is adopted in commercial buildings. This requires conditions where the ground temperature is well below the air temperature so that sufficient cooling can be achieved. Since the ground is the heat sink, in conditions where ground temperatures are close to air temperatures the ground surrounding the tube will quickly equalize with the air temperature and the effect of the tube will be lost.
- Water sprays (to roofs) are an effective way of reducing solar heat by evaporative cooling; but a more active approach to cooling the rooms is the roof-mounted radiant cooling system using flowing water. This system makes use of water running down the roof in a thin film; through evaporative cooling, it produces cool water for radiant cooling during the night and it also cuts down the solar heat gain during the day. This system uses electric power only for the pump, so power consumption is less than one tenth of that of an air conditioner. If groundwater, which has a low temperature, is used for this type of cooling, the internal surface temperatures of the building may be reduced below the dew point, creating condensation.
- Thermal mass and air conditioning: thermal mass can be used with air conditioning to purge heat during the day and to allow natural ventilation at night. If combined with ceiling fans, this approach provides a system that is favoured by occupants who do not sleep well in air-conditioned rooms. Trickle ventilation is used to provide fresh air. This system is not advised for night-time air temperatures above 30° C.

Within the range of life systems, it is important to separate out the resource-producing systems from the climate control systems.

An important issue is emerging regarding the integration of new technologies within buildings. If resource-producing systems, such as photovoltaics and rainwater harvesting, and recycling systems, such as grey-water waste recycling, are used, how are these best integrated within the site and fabric of the buildings?

In many cities, water, energy and waste disposal are centralized. While there are many advantages to using this approach, there are also many benefits from additional decentralized site-specific systems. Water harvesting through rainwater tanks and organic waste treatment through worm farms and composting are examples of how pressure on centralized systems can be reduced.

Resource-producing systems

Resource-producing systems, which use a combination of passive and active systems, include the following:

(a)

(b)

Source: Richard Hyde

0.11 (a) Integration of new green technologies, allowing space for technological change in housing, is a future-proofing design strategy that embodies life-cycle thinking, such as the Healthy Home Project utilization of raised floors for future service flexibility; (b) 'flat' (approximately 1.5m deep) rainwater storage tanks fit under the living room of this Agnes Water Home, Queensland, Australia: the local council mandates the use of water tanks to reduce pressure on centralized services

- Solar thermal systems for water heating: there are two types of solar collectors – a water type and an air type. In the water type, the flat plate collector is popular. Its design temperature is 60°C and it is used for heating and hot water supply. The vacuum tube collector heats water to about 90°C and is used not only for heating and hot water supply, but also for cooling. The air-type collector is used primarily for heating, but is, at the same time, employed for providing a hot water supply by using a heat exchanger.
- Photovoltaic systems (PV cells): there are two types of PV cells – a crystal type and an amorphous type. The crystal type has a higher rate of power generation, but its cell area is small. The amorphous type has a lower rate, but has a larger cell area, and some products are bonded to a metal roof. The system should be connected to the usual power supply system because the storage battery is expensive and features a low performance. In the design process, the choice of the type and the area of collector/PV cells is determined by considering the design's overall purposes, the heat/power demand and the architectural design.

Material systems

Advances in material systems have reduced their environmental impact. The dominant factor is the energy embodied in their materials; but the selection of materials is complex, and single-criterion systems have been replaced by a holistic approach. A framework for assessing these materials is provided in this book.

Lessons from the case studies demonstrate a shift in the balance between passive and active systems. The current practice for climate control using a low performance building envelope and air conditioning should be replaced if improved life-cycle performance is to be achieved. The possibilities of using microclimate enhancement of the site, a high performance envelope with high efficiency active systems, and a mixed mode (provision of both passive cooling and air conditioning that can be selected in appropriate conditions) should be considered.

A high performance envelope, which comprises elements for reducing heat gains, can be a major element of the solution set. This includes the use of light-coloured external surfaces, shading, lower U values for transparent elements, such as windows, appropriate glazing ratios and higher levels of insulation, including radiant barriers (critical for rejecting heat). The capital cost of improving the performance is offset against reduced life-cycle costs, including energy costs and maintenance.

REFERENCES

Givoni, B (1998), *Climate Considerations in Building and Urban Design*, Van Nostrand Reinhold, New York

Greenland, J. (1991) *Foundations of Architectural Science: Heat, Light, Sound*, University of Technology Sydney, Faculty of Design Architecture and Building, New South Wales

Olgyay, V. (1963) *Design with Climate: Bioclimatic Approach to Architectural Regionalism*, Princeton University Press, Princeton, NJ

Overbay, M. (1999) 'Ecological foot printing', *Yes Magazine*, www.yesmagazine.org/article.asp?ID=760 (accessed 10 November 2005)

Price & Myers (2005) www.pricemyers.com/sustainability/efficiency.htm (accessed 10 November 2005)

Rapoport, A. (1969) *House Form and Culture*, Prentice Hall, Englewood Cliffs, NJ

Rees, W. E. and Wackernagel, M. (1996) *Our Ecological Footprint: Reducing Human Impact on the Earth*, New Society, Gabriola Island

Roaf, S. (2003) *Ecohouse 2: A Design Guide*, Architectural Press, Oxford; Burlington, MA

Solar Cities EHT (2005) *Defining Solar Cities*, http://sc.ises.org/ (accessed 10 November 2005)

Part I
Redefining Bioclimatic Housing

Chapter 1
Definitions, Concepts
and Principles

Richard Hyde and Harald Røstvik

Source: Richard Hyde

1.1 Redefining bioclimatic housing can start from studies of built works: Mapleton House, subtropical Queensland, Australia. Richard Leplastrier is one of the contemporary Australian architects whose architectural ideas have had a significant impact on its architecture. Mapleton House, located in subtropical Queensland, provides insight into new definitions, concepts and principles (architect: Richard Leplastrier)

INTRODUCTION

The aim of this chapter is to discuss some of the current definitions, concepts and principles for bioclimatic houses and housing.

First, an examination of bioclimatic architecture is given. This demonstrates the increasing importance and relevance of bioclimatic design to solving some of the emerging issues of the 21st century – namely, the concern for the state of the environment. It suggests that bioclimatic design can be a means of implementing international policy, such as the Kyoto Protocol, through a reduction of energy use and other environmental impacts. Hence, if bioclimatic design is the means, then sustainability is the outcome. New definitions and standards are emerging, which are now called sustainable development; therefore a discussion is provided in this chapter of some of the assumptions about the notion of sustainability in the context of sustainable development for buildings.

Generally, because of the multidimensional nature of sustainability, it is often difficult to conceptualize what sustainability means unless it is described with reference to a particular context. This is because being sustainable brings to a phenomenon the idea of being able to maintain its condition or state. The sense of the temporal and the future are an important dimension to the discussion of sustainability. More commonly, sustainability is discussed with reference to the operation of natural systems, with particular reference to the way in which natural resources are used and managed. Exploiting natural resources without destroying the ecological balance of a particular area is a key requirement of being sustainable. Currently, international policy has described the need for sustainable development as a means of achieving this end.

From these definitions have arisen new concepts for houses. New terminology has developed to describe these concepts. Countries such as the UK refer to eco-houses and carbon neutral housing, while from Australia there is SMART housing. All seem to have a similar but different interpretation of the relationship between the form and fabric of housing and living organisms and climate. What can be drawn from this diversity helps with the task of redefinition. Interesting aspects from this rich variety of concepts bring new ideas and priorities. Some argue that to be 'green', housing should be autonomous in terms of its servicing – that is, the building should not be connected to mains services. Other concepts emphasize the quality of life provided by these types of homes – that is, homes that are healthy for the occupants and healthy for the environment.

Principles have arisen which are useful to the building professional. The work of Brenda and Robert Vale is used as a basis for describing a range of principles that are generally applicable to green design but can be modified to the specific needs of housing in warm climates. The objective of this book is to assist with redefining principles for bioclimatic housing in warmer climates.

Finally, an example of how these principles have been applied to a domestic building in a developing country (Sri Lanka) is provided in Case study 1.1. This case study is of importance because it shows the need to view bioclimatic housing in its social context. To state the obvious, housing is primarily a social challenge; yet, often its relationship with climate and living organisms is lost, particularly in developing countries. Achieving a sustainable future in these countries is as much a bioclimatic issue as a social issue.

DEFINITIONS

Bioclimatic architecture

Linking international policy and industry transformation and creating local actions

During recent years, initiatives have been developed to create an international policy framework of environmental improvement leading to sustainable development. This policy framework can be seen in action in many countries. For example, in Australia, there has been a move to map the environmental impacts of buildings and to use this as a target to assist with the transformation of industry.

According to the Australian Greenhouse Office report (AGO, 1999), building energy consumption accounts for nearly 27 per cent of all energy-related greenhouse gas emissions. According to the same report, it was also predicted that by 2010, emissions from buildings as a result of energy consumption are estimated to increase by 48 per cent above 1990 levels. This projected trend is alarming given the fact that Australia's obligation under the Kyoto Protocol is to limit greenhouse emissions to only 8 per cent above 1990 levels.

In the near future, residential buildings will be the main outputs of the building sector. This is because, as predicted by the Australian Greenhouse Office (AGO, 1999), in 2010 80 per cent of the homes in Australia will already

(a)

(b)

Source: Richard Hyde

1.2 Mapleton House, Queensland, Australia. Leplastrier's architecture explores an intimate relationship with the environment and the people who dwell in it. This derives from a belief that 'architecture stems from the life within and the structure of the landscape without' and that one has to start with respecting 'the land and its original people' (Stutchbury, 1999). The starting point for his architecture comes from establishing key points of interaction. The deck is one of these points

have been built. The booming housing market during recent years has, of course, contributed significantly to this. It is argued that new buildings and renovation should be directed more towards utilizing bioclimatic principles to make better use of available natural energy and energy sinks – hence, reducing energy use. The use of bioclimatic principles, strategies and best practice examples will assist in transforming the building industry towards this goal.

Bioclimatic design

Bioclimatic issues in building were identified by Olgyay during the 1950s and were developed as a process of design during the 1960s (Olgyay 1963). The design process brings together disciplines of human physiology, climatology and building physics. It has been integrated within the building design professions in terms of regionalism in architecture and, during recent years, has been seen as a cornerstone to achieving more sustainable buildings (Szokolay, 2004). Research into bioclimatic issues has taken the form of passive low energy architecture research and is carried out worldwide, with a well-developed field, as is evidenced in the passive and low energy architecture (PLEA) conferences. PLEA is committed to 'the development, documentation and diffusion of the principles of bioclimatic design and the application of natural and innovative techniques for heating, cooling and lighting' (see www.plea-arch.org/).

This research has led to the development of bioclimatic design principles, which are used by design professionals as a starting point for designing with climate in mind. These have been developed primarily for low- and medium-scale buildings; the rationale for this is that these types of buildings are relatively easy to make bioclimatically interactive. That is, the form and fabric of the building can be matched to human and climate factors in order to optimize climate response (Hyde, 2000).

Larger-scale developments have generally escaped attention due to issues of programme complexity, the density in urban context in which these buildings are located and the availability of cheap energy for cooling and providing comfort. Their design principles have largely been to exclude bioclimatic influences and to achieve an adequate internal environment by adding comfort through mechanical energetic systems. Exceptions to this approach are increasing and can be found in a number of buildings, such as the pioneering high-rise buildings of Yeang and the large floor-plate buildings of Bligh Voller Neild (Rajapaksha and Hyde, 2002).

Yet, evidence of a set of bioclimatic principles, strategies and best practice solutions for buildings is still to be fully researched and acknowledged within the field (Yeang, 1999). While a number of case studies on bioclimatic design of large-scale buildings have been written, the extent to which general principles can be advanced from these examples is limited. Jones (1998) has developed some bioclimatic principles for large buildings. These are concerned with the

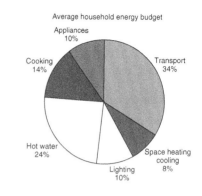

Average household energy budget

Mapleton house

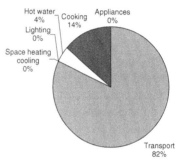

Source: Richard Hyde

1.3 Mapleton House in Queensland, Australia, is an autonomous house in a remote location, with no mains services; through its use of renewable energy and passive design, it uses a fraction of the energy of a conventional house. The significant energy use is transport energy. In warm climates, hot water heating and transport are the largest items in the energy budget provided that space heating and cooling can be provided by passive means

form of energy used – renewable or non-renewable – the use of energy efficiency, conservation, human well-being, and comfort and amenity. These elements are largely unconnected to building strategies in a building science context and are rarely fully demonstrated in best practice solutions of large-scale buildings.

Redefining bioclimatic housing: The issues of synergy

Design professionals often remain sceptical of the bioclimatic approach to large-scale projects. This is due to the lack of workable models of bioclimatic large-scale buildings and the cost of additional design work, such as simulation modelling, to demonstrate proof of concept, cost effectiveness and the comfort of these types of buildings (Pedrini, 2003). Energy efficiency has been seen to centre on the design of more efficient mechanical systems, rather than examining the use factors, such as the passive elements of the building, to engage in synergies that lead to an integrated solution. An examination of modelling tools is thus provided in subsequent sections.

With regard to synergy in terms of housing, many housing projects are now large scale, either in terms of whole neighbourhoods or precincts of housing. New trends have brought about the need to redefine bioclimatic housing in terms of this broader scale. Combining new forms of technology at the larger scale is providing a means of achieving zero energy targets. For example, the Beddington Zero Project in the UK, a combination of passive and active design, has made possible the achievement of 'a carbon neutral project' (EST, 2007)

The carbon neutral concept aims to create sufficient power from renewable sources onsite, which can balance power drawn from non-renewable sources supplied by the electricity grid (the grid it largely a backup system). The use of bioclimatic principles is crucial to the workability of this zero carbon target.

The importance of passive elements:
The building's microclimate, form and fabric

Research suggests that bioclimatic buildings will use five to six times less energy than conventional buildings over their lifetime (Jones, 1998, p45). This is achieved primarily through the use of the buildings' microclimate, form and fabric, rather than through the use of efficient mechanical equipment. For example, in warm climates where cooling is needed most of the year, 34 per cent of energy is used to address the need for cooling to mitigate heat gain from solar radiation. In many cases this is due to poor design of the building envelope' (Parlour, 2000, p94).

This is normally achieved through air conditioning and, as a consequence, incurs large environmental penalties, such as high energy use and high greenhouse gas emissions. Bioclimatic design refocuses on providing high-quality passive design of buildings through new technologies in the building envelope and in its form and fabric. Pioneering work by Yeang has defined a number of passive elements using a range of biophysical elements:

Source: Richard Hyde

1.4 Mapleton House, Queensland, Australia: Leplastrier worked with Jorn Utzon, whose interest in the ancient and vernacular architectural forms and their relation to the landscape made a profound impact on forming Leplastrier's views on architecture. Through Utzon, he also learned to appreciate the beauty of constructional expression in architecture (Frampton, 1995)

- thermohydronic – thermal, humidity and water sinks;
- kinetic – adaptive thermal defences;
- organics – use of flora as heat sinks;
- aerodynamics – adaptive wind defences;
- materials – phase change, heat storage and radiant defences;
- ground effects – heat storage (Yeang, 1998, 1999; Law, 2001).

Indeed, in order to achieve 'net zero-energy buildings', ongoing research into this area is needed.

Developing environmental targets: Evolving concepts of the net zero-energy building to a 'net carbon building'

The 'zero energy building' is an ideal concept where no fossil fuels are used and sufficient electricity is generated from natural sources to meet the service needs of the occupants (Gilijamse, 1995). This idea sets the optimal design target for a building in terms of minimizing the environmental effect and minimum energy cost of the dwelling. The appealing side of this idea is the availability of 'free' energy coming from nature, such as solar and wind energy, and energy sinks such as air, water and earth. Ideally, this 'free' energy can be utilized to achieve the zero energy target by employing passive and active solar technologies.

As an ideal concept, the 'net zero-energy building' serves as a guide to direct all possible technologies available to approach this target. The real target is then to define and set the criteria for achievable minimum energy buildings in terms of environmental benefits, economic feasibility, thermal comfort and energy sustainability. The role of innovative green technologies in achieving this target is crucial. Since the technologies come in various forms and state of the art, it is crucial to explore how they can be synergized to achieve the goal. Hence, as seen previously, the zero energy concept has evolved into a carbon neutral target to account for the importance of the use of renewable energy in this approach.

Emerging and proven technologies

Theoretically, the zero-energy dwelling concept can be achieved by simultaneous actions of:

- reducing the energy demand (increasing energy efficiency) of dwellings through various energy conservation measures; and
- utilizing the solar energy incident on the wall surface, roof and ground surfaces surrounding the house for electricity generation and for satisfying the dwelling's heating requirement.

The implementation of these actions entails the careful assessment of economic feasibility, benefits for the environment and human comfort.

(a)

(b)

Source: Richard Hyde

1.5 Mapleton House, Queensland, Australia: (a) the use of liquid petroleum gas supports a gas cooker, which also acts as a thermal storage unit; the waste heat in winter provides heating and drives off humidity in summer; (b) one of the important decisions regarding achieving a zero energy house is the choice of energy source: in subtropical climates, water heating is the major demand for energy in installing solar thermal systems, and photovoltaic systems support by renewable energy is crucial to achieving a zero energy home

Various technologies for energy conservation and solar energy utilization are already available or are being developed. Energy conservation measures available for reducing energy demand include proper insulation of the building envelope and selection of high efficiency (high energy rating) appliances.

Solar technologies can be divided into two groups – namely, passive technology and active technology. Passive solar technologies include passive solar heating, natural ventilation, daylighting, thermal mass storage and ground cooling. Passive solar heating employs the structural elements of a building to collect, store and distribute solar energy without or with minimal use of mechanical equipment. In this scheme, sunlight is collected through a large glazing area of the building wall and is absorbed in the wall, floor or partition. The distribution of this stored heat is through the heat transfer modes of conduction, convection and radiation. Fan and ducting may be used for specific applications.

Trombe walls (or solar walls) are among the popular technologies to catch solar energy. Solar energy transmitted through the transparent cover of the wall is absorbed by the outer surface of the wall. This absorbed energy is either conducted through the wall and reaches the inner surface several hours later or is conveyed by the air flowing through the space between the cover and the wall outer surface (Wittchen, 1993). Recent research works by Manz and Egolf (1995) and Stritih and Novak (2002) explored the use of phase change materials (PCMs) as the thermal storage in the solar (trombe) wall. This technique reduces the energy losses and energy requirements of the room.

Integrated photovoltaic systems have been the subject of intensive investigation during recent years. According to this concept, the roof or façade of a building is designed to accommodate the solar panels (or cells) and thereby minimize the cost of the system's support structure (Clarke et al, 1997). A similar concept has been applied for solar thermal systems, where the steel roofing of a house is utilized as a thermal absorber. To improve efficiency of the system, PCM thermal storage is introduced into the system (Halawa, 2005).

The idea of combining a photovoltaic and thermal (PV/T) system with thermal storage has achieved research interest recently. An interesting feature of this system is that it stores thermal energy and, at the same time, reduces the operating temperature of the photovoltaic (PV) cells, which results in increased efficiency of the PV modules. Sandnes and Rekstad (2002) designed and tested experimentally a hybrid PV/T collector consisting of a polymer solar heat collector and single-crystal silicon PV cells. The thermal efficiency of the PV/T was reduced compared to a pure thermal absorber, which, according to them, was due to:

- reduced fraction of thermal energy by electricity conversion in PV cells;
- lower optical absorption in PV cells compared to black absorber plate; and
- increased heat transfer resistance in the interface of the PV cell and absorber.

Sandnes and Rekstad (2002) found, however, that PV efficiency improved due to a lower operating temperature.

Similar to the work of Sandnes and Rekstad (2002) was the work carried out by Huang et al (2002), who incorporated PCM into the PV system to reduce the PV module operating temperature and to store some thermal energy for later use. Three configurations were studied – namely, a single flat aluminium plate system; a PV/PCM system with no fins; and a PV/PCM system with fins.

These technologies are only a few of the many emerging technologies relating to the goal of realizing extremely low energy-demand houses. Given these technologies, the concept of zero energy dwellings can theoretically be realized. The realization of this concept, however, depends upon whether it satisfies the three criteria previously mentioned: environmental benefits, human comfort improvement and economic feasibility.

The first and the second criteria can easily be delivered by solar technologies; in fact, these have been used by solar advocates to back up their 'switch to solar' movement. The third criterion – namely, economic feasibility – is equally important and may not be separated from the other criteria.

Until now, there has been no significant research aimed at exploring the synergies of these emerging green technologies to bring about new and indisputable sets of solutions to the environmental problems and energy issues relating to residential buildings. Research efforts so far have been fragmented, with researchers/research groups focusing on particular technologies without looking at the whole problem and exploring the best solution.

The role of demonstration projects

Part of the strategy for transforming industry rests on exemplars for new building work. Many examples in the form of case studies are provided throughout this chapter.

An example of sustainable housing is the Healthy Home Project currently implemented with support from the state of Queensland, Australia, and through the Centre for Sustainable Design, University of Queensland (see www.econnect.com.au/pdf/hh1.pdf).

This example shows how solar and energy conservation technologies can play an important role in bringing about healthy, sustainable and environmentally friendly human dwellings. The importance of ecological housing with low energy demand is being addressed by collaborative research work among 12 countries in Europe, North and South America, Asia and Australia, in which the Centre for Sustainable Design of the Architecture Programme, University of Queensland (see www.csdesign.epsa.uq.edu.au), is an active participant. The significance of this work is that it is linked to industry

(a)

(b)

(c)

Source: Richard Hyde

1.6 Current housing: 19th-century principles are used to solve 20th-century problems. The need for small passive design measures and proven technology is evident in making these homes more suitable. Light-coloured roofs, more insulation, better shading and mass on the inside of the light frame construction can reduce internal temperatures to ambient. Reduction of waste to landfill can be achieved with more prefabrication

groups who provide demonstration projects, which act as 'blueprints' for future sustainable ecological development.

Sustainable development and building

Sustainable development has commonly come to be understood as economic and social development that maintains growth within acceptable levels of global resource depletion and environmental pollution. Buildings, in general, and housing, in particular, which claim to be sustainable should address these problems through design and construction.

There is debate as to how to achieve this goal within the current context. One view of sustainability is called 'balance theory', which views sustainability as a matter of trade-offs in social, economic and environmental factors within organizations. The argument is that simply placing priority only on environmental factors without considering the economic and social context is likely to frustrate any attempts to achieve a realistic level of sustainable development. One prevailing opinion is therefore to balance these factors.

This view of sustainable development in buildings has led to a number of strategies and approaches to designing buildings that assist in reducing pollution and resource consumption. There are a number of new approaches that are fundamental to the design of sustainable buildings:

- holism;
- selecting environmental criteria; and
- ethical frameworks.

Holism

The term holism has been used to describe the view that a whole system must be considered, rather than simply its individual components. The Vales have addressed this point in their book *Green Architecture*, suggesting that a building should attempt to address all of the principles of green design in a holistic manner (Vale and Vale, 1991).

Holism emphasizes the relationship between the parts and the whole; the effects of the whole are more than the simple sum of the parts. The first order combination of elements leads to second order combinations to create greater effects and influences. For example, in a house, there are many physical components and systems; often, these are designed as individual objects. Walls, for example, are designed to take a particular wind load for structural purposes and to resist a particular heat load for energy purposes.

Holism relates to the way in which these elements and their associated performance attributes combine to create a whole. The concept of holism addresses both the sense of dwelling within a home and the home's environmental performance. This can be viewed as a home for living, as well as a machine for reducing environmental benefits. The sense of holism from an experiential point of view is as important as the holistic performance. Another dimension of holism comes from an engineering perspective and relates to the synergy between systems, which is a critical part of the physicality of a house. Holism can mean the way in which systems are integrated within the home – for example, the way in which the solar system is integrated with the roof or the water recycling system is integrated with the irrigation system.

In summary, the outcome holism can create leads to greater improvements in environmental performance where the effect of the whole is greater than the sum of the parts. It can also mean the balance between tangible and intangible elements in a building to create a sense of place: a home for occupants that produces a sense of well-being and comfort. Yet, while eco-homes can have the potential to far outreach traditional housing in terms of quality-of-life issues, in theory, the reality is quite different.

Selecting environmental criteria

It is common to attempt to minimize the environmental impacts of buildings by selecting environmental design criteria. This approach involves defining a set of environmental criteria that are used to inform the design process. Yet, the question arises as to which criteria are important.

Conventional wisdom suggests that energy is the most important criterion. The oil shocks of the 1970s and the current volatility of oil prices have demonstrated the vulnerability of our dependence upon non-renewable energy sources. In recent years, however, another argument has arisen for moving away from fossil fuels: environmental pollution.

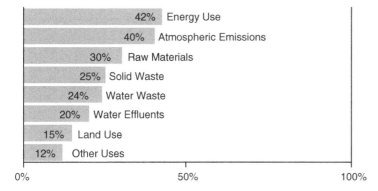

Source: Levin (1997)

1.7 The environmental impacts of buildings as a percentage of overall environmental impacts for the US

It is for this reason that a convincing argument has been advanced that energy efficiency and conservation should be the main criteria when designing sustainable buildings. Edwards (1999, p7) makes the argument that energy sources – in particular, coal and oil – are not only finite, but their associated pollution is most damaging to the ecosphere and its biodiversity. Therefore, two main design issues are important:

1   energy sources – priority for solar energy as a source and the use of other renewable energy sources;
2   energy efficiency and energy conservation in construction, operations and resources use (Edwards 1999, p7).

Yet, during recent years, rather than subscribe to a single criteria approach, multiple environmental criteria are used. This is viewed as a more holistic approach and is more directly aligned with an ecological view.

Box 1.1 Royal Australian Institute of Architects (RAIA) Ethical design framework

---

Key points:

- Maintain and/or restore biodiversity and biophysical ecology (air, land, water, flora and fauna).
- Maximize resource-use efficiency and reduce or eliminate unsustainable consumption of resources.
- Minimize adverse emissions to air, soil and water.
- Maximize the health, safety, comfort and productivity of people and places.
- Leverage multidisciplinary teams and skills and incorporate stakeholder and community input.
- Integrate ecological sustainable development (ESD) objectives within all projects and project design, development, delivery and commissioning processes.
- Minimize negative cumulative impacts and generate positive cumulative impacts within and between human-made and biophysical environments over time.
- Reflect and respond to global, regional and local impacts of local actions and practices.
- Promote equity within and between generations.
- Embody a precautionary approach to the environmental consequences of decision-making.
- Precautionary principle: lack of full scientific certainty shall not be used as a reason to postpone measures to prevent environmental degradation.

Source: RAIA (2004)

Box 1.2 Rationale for the RAIA Ethical Framework

Supporting background statement: *BDP Environment Design Guide*
Planning and Development Committee

The rationale of the *BDP Environment Design Guide* (RAIA, 2004) is as follows.

The professions which make up the building design professions (BDP) are individually, and collectively, major contributors to shaping the environments of the future. Increasingly, the environments in which we live are made ones – made up of built structures, technologies and systems of use that interact with one another to care (or fail to care) for the sustainability of the made and biophysical systems upon which we rely for life.

BDP professions each enjoy a duty of care to look after the environments in which they work and to respect the needs of the communities whom they serve. Increasingly, they are called upon to work together in often complex and multidisciplinary teams to create built environments, which in their construction, operation and reuse (disassembly, refurbishment and redevelopment) will foster more ecologically sustainable outcomes.

This involves the development and application of more holistic design approaches through which greater care for biophysical impacts can be incorporated, along with a reorientation towards the materials, products and built structures which we use and create so that they are treated as ecological assets not only throughout their economic life, but also throughout their material life.

The rationale of creating the *BDP Environment Design Guide* is predicated upon a recognition that producing more sustainable outcomes requires multidisciplinary skills and teamwork.

The guide aims to assist professionals in each discipline to identify design skills and resources that are directly pertinent to their respective area of practice and to assist them to better identify and understand many of the sustainable design tools, issues and approaches of other BDP disciplines, which, collectively, may be mobilized to design and create more sustainable outcomes.

Source: RAIA (2004)

Ethical frameworks

Bound up with the idea of addressing environmental impacts through building design is the drive and motivation to move forward with what is a challenging goal.

It is often argued that the ethic constructs concerning the environment that drive design are as important as responding to the pragmatic criteria provided in environmental design guidelines. It is for this reason that many professional organizations have developed environmental ethical frameworks, which guide and inform design. An example of this is the Royal Australian Institute of Architects background statement, which supports their *Environment Design Guide*. These have led to a number of working principles (RAIA, 2004), which have been developed through a review of each of the BDP member ecological sustainable development policies and statements and which provide the editorial and design direction for the ongoing development of the guide.

The sustainable home: Diversity of approach

During recent years, the concept of the bioclimatic house has evolved to what can now be considered a 'sustainable home'. Yet, what is a sustainable house; how does it differ from a conventional home; what are the particular issues addressed in the definition?

Currently, sustainable housing appears to place emphasis on energy as the main environmental impact that should be reduced. This definition can be found from the aims of the International Energy Agency (IEA) Solar Heating and Cooling (SHC) Programme Task 28 on Solar Sustainable Housing.

Until now, such housing achieved this high performance primarily by reducing heat losses through compact building form, thick insulation and ventilation heat recovery. While the number of such buildings is growing, market penetration would be enhanced by a complementary approach – namely, in very well-insulated housing, increasing energy gains. This can be achieved by:

- passive solar design;
- active solar systems for domestic hot water and space heating;
- PV electric supply systems;
- improved daylighting (for improved living quality); and
- natural cooling and solar/glare control (IEA, 2004).

In the context of warm climates, a different configuration of these parameters is used to achieve solar sustainable housing. This is particular to climates and countries where differing social and economic factors are found.

At present, there appears to be no precise definition of sustainable housing or, indeed, what comprises sustainable development. Rather, there is a diversity of design approaches and concepts. Attempts to define sustainable housing can be found in the following areas:

- built projects – demonstration projects;
- emerging standards; and
- various concepts for sustainable housing.

Built projects

First, the definitions for sustainable housing come from built works, including a range of demonstration projects. The following are some that can be looked at in more detail; the first three are from cool climates where heating is needed, the last is from warm climates where cooling is required:

1   Autonomous House (Vale and Vale, 2000);
2   Integer Millennium House, Watford (Cole Thomson Associates, 2004);
3   Oxford House (Roaf, 2003);
4   Healthy Home (Hyde, 2000).

What can be learned from these buildings is the breadth of environmental issues addressed. The Autonomous House seeks to include issues such as the impact of water and material use. In addition, the houses incorporates innovative servicing, reducing the dependency of the 'big systems of servicing', such as the grid or the sewer. The approach originated during the 1970s from the philosophy of Schumaker and Lovins, who argued against large-scale 'servicing' centralized systems of buildings in favour of decentralized, site-specific solutions for energy, water and waste (Lovins, 1997). The reasoning for this comes from the view of life-cycle thinking: 'What comes to the site stays on the site' – hence the current philosophy of using the site as a transition space for resources so that waste is avoided. The Integer Millennium House defines a home as a machine metaphor, drawing from intelligent computer technology to accommodate the complexity of the human–environment interface. The Oxford House, on the other hand, draws on vernacular architecture. The traditional building typology is maintained and adapted with passive and active systems to improve environmental performance. This creates a building that provides future asset protection and very low carbon dioxide (CO_2) production. Finally, the Healthy Home follows a similar path as the Oxford House, but for warm climates. The focus is on providing a home that increases the quality of life of its occupants through comfort, health and well-being.

Attempts to define sustainable housing more closely can be found in the plethora of design advice, rating tools and standards.

Standards: Performance not prescription

Emerging environmental standards for housing are evolving. This process attempts to link environmental principles, design criteria or indicators with a standard for 'best practice' performance. Thus, if the Oxford House is taken as 'best practice', the environmental principles and design criteria from this house can be used to form the standard.

Some emerging voluntary standards, such as the Building Research Establishment's *EcoHomes* rating tool, have emerged to help define sustainable

Source: Rao et al (2000)

1.8 *EcoHomes* standard, Building Research Establishment (BRE)

housing. These give a useful set of environmental criteria for what could comprise sustainable housing. Hence, the definition for sustainable solar housing can utilize the *EcoHomes* criteria. At present this standard is based on cold climate UK practice. One of the goals of redefining sustainable houses is to investigate the set of environmental criteria appropriate for warm climates (Rao et al, 2000). The inherent problem underlying this approach is that the criteria tend to apply to just the environmental issues. Housing, by definition, arises out of social needs and how these fundamentally drive the current design, marketing and process of building houses. Hence, any eco-house standard would need to apply across a broader set of criteria, including social, economic and environmental factors.

New standards are emerging that not only relate to the design of houses, but also to the planning design of housing at the neighbourhood level. One tool developed for Green Globe 21 is called the Precinct Planning and Design Standard (PPDS) (Hyde et al, 2005).

By way of introduction, the PPDS tool, in concept, is similar to other building environmental assessment (BEA) tools, such as Green Star, LEED and the Green Globe 21 Design and Construct Standard (Hyde et al, 2005). The difference is in the scale and character of development that the tool assesses. Precincts or neighbourhoods are usually clusters of buildings, often of different building types, created as a social unit or community within a rural area or city. The scale of this type of development offers significant opportunities for creating environmental benefits compared to single buildings. This is evident from a number of recent projects, such as BedZED–Beddington Zero Energy Development Sutton (EST, 2007) in the UK.

This development demonstrates the advantages of making areas of a city more autonomous in terms of their services, and also makes better use of the resources available to the community. For example, Beddington Zero is a zero energy project – that is, the onsite renewable energy generation is balanced with the energy drawn from the grid to create net zero carbon emissions. Beddington Zero also addresses social and economic issues, what Yencken and Wilkinson (2000) call 'the three pillars of sustainability: environmental, social and economic parameters'. They also recognize a fourth pillar: culture. The four-pillar approach underpins much of the work at BedZED and has led to an operation theory for sustainability called 'balance theory' (Mawhinney, 2002).

Balance theory proposes that to achieve sustainability of developments, such as BedZED, optimization of the four pillars is needed. The past emphasis on environmental issues has often meant that sustainable objectives have been unworkable in the design of a development because it has entailed trading off environmental parameters for social, economic and cultural parameters. It is argued that optimization can be achieved with balance theory. Hence, in BedZED, economic parameters support environmental parameters, which, in turn, support social progress. For example, the use of retail units within housing

Table 1.1 Matching indicators to principles

Principles	Precinct Planning and Design Standard (PPDS) indicators
1 Create an improving quality of life for the occupants and users.	Indicator 1: Sustainable master planning approach
2 Protect and conserve ecosystems through respect for the site.	Indicator 2: Precinct location and siting planning
3 Reduce environmental global and local impacts through conservation of energy and resources.	Indicator 4: Energy efficiency and conservation
	Indicator 5: Water conservation and management
	Indicator 6: Solid and other waste management
	Indicator 8: Chemical use
	Indicator 9: Storm water management
4 Support local communities and economies.	Indicator 3: Social commitment
	Indicator 10: Economic commitment
5 Source materials and energy within the local bioregion.	Indicator 7: Resource conservation (materials)

Source: Richard Hyde

development is aimed at subsidizing housing rental, and the multi-use, retail and housing strategy creates a scale effect that supports the cogeneration energy strategy, leading to the zero energy capability of the scheme. While there are many theories and much debate about sustainable development (Mawhinney, 2002), the development of PPDS has followed the balance theory paradigm – hence, its development attempts to work from a broad base of issues rather than confining its scope largely to environmental issues.

In fact, research into PPDS has been initiated and supported, in part, by a developer in the travel and tourism industry who wished to plan, design and develop a sustainable precinct. The purpose of this initiative was to assist with demonstrating that the developer had adopted environmental measures of a higher standard than normally found. This early finding led to research into the changing landscape of the building development process to identify the drivers of this change (Hyde et al, 2005) and to the development of the research

Source: Energy Saving Trust (2007)

1.9 BedZED–Beddington Zero Energy Development, Sutton, UK

hypothesis that a tool would provide a means of demonstrating 'best practice' precinct planning and design for developers. Five main principles were distilled from the review of principles that were relevant to the scale and scope of the PPDS tool:

1 Create an improving quality of life for the occupants and users.
2 Protect and conserve ecosystems through a respect for the site.
3 Reduce environmental global and local impacts through conservation of energy and resources.
4 Support local communities and economies.
5 Source materials and energy within the local bioregion.

The PPDS is a key standard that aims to define sustainable housing in a comprehensive approach. The inclusion of both social and economic principles complements environmental issues. A benchmarking system supports the indicators to provide a mechanism for industry to rigorously apply the standard. Local government has also been interested in this tool since it provides a metric for evidence-based policy planning. Further advantages for developers are the use of a standard by which to gauge the sustainability of proposals at the master planning point in the development process (Hyde et al, 2005).

The important issue that derives from these emerging standards is that they should be performance based (that is, they should establish an objective for quantitative reduction in energy, rather than prescribe solutions for energy reduction). The reason for this is that it allows some flexibility in how the standards are met in a particular context.

CONCEPTS

There are some important concepts concerning sustainable housing that help to assist designers in creating sustainable houses and sustainable housing. These can be found from examining some of the ideas behind eco-homes and eco-villages. Other concepts, such as SMART housing, also exist. These concepts have evolved depending upon how the authors have drawn the boundary around phenomena. In some cases, the boundaries are intellectual and relate to the domains of knowledge – social, environmental and economic issues. In other cases, these are physical and relate to the site or subdivision.

Autonomous House

The idea behind the Autonomous House is that it is a house operating as a semi-closed system with regard to servicing. It is not linked to any city services, such as gas, electricity, water and drainage. It seeks to harvest the natural resources from the site: water, sun and wind. These are processed onsite, reused and/or recycled and finally disposed of through the natural system (Vale and Vale, 2000, p4).

The Autonomous House should not be confused with semi-autonomous houses, which are connected to city services but utilize the natural resources of the site to reduce the demands from city services, such as energy and water and city waste disposal systems.

At the centre of this debate are a number of issues concerning health and management and risk. While it may be argued that the Autonomous House is simply city planning since it does not require the extensive servicing infrastructure that currently makes up our cities, it increases the risks of management and health while requiring owners to operate and augment their level of service infrastructure onsite, complicating the management of onsite services.

Autonomous Housing

This is an extension of the Autonomous House. The boundary, in this case, is physical and is concerned with the area around a cluster of houses. In this case, the housing should be completely independent in terms of energy, water and waste services. Many would argue that while this is theoretically possible, it is practically impossible given the level of servicing required to facilitate today's lifestyle.

Source: Vale and Vale (1991)

1.10 One of the key texts on green architecture

> ## Box 1.3 Integer Millennium House: Intelligent technologies
>
> Intelligent technologies for the Integer Millennium House include the following:
>
> - A sophisticated, yet simple to use, building management system ensures that the performance of the heating system is optimized, only operating when heat is required.
> - Soil humidity is monitored to ensure that the automatic garden irrigation system only waters those plants that need it.
> - An intelligent security system not only picks up intruders, but also interacts with the lighting, heating and door control systems to ensure that no energy is wasted while the house is unoccupied. When owners return home, all of the systems within the house 'wake up' automatically to provide immediate safety and comfort.
> - Lighting can be set to four predefined moods at the touch of a button or by infrared remote control.
> - Door keys equipped with a microchip allow the key to be programmed to open any door in the house – or keys can be restricted to only give access to certain areas. For example, a parent might not want their child to have access to the study, and the key could be programmed to achieve this. This means that occupants need only one key for everything.
> - Telephony is distributed around the house via a local building exchange. The ISDN line carrying this allows multiple numbers to be allocated, so each person in the house can have their own personal phone number.
> - Digital satellite and terrestrial television is distributed to every room around the house providing freedom of choice for both programme and location.
> - WebTV is available on the Philips wall-hung, flat-screen television in the lounge, allowing people to browse the internet together and in comfort. CCTV cameras at the front and rear of the house are broadcast on a spare analogue television channel around the house so that they can be viewed on every TV screen.
>
> *Source:* Integer Millennium House (2004)

Integer house

The integer house concept in its original form, found at the Integer Millennium House in Watford, UK, adopts a similar approach to the Autonomous House. The difference is that it aims to reduce environmental impacts in four areas:

1. social sustainability;
2. environmental sustainability;
3. economic sustainability; and
4. informational sustainability (Integer Millennium House, 2004).

Source: Integer Millennium House (2004)

1.11 Integer Millennium House, Cole Thomson Associates

The house development was linked to the Echelon Group, which focuses on using information technology (IT) to create an interface between the user and the building infrastructure (Integer Millennium House, 2004).

The IT approach has the potential of not only managing the environmental servicing systems effectively, but reducing the risk of malfunction, which results in health risks and other risks of onsite servicing. In this case, boundaries are multilayered: there is the physical boundary of the site and the virtual boundaries of the IT domain.

Eco-homes

The eco-home concept is similar to the semi-autonomous house in that it involves a number of ideas. Work by Roaf (2003) in her book *Ecohouse 2* provides design guidance for this type of building concept. Although the title focuses on the house, there are important principles for housing, such as site planning.

Dimension of eco-homes

The meaning of eco-homes comes from the stem 'eco', which derives from the Greek root '*oikos*', meaning 'household'. The Greek root, as with many Greek words, has two meanings: the sense of 'ecological' relationships between organisms in nature, and 'economics' – relationships concerned with the use of 'resources', Hence, 'eco' has as its basis two dimensions that need to operate within an 'ecological' philosophy – that is, its relation in design and use should follow a natural order, while it also has the dimension of using resources in an economical and efficient manner.

The inherent problem with eco-homes is that while building design professionals can embrace an ecological view, the social and economic dimensions may be missing or totally absent.

The discussion on world and country trends examines the question of how eco-homes or, indeed, sustainable housing are reaching acceptance within the local context and how the economic models within society accept eco-home design. The question that comes from this analysis is: do we need eco-homes (Hay, 2002, p205)?

Need for eco-homes

The need for eco-homes comes from concern about the current excessive use of resources to make houses, the waste that is created and its impact on the environment.

Edwards (1999) argues that one of the key issues with regard to environmental impact is *energy,* since the present use of fossil fuels as a primary energy source is driving climate change and this, in turn, radically affects the biodiversity of the planet – the threat to other life systems. This primary linkage in what is a complex web of interactions singles out a tangible goal, which is universal to all eco-homes. Hence, the term sustainable solar housing may be a better description since it focuses on 'energy' as a key performance area to be addressed in any low impact housing.

Eco-villages ('carbon neutral' developments)

Some argue that eco-homes by themselves cannot address the social and economic needs for housing effectively, and that there are benefits from making more comprehensive developments that are environmentally friendly. Eco-villages address, in particular, the social and economic needs of a community.

The BedZED project in the UK has become an example of what can be achieved in the area of precinct design. Although this project does not qualify as autonomous housing, it significantly reduces impacts – particularly in the energy area – through careful selection of fuel sources. Carbon neutral developments have zero carbon emissions from fossil fuels such as coal, gas and oil. This is achieved through many of the strategies found in autonomous housing, but primarily relies on using renewable energy sources, in this case solar thermal, solar electric and wood fuel from waste from the surrounding area (BRESCU, 2002, p11).

SMART housing

The current SMART housing initiative and the developing sustainability code for housing in Brisbane, Australia, addresses a broader set of issues:

- accessibility (for the socially disadvantaged);
- affordability (cost);
- environmental responsibility (environmental impact); and
- liveability (lifestyle and image).

SMART housing addresses the triple bottom line (TBL) of sustainability through balancing social, economic and environmental criteria. TBL methodologies have been imported into the design and procurement process of buildings from accounting methods used in sustainable business management.

Setting this aside, SMART housing, in simple terms, means taking a more intelligent view of people's needs. In this case the boundary is temporal, examining the varying human needs through life and designing accordingly. Universal design, as it is called, creates an environment for a wide range of human abilities and stages in life. This is set in a context of resource and economic efficiency (Queensland Government, 2004).

Healthy homes

Healthy homes (similar to SMART housing) focus on the impacts of buildings on people and the environment. The idea has some similarity to concepts presented in Pearson's book *The Natural House* (Pearson, 1989). The natural house is constructed as part of its local ecosystem and ensures its health and sustainability.

Pearson identifies the modern house as a 'dangerous house': the very technologies that support our lifestyle are bringing about its demise through pollution. Much of Pearson's focus is on selecting materials and systems that are benign to humans and to the environment (Pearson, 1989, p25). He cites the new era of sick building syndrome, which plagues many buildings, creating what he calls the 'alienating house': one that destroys cognitive links between the individual and the environment (Pearson, 1989, p60).

PRINCIPLES AND APPLICATIONS

Redefining principles

The notion of bioclimatic housing can be defined not only in terms of environmental criteria, but also in terms of design principles. Pulling together bioclimatic principles from the overview and concepts previously discussed, a number of new principles can be examined. The use of principles is an important way of translating theory into practice. These principles are:

- creating user health and well-being;
- using passive systems;
- restoring ecological value;
- utilizing renewable energy;

- utilizing sustainable materials; and
- applying life-cycle thinking, assessment and costing.

Creating user health and well-being

There are three main dimensions to creating a sense of health and well-being among users with regard to houses and housing:

1   working with owner perceptions and values;
2   creating a feeling of comfort; and
3   developing within the owner a state of health.

These dimensions are highly subjective and specific to individuals and family groups. Very often, the dimensions are subject to the home's externalities and to the individual's economic and social circumstances. Increasingly, as a result of complex social circumstances, users are applying new initiatives with regard to building homes. At one level, the home is seen as a 'natural' phenomenon. David Lee (Pearson, 1989) remarks on the desire for the 'self-built home' and Hassam Fathy, the Indian architect, reports: 'people build houses like birds build nests' (Fathy, 1972). This phenomenon of the owner home has meant that architects are becoming less involved in designing houses. Rather, housing is now systematized and commercialized. People can now buy houses much as they buy cars, selecting a model that suits their needs and tastes. Creating awareness of the improvements of eco-housing may thus help drive the 'market' to a better alternative.

Addressing the core issue of how eco-homes create a more comfortable environment, leading to a state of well-being and health, is central to addressing the mass 'marketing' approach. Increasingly, this 'market' is becoming sensitive to performance issues. This trend began during the 1960s with concern for maintenance and durability, and now encompasses other issues, such as value adding and running costs (energy and water). Value adding is increasingly becoming important as a way of maximizing return on investment in property. It is indicative of a rising duality in housing – the need to use the 'house' as an investment as well as a 'home'. What may be appropriate for a home is not appropriate as an investment.

Perceptions are also changing. At one level, a house owner appears to be concerned about the environment; yet the use of environmentally friendly strategies in homes is often problematic. This is either because the strategies often bring little economic benefits or add little value. So, a new bathroom or kitchen is perceived to add more value to a home than placing solar hot water or PVs on the roof (for a similar cost).

Yet, as can be seen, the economic context to this phenomenon is changing. Comfort is more of a key issue that is related to the fabric of the building. Comfort has, since the 1960s, been expected in housing by virtue of cheap energy. No matter if the building's envelope is poor: turning up and down the thermostat is a remedy. The consequences of this have been staggering from an environmental point of view.

However, new standards of comfort are emerging that promote variability in temperature as an important criterion and the adaptability of human response. The way forward is to return to making more climate responsive buildings, rather than energy-intensive buildings.

Utilization of passive systems

Climate responsive design is a central part of sustainable solar housing and eco-homes. Science has provided tools such as the bioclimatic chart and the psychometric chart. These tools enable the designer to select concepts, strategies, elements and systems to suit a particular building type and climate. These tools provide a framework of criteria for existing house types and enable the evolution of new forms and types of houses to better match the climate and the social context.

'Form match', which initially was seen as a form of 'climate determinate' where the building needs to face a particular orientation and a particular geometry to be truly climate responsive, has been replaced. The solution sets that this form of analysis created are highly introspective and hold that complicated design problems should match building programme with climate. Hence, the solution sets presented in this chapter represent prescriptive advocacy for plan form, orientation and section, but also describe broader performance requirements that are readily translated to specific projects that do not have an ideal orientation.

Maintaining and restoring ecological value

Issues concerning the building site revolve around the need to re-evaluate the relationship between the building and its environment. Many cultures view the earth as 'sacred' – the earth and the creatures that live in the subsoil are as important as those that live above it. This symbolic link with the land means that decisions concerning foundations and the 'fit' with the building and ground plane should not be taken lightly. Respecting the site therefore takes on a pragmatic meaning and means addressing three main issues:

1   'touching the ground lightly';
2   building foot-printing; and
3   consideration of the ecological value of the site.

The term 'touching the ground lightly' was a prosaic term attributed to architect Glenn Murchutt to emphasize his concern for the way in which buildings interface with the ground (Fromonot, 2003). Many designers take this literally to

Source: Richard Hyde

1.12 The Queenslander house: passive systems of climate control are used and the veranda is ideally suited to provide shade

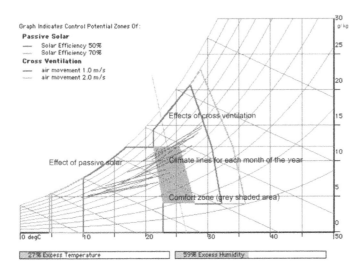

Graph Indicates Control Potential Zones Of:
Passive Solar
— Solar Efficiency 50%
— Solar Efficiency 70%
Cross Ventilation
— air movement 1.0 m/s
— air movement 2.0 m/s

Effects of cross ventilation

Effect of passive solar

Climate lines for each month of the year

Comfort zone (grey shaded area)

27% Excess Temperature   59% Excess Humidity

Source: Hall and Blakay (1996)

1.13 The psychometric chart can be used as a design tool, with the comfort zone in the centre. Temperature forms the horizontal axis, humidity the vertical axis. Warm climates normally sit to the right of the comfort zone

mean building on piles or piers. More generally, the term 'building footprint' has been used – that is, the area occupied by the building on which things cannot live. Essentially, this is developed as a ratio of the building area to the site area. Hence the footprint is also a measure of loss of habitat and reduction in the ecological value of the site. Most buildings change the ecological value (the biodiversity) of their site in some ways. A definition for ecological value is complex and relates to the number of species of flora and fauna on the site. Assessing ecological value is therefore a specialist task, but has been simplified for the purposes of design to give some simpler measures such as the concept of building foot-printing. In some instances, brownfield sites (that is, sites that have been abandoned or contaminated) have little ecological value and require regeneration, hence providing a dilemma for sustainable construction. Remediation work on degraded sites often requires more environmental impact than the construction of the building. Alternatively, the impact of greenfield site construction is more of an issue of protecting the existing environment. Hence, different strategies are needed for different sites.

The issue of respecting sites adds another dimension to building design since it is no longer sufficient to view sites as the sole territory of the building form – rather, it can be argued that sites are only one part of the ecological system of the area. Viewing buildings from a broader macro-ecological perspective is needed if these principles are to be adopted.

Utilizing renewable energy sources

Renewable energy is naturally regenerated over a short timescale and is derived either directly or indirectly from the sun. Direct sources are thermal, photochemical and photoelectric energy, and indirect sources include wind, hydropower and photosynthetic energy stored in biomass.

Sources of renewable energy include natural movements and mechanisms of the environment (such as geothermal and tidal energy). Renewable energy does not include energy resources derived from fossil fuels, waste products from fossil sources or waste products from inorganic sources. It is necessary to now define this in legislation to move forward with developing the integration of these systems in buildings (TREIA, 2007).

Concern for the environmental impact of fossil fuels, which are non-renewable, has brought about the need for energy conservation. Design principles for conserving energy involve three main approaches:

1   Energy conservation involves the reduction of energy demand in houses. 'Houses without heating systems' are an approach to housing that reduces the demand for heating through improvements in the quality of the building envelope. The premise for the Solar Heating and Cooling Programme Task 28 on Solar Sustainable Housing is that by utilizing high-quality envelopes in houses, in addition to passive solar heating, significant energy conservation can be achieved.

2   Energy efficiency involves the use of mechanical equipment and appliances, the level of service provided and energy use. Ideally the objective is to choose mechanical equipment or appliances which use less energy to give better service, if possible. Service embraces such aspects as providing comfort and labour saving. Thus, an energy efficient house could be defined as one that creates a higher level of comfort with less energy than a traditional house. Increasingly, it is becoming common to discuss energy efficiency in terms of the 'solution set' used to achieve service objective. Hence research examines the way particular combinations of appliance and mechanical equipment can be created to achieve efficiency. For example, with air-conditioning systems, more efficient systems normally have a higher capital cost but save operational costs. Life-costing exercises are needed to evaluate systems like this in the context of the total building project.

3   Fossil fuel energy abatement involves the reduction in use of energy from fossil fuels. This can mean simply avoiding use of energy from fossil fuel sources or by using renewable energy.

An area for design research is emerging that examines the 'solution sets' for energy efficient mechanical systems and appliances in the context of the energy conservation strategies used in the building envelope. In this way, solar sustainable housing can evolve as a new building type addressing environmental imperatives

Generally, solar energy, compared to other forms of energy such as wind or water, is seen as being a readily available natural resource and has for many

Source: Walker (1998)

1.14 Gabriel Poole, North Stradbroke Island, Queensland, Australia: use of lightweight construction to touch the ground lightly and reduce the building footprint has considerable environmental benefits for maintaining the ecology of the site

years been viewed as a method of saving fossil fuel energy. Powered by the sun, the solar house can reduce dependence upon fossil fuels.

The problem with solar energy is that it is dispersed widely across the globe and is variable in availability due to seasonal effects and conditions in the atmosphere. Solar energy is also low-grade heat (in the form of photometric or thermal energy) and requires upgrading to higher grades of energy, such as electricity, which have a wide variety of uses.

Utilizing sustainable materials

As with minimizing energy resource use, the use of water and other materials is a key issue for eco-homes. The design of eco-homes features careful selection and management of resources within the project. The underlying principle behind resource conservation is to design closed systems in the house. Conceptually, this means departing from the present open-system approach, where resources enter the home from outside its boundary, are used and then the waste is exported, with its consequent impacts upon the wider environment. Instead, eco-homes are characterized by a closed system, with minimal resources entering the home, recycling and reusing of materials, and disposing of waste on the site.

Suggested strategies include:

- reducing resource intensity and impacts;
- recycling and reusing materials; and
- harvesting resources from the site.

Life-cycle thinking

Life-cycle thinking is part of the green design approach and combines two constructs. First, it concerns whole-of-life issues, and uses as its frame of reference the process of change in the hard and soft systems of a building's 'life'. A conventional view of design is to think of the building's first use and current purpose, with only little regard of subsequent uses. Hence, life-cycle thinking is about designing from the standpoint of a whole-of-life scenario – called a cradle-to-grave approach.

A second construct examine the cycles within the overall life cycle of the building, This cycle within cycles parallels how natural systems operate and uses this as a model for buildings. It focuses on the need to create buildings as closed or semi-closed systems that mirror the ecological systems on Earth. The Earth, with the exception of solar energy and other inputs, relies on the energy and resources within the specific boundaries of its ecosphere and lithosphere. In a similar way, a building can have a boundary placed around its site, and strategies may be adopted to only use the energy and resources available to that site, where practicable. The 'closed' system developed by this approach is

Source: Hyde et al (2003)

1.15 Transport energy is a crucial element in some housing developments

Source: Hyde et al (2003)

1.16 Redevelopment of the Heron Island Research Station, Queensland, Australia: resource conservation and construction waste monitoring to examine sustainability of the construction systems used

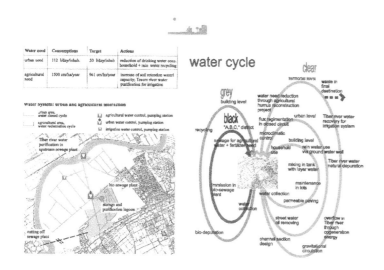

Source: Sartogo (1999)

1.17 Life-cycle mapping of water systems used in demonstrating the application of ecological principles. The qualitative and quantitative approach involves setting targets and then selecting strategies to meet these targets

essentially cyclic and relates to the life of the building. The term 'life cycle' is therefore used to define buildings designed in such a manner.

This contrasts with the current approach of building development, which is linear and open. Resources and energy enter the site, are used by the building and its occupants, and waste is exported. This approach is seen as being sustainable because the pollution and resources drawn into the system are expanding at a rate that cannot be supported. The challenge for designers is to create closed-system buildings. Although, in theory, this sounds simple, in practice, it can be complicated.

Useful cycles within the overall life cycle of a building are waste, water and heating and cooling cycles. Figure 1.17 shows the life-cycle mapping of

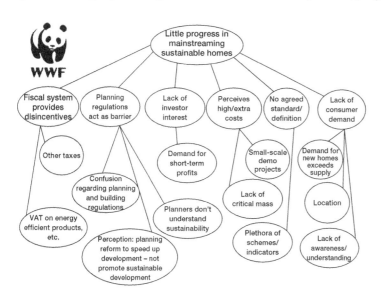

Source: Wheeler (2007)

1.18 World Wide Fund for Nature–UK study of sustainable housing

building materials. Within the cradle-to-grave system are sub-cycles of disposal, recycling and reuse.

Life-cycle assessment

Methods have been developed to assist with mapping and quantifying these cycles through examining the outcomes of life-cycle thinking. Life-cycle assessment (LCA), as described by the International Organization for Standardization's ISO 14040-43, provides a framework for examining the environmental impacts of products, including buildings.

The intention is to examine the predicted environmental loads created by a product or process with the purpose of avoiding or reducing environmental harm. This can be a highly rigorous and time-consuming process and is often carried out as a research exercise on prototype buildings. The results of LCA work are included in a number of building environmental assessment (BEA) tools and standards. For example, LCA data has assisted in creating the Green Specification developed by the Building Research Establishment (BRE) for comparing the environmental performance of sustainable materials, and this underpins the Eco-home Standard (Anderson and Shiers, 2002).

While there has been software development to carry out full LCAs of buildings, its application at the design stage still remains on the horizon. At present, BEA tools offer a partial assessment.

In summary, this chapter has discussed the development of principles to redefine bioclimatic housing. However, applying these principles in practice remains a challenge. This is elaborated upon in the following section.

Applications

Work in the UK by the World Wide Fund for Nature (WWF) has suggested a number of barriers to an environmental approach to housing, ranging from factors concerning social issues, to economic and political issues (see www.wwf.org.uk/sustainable homes/).

Important barriers emerge during the planning process. WWF cites confusion regarding the way in which planning authorities view sustainability – in particular, the need to enhance training and education to improve understanding among planning staff. In addition, the aim of reform in planning has been to improve the speed of the planning process, rather than to reduce environmental impacts. WWF suggests that part of the problem lies in the lack of principles and standards. Furthermore, while the forces of government lack a mechanism for change, the application of bioclimatic principles in consumer culture is still not a mainstream consumer issue. Hence, the development of planning standards such as PPDS can be a way of addressing these problems.

Raising consumer awareness by governments and other groups has thus become a priority. For example, marketing strategies by the Australian government called *Your Home* have been published on an extensive website (see www.greenhouse.gov.au/yourhome/). These strategies emphasize the benefits of eco-homes to owners and builders. This market-oriented approach places less emphasis on sustainability and more on the benefits of bioclimatic homes, particularly from the value-adding, health and well-being perspectives. 'Building green is building smart' is the prevailing philosophy and will perhaps break down the barriers to enable an uptake of sound environmental principles (Wheeler, 2007).

Case study 1.1 Media Centre, Colombo, Sri Lanka

Harald N. Røstvik, Architect

PROFILE

Table 1.2 Profile of the Colombo Media Centre, Sri Lanka

Country	Sri Lanka
City	Colombo
Building type	TV studios and media centre
Year of construction	Completed autumn 2001
Project name	WGM
Architect	Sivilarkitekt, Harald N. Røstvik in association with Kahawita de Silva Ltd

Source: Harald N. Røstvik

1.19 Media Centre, Colombo, Sri Lanka: front elevation

PORTRAIT

Buildings account for close to 50 per cent of global carbon dioxide (CO_2) emissions. Instead of being polluters, can buildings and even whole cities become solar power stations? Leading ecological designers now suggest that they can and must.

In time, can a sun-blessed country such as Sri Lanka be connected to the sun and be free of dependency upon imported fossil fuels? With the help of well-designed ecological architecture it possibly can.

What, then, is sustainable design? The answer will be different in different contexts. The approach must be determined by local factors: climatic, economic, technological and, not least, cultural.

The Media Centre in Colombo contains a range of ecological features. It is an example of climatic adaptation and innovative sustainable design. It also addresses the growing trend of putting up slick glass buildings regardless of the climatic and cultural context.

CONTEXT

The design of the 3000 square metre building was developed with a Norwegian firm as lead architects in collaboration with a Colombo firm. Most of the consultants were Sri Lankan; but the specialist energy consultant, Max Fordham & Associates in London, supported the local know-how base and the heating, ventilating and air conditioning (HVAC) consultant Koelmeyer. All contractors, with the exception of the solar system contractor (Engotec GmbH in Germany), were local. The local University of Moratuwa provided input studies. Transfer of knowledge and awareness-building played a key role in this North–South collaboration.

ECONOMIC CONTEXT

The building contains workspace for up to 450 people, including visitors. The several private companies and non-governmental organizations (NGOs) that occupy it and built it were previously scattered throughout Colombo in five or six different buildings, resulting in communication problems and unnecessary traffic. Each had their own canteen, stores and toilets.

By co-locating these different units, a more rational use of space was made possible. This has resulted in a space reduction of up to 30 per cent and, hence, a reduction of material use as well as of energy needs. The Norwegian Agency for Development Cooperation (NORAD) supported part of the 'green package' of technologies; but most of the funding came from the visionary client itself. The 'green package' has had an extremely tight budget, which considerably restricted the project.

SITE DESCRIPTION

The site in Battaramulla, close to the new parliament, on the outskirts of Colombo was chosen to move the occupants out of the polluted city centre of Colombo. The site was sloping and the building cut right into the slope. All

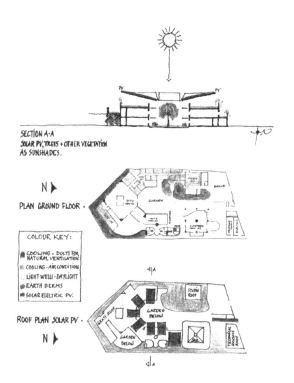

SECTION A-A
SOLAR PV, TREES + OTHER VEGETATION
AS SUNSHADES.

N ▶
PLAN GROUND FLOOR .

COLOUR KEY:
■ COOLING · DUCTS FOR NATURAL VENTILATION
▨ COOLING · AIR CONDITION
░ LIGHT WELL · DAYLIGHT
▨ EARTH BERMS
■ SOLAR ELECTRIC PV.

ROOF PLAN SOLAR PV .

N ▶

Source: Harald N. Røstvik

1.20 Media Centre: section showing solar photovoltaic (PV) modules and vegetation as roofs and shade. Plans show the location of solar PV modules. The coloured 'key plan' shows natural ventilation ducts, locating which spaces use air conditioning and which use natural ventilation, as well as light shafts that allow daylight into the offices

trees were preserved in order to shade the building. The concept is that of a lush garden with no vehicle access, thus greatly reducing vehicle noise and airborne pollution in the working environment. Vegetation and external window shades, along with solar photovoltaic (PV) modules that 'float' over the building as flakes of shading devices, are attempts to reduce the cooling load. Air entering the building is filtered through vegetation for cooling and cleaning purposes.

BUILDING STRUCTURE

Most of the building is three storeys high, while part of it comprises two storeys. The roofs are shaded with great sails or solar modules, and there is the possibility of incorporating working spaces beneath the roofs as long as there is no rain. The building contains several split-up divisions, all shaded or covered but in the open air. Rooms within each division are arranged mostly as an open plan or, where closed rooms are necessary (editing rooms, etc.), along a corridor. There is no basement; but since the site originally was sloped, parts of the building have their 'back' towards the earth shelter and are situated underground towards the neighbouring sites. Vertical light shafts are used to bring daylight deep into the building, where possible.

Most of the building is naturally ventilated and lit. The parts that must have air conditioning due to the need for a stable temperature are the technical rooms (studios and editing rooms). These have split-unit air

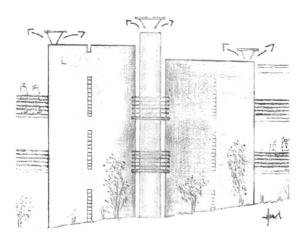

Source: Harald N. Røstvik

1.21 Media Centre: architecturally visible natural ventilation ducts draw air from interior spaces – cooling is helped by solar PV extract fans

conditioning with an economy cycle on the cooling system to improve energy efficiency.

BUILDING CONSTRUCTION

For fire regulation and mass (cooling) purposes, the main structure is concrete and plastered brick. Part of the top floor roof is earth sheltered. The remaining solid roofs are tiled.

The majority of the building was constructed *in situ* in order to create local employment. Shading devices comprising solar photovoltaic panels and other structures, such as the sails and external window shades, are prefabricated.

VENTILATION SYSTEM

The major parts of the building are naturally ventilated. Huge, wide and vertical (mostly round) ducts project through the building as architectural elements. Inside the ducts are huge ceiling fans. The fans are run directly by low-voltage DC electricity from solar PV panels, which function as a shade and water barrier on top of the duct. These ducts suck air out and naturally cool the building interior through a high proportion of air changes.

APPLIANCES

Throughout the building, energy efficient light bulbs are used. Some of the new computer equipment and studio mixing and recording equipment are chosen with energy efficiency in mind. There is cooling recovery on the AC system in the studio and editing rooms.

ENERGY SUPPLY SYSTEM

The overall concept of the building is influenced by the aim of reducing energy needs. The building was the nation's first to have a grid-connected solar PV system, with a calibre of 25kWp (kilowatt peak: the current/voltage curve for a solar cell; so a 25kWp solar system creates 25 kilowatts of power at peak solar conditions). The base load in the building is, at times, smaller than that and power is exported to the grid. The grid connection of the solar PV eliminates the need for batteries. The hot water supply is intended to be solar thermal, but is not yet installed, nor is the cooking water supply, also planned to be solar thermal assisted.

SOLAR ENERGY UTILIZATION

To reduce electricity demand, the building is designed to allow maximum indirect daylight penetration into rooms via courtyards, light wells and patios, cutting vertically right through the building mass.

The common design of 'deep', badly lit offices has been avoided. Shading by vegetation, sail cloth and solar PV modules (which thus have a double function) protects against direct sunlight and overheating. Windows have horizontal external shading fins that project 1m out of the wall.

BUILDING HEALTH AND WELL-BEING

Most rooms are naturally ventilated so that fresh air flows through large window openings and vertical extract ducts.

In all rooms (especially important in air-conditioned spaces where air changes are less than in naturally ventilated spaces) all interior walls and ceilings are painted with natural paints produced from plants. The lush garden, which excludes cars, ensures fresh air surrounding the building.

The issue of electromagnetic fields has been addressed by gathering most cables in under-floor ducts away from people.

Fittings and furniture were selected on the basis of their environmental qualities. Bulky material-consuming furniture was passed up, health-damaging glues and paints were avoided, rational short-run cables were chosen, and timber from eco-forestry was preferred.

WATER RECYCLING AND CONSERVATION

No public sewage system exists near the site. All sewage from the 450 occupants is therefore treated onsite using biological methods in an extended aeration underground-type sewage disposal unit.

Water supply is provided locally onsite. The wastewater treatment systems enable onsite recycling and they 'clean' grey water for the main secondary uses – for example, to water the gardens and to flush toilets. Water-saving toilets reduce water consumption by 60 per cent (5 litres per flush as opposed to normally 13 litres per flush in Sri Lanka). All basin taps are the automatic water-saving, self-closing type push taps.

Sorting and recycling of paper, metals and other solid wastes, seldom done in Sri Lanka, will gradually occur within the building.

PLANNING TOOLS APPLIED

Natural ventilation duct dimensioning, fan size, speed and regulations have been simulated in London based on input from Colombo. The solar system has been simulated in collaboration with a German supplier. Cooling loads

Source: Harald N. Røstvik

1.22 Media Centre: solar PV modules replace traditional roofs as shade and water catchers. Large overhangs stop the blazing sun from reaching the building walls. White tent-like structures are used as shades and more are currently being installed. Large ducts, extending above the tents, extract air from rooms to cool the building interior through a natural ventilation system that is assisted by solar PV-driven fans. In addition to this vertical ventilation and cooling, horizontal cross-ventilation is extensively used and shading issues have been calculated and designed in collaboration with consultants and architects. Waste systems have been designed and simulated by Colombo consultants.

PRINCIPLES, CONCEPTS AND TERMS

Additional features related to the building are also of interest. The building process follows a 'clean worksite' programme that minimizes all material waste and construction pollution. The programme was also a learning process, with the goal of keeping the number of accidents down through organizing site layout and hygiene to protect workers.

Accessibility for the disabled has been ensured – something that is still rare here. Ramps and level design ensure access to most, but not all, levels. All doors are of a wheelchair width and there are toilets for wheelchair access.

INFORMATION

For further information see Sivilarkitekt, www.sunlab.no.

Source: Harald N. Røstvik

1.23 Media Centre: view from the access road, showing the tent structure and roof terrace, which is planted to provide vegetation overhanging the roof. There is a large column at the centre and some of the side columns are hollow ducts for natural ventilation, bringing stale, warm air out of the interior spaces, assisted by solar PV-powered fans set in the top of the ducts

SUMMARY

Bioclimatic principles form the cornerstone of sustainable housing. New definitions and concepts can be found in a range of built projects and are typified by a wide diversity of approaches, with differing names and priorities. Some houses focus on the social issues of sustainability, while others have an ecological and economic focus. Supporting the drive for new concepts are various new standards for houses and housing. The object is to improve performance across a range of social, environmental and economic conditions. The challenge for new building, such as that found in Case study 1.1, is to adequately integrate these diverse measures within a building project. The following chapter examines how this challenge has been addressed in a number of countries.

REFERENCES

AGO (Australian Greenhouse Office) (1999) *Australian Residential Building Sector Greenhouse Gas Emissions 1990–2010*, Department of the Environment and Heritage, Australia

Anderson, J. and Shiers, D. (2002) *The Green Guide to Specification*, Blackwell Science, UK

EST (Energy Saving Trust) (2007) *BedZED–Beddington Zero Energy Development Sutton,* General Information Report 89, Energy and Efficiency Best Practice in Housing, www.est.org.uk. (accessed 1 Jan 2007)

Clarke, J. A., Johnstone, C., Kelly, N. and Strachan, P. A. (1997) 'The simulation of photovoltaic-integrated building façades', *Proceedings of the International Building Performance Simulation Association Conference 1997,* volume 2, Prague, Czech Republic, pp189–195

Cole Thomson Associates (2004) *The Integer Millenium House, Watford*, www.colethompson.co.uk/w_housing.html (accessed 8 December 2004)

Edwards, B. (1999) *Sustainable Architecture*, Spon Press, UK, p7

Fathy, H. (1972) *Architecture for the Poor: An Experiment in Rural Egypt*, University of Chicago Press, Chicago, USA

Frampton, K. (1995) *Studies in Tectonic Culture,* MIT Press, Boston, MA

Fromonot, F. (2003) *Glenn Murcutt: Buildings and Projects, 1962–2003*, Thames and Hudson, London

Gilijamse, W. (1995) 'Zero-energy houses in The Netherlands', *Proceedings of of the International Building Performance Simulation Association Conference 1995,* Madison, WI, pp276–283

Halawa, E. (2005) *Modelling and Thermal Performance Evaluation of Roof Integrated Heating System with PCM Thermal Storage*, PhD thesis, University of South Australia, Australia

Hall, J. D. M. and Blakay, I. (1996), *DA Sketchpad*, computer software, University of Tasmania, Launceston, Australia,

Hay, P. (2002) *Main Currents in Western Environmental Thought*, University of New South Wales Press, Australia, p205

Huang, M. J., Eames, P. C., Norton, B. and Griffiths, P. (2002) 'Experimental validation of a numerical model for the prediction of thermal regulation of building-integrate photovoltaics using phase change materials', in *World Renewable Energy Congress VII (WREC 2002),* Cologne, Germany

Hyde, R. A. (2000) *Climate Responsive Design*, E. F. & N. Spons, London and New York

Hyde, R. A., Law. J. and Bridges, S. (2003) *Benchmarking Report: Heron Island Research Station Redevelopment: Dining Facility and Teaching Facility*, The Collaborative Research Centre for Sustainable Tourism, Queensland, Australia, p28

Hyde, R. A., Moore, R., Kavanagh, L., Watt, M., Prasad, D. and Blair, J. (2005) 'Development of a planning and design tool for assessing the sustainability of precincts', in *Proceedings of the Australian New Zealand Architectural Science Association (ANZAScA) Conference*, Wellington, New Zealand, p29.

IEA (2004) *Solar Heating and Cooling Task 28*, International Energy Agency, www.iea-shc.org/task28/ (accessed 29 September 2004)

Integer Millennium House (2004) *The Integer Millennium House, Watford*, www.integerproject.co.uk/watford.html (accessed 4 December 2004)

Jones, D. L. (1998) *Architecture and the Environment: Bioclimatic Building Design*, Laurence King, London

Law, J. H. Y. (2001) *The Bioclimatic Approach to High-Rise Building Design: An Evaluation of Ken Yeang's Bioclimatic Principles and Responses in Practice to Energy Saving and Human Well-Being*, BArch. thesis, University of Queensland, St Lucia, Queensland. Australia

Lawson, B. (1996) *Buildng Materials, Energy and the Environment*, RAIA, Red Hill, ACT, Australia

Levin, H. (1997) 'Systematic Evaluation and Assessment of Building Environmental Performance (SEABEP)', paper presentation to Buildings and Environment, Paris, 9–12 June, available online at www.wbdg.org/design/sustainable.php

Lovins, A. (1997) *Soft Energy Paths: Toward a Durable Peace*, Friends of the Earth International, Cambridge, MA

Manz, H. and Egolf, P. W. (1995) 'Simulation of radiation induced melting and solidification in the bulk of a translucent building façade', in *Proceedings of Building Simulation 1995*, pp252–258

Mawhinney, M. (2002) *Sustainable Development: Understanding the Green Debates*, Blackwell Science, Oxford

Olgyay, V. (1963) *Design with Climate: Bioclimatic Approach to Architectural Regionalism*, Princeton University Press, Princeton, NJ

Parlour, R. P. (2000) *Building Services: A Guide to Integrated Design – Engineering for Architects*, Integral Publishing, Pymble, New South Wales

Pearson, D. (1989) *The Natural House Book*, Conran Octopus, London

Pedrini, A. (2003) *Integration of Low Energy Strategies to the Early Stages of Design Process of Office Buildings in Warm Climate*, PhD thesis, University of Queensland, St Lucia, Australia

Queensland Government (2004) *Smart Housing Objectives*, Department of Housing, Queensland, p1

RAIA (2004) *BDP Environment Design Guide*, Royal Australian Institute of Architects, www.architecture.com.au/ (accessed 29 September 2004)

Rajapaksha, U. and Hyde, R. A. (2002) 'Passive modification of air temperature for thermal comfort in a courtyard building for Queensland', *Proceedings of the International Conference Indoor Air,* Monterey, US

Rao, S., Yates, A., Brownhill, D. and Howard, N. (2000) *EcoHomes: The Environmental Rating Tool for Homes*, BRE Bookshop, Watford, UK

Roaf, S. (2003) *Ecohouse 2: A Design Guide, Architectural Press*, Burlington, Massachusetts

Sandnes, B. and Rekstad, J. (2002) 'A photovoltaic/thermal (PV/T) collector with a polymer absorber plate: Experimental study and analytical model', *Solar Energy*, vol 72, no 1, pp63–73

Sartogo, F. (1999) *Saline Ostia Antica*, Alinea, Italy

Stritih, U. and Novak, P. (2002) *Thermal Storage of Solar Energy in the Wall for Building Ventilation*, IEA, ECES IA Annex 17, Advanced Thermal Energy Storage Techniques – Feasibility Studies and Demonstration Projects Second Workshop, 3–5 April, Ljubljana, Slovenia

Stutchbury, P. (1999) 'Architecture and place', *Architecture Australia*, vol 88, no 1, January/February, p58

Szokolay, S. V. (2004) *Introduction to Architectural Science: The Basis of Sustainable Design*, Architectural Press, UK

TREIA (Texas Renewable Energy Industry Association) (2007) *Renewable Energy Definition,* www.treia.org/backup/definition.htm (accessed 1 Jan 2007)

Vale, B. and Vale, R. (1991) *Green Architecture,* Thames and Hudson, London

Vale, B. and Vale, R. (2000) *The New Autonomous House,* Thames and Hudson, London

Walker, B. (1998) *Gabriel Poole: Space in which the Soul Can Play*, Visionary Press, Noosa, Queensland

Wheeler, J. (2007) 'One million sustainable homes', www.wwf.org.uk/sustainablehomes/

Wittchen, K. (1993) *Proceedings of International Building Performance Simulation Conference 1993*, Adelaide, Australia, pp377–383, www.ibpsa.org/

Yeang, K. (1998) *T. R. Hamzah and Yeang: Selected Works*, Images Publishing, Mulgrave, Victoria, Australia

Yeang, K. (1999) *The Green Skyscraper: The Basis for Designing Sustainable Intensive Buildings*, Prestel, Munich

Yencken, D. and Wilkinson. D. (2000) *Resetting the Compass: Australia's Journey Towards Sustainability,* CSIRO Publishing, Collingwood, Victoria, Australia

Chapter 2
Trends, Promotion and Performance

Peter Woods, Richard Hyde, Motoya Hayashi, Marcia Agostini Ribeiro, Francesca Sotogo, Valario Calderaro, Veronica Soebarto, Indrika Rajapaksha, Upendra Rajapaksha and Vahid Ghobadian

INTRODUCTION

This chapter discusses trends in the design of solar sustainable housing for warm and hot climates. The first section, 'Trends towards sustainable housing', examines these trends for a number of warm climate countries in terms of the following issues:

- trends in solar sustainable housing;
- government policy, legislation and structure (which part of government is responsible for sustainability);
- standards of energy consumption for housing;
- initiatives to promote solar housing; and
- marketing.

The trend is towards improving the environmental performance of houses, so this chapter also discusses techniques that aid the building professional in achieving these improvements. The focus is on analytical tools and methodologies used both in the design phase and post-construction, including simulation and monitoring approaches. These techniques are described in the case studies that follow. Finally, the section on 'Performance assessment' illustrates the way in which solar sustainable housing has been promoted through design guidance, demonstration projects and sustainable development codes.

TRENDS TOWARDS SOLAR SUSTAINABLE HOUSING
World energy and sustainability trends: 'The energy squeeze'
Peter Woods

This introduction is concerned with world energy trends as they pertain to sustainable housing in climates where cooling of housing is a dominant need – 'cooling-dominated climates'. The worldwide energy consumption projections are drawn from *International Energy Outlook 2004* (IEA, 2004).

Globally, energy consumption is projected to increase by 54 per cent during the period of 2004 to 2025, with developing Asia accounting for 40 per cent of the total projected world increase (IEA, 2004). Developing Asia includes a large proportion of the world population who live in cooling-dominated climates. Figure 2.1 shows the world marketed energy consumption by region projection for 1970 to 2025. The projection for the developing world reflects both its lower starting level and its requirements for industrialization and transportation.

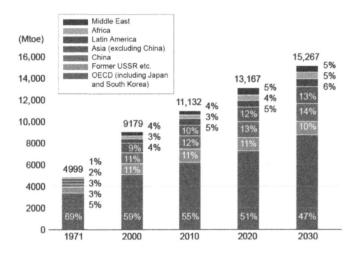

Source: Clini (2004)

2.1 World marketed energy consumption by region projection, 1970–2030

The projection shows that use of oil will continue to dominate primary energy consumption, at about 39 per cent of the total. Indeed, the projection (see Figure 2.2) suggests that the proportion of major energy sources will remain fairly constant, although there may be major downward adjustments in the proportion of oil versus natural gas for electricity generation. The change will, however, be

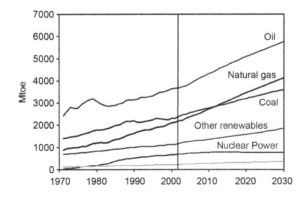

Source: Clini (2004)

2.2 Projections of primary energy consumption

compensated by the increasing demand for petroleum products for transportation. Marketed renewable (non-nuclear) energy sources will increase in line with the general trend of the projection, maintaining their share of about 8 per cent of global energy supply.

In countries with cooling climates, predominantly developing tropical and equatorial countries, the marketed renewable energy increases will come primarily from large-scale hydroelectric schemes or power stations using biomass. There are complexities in determining the extent of non-marketed passively renewable energy consumption, both indigenous (such as wood or charcoal burning) and modern (such as photovoltaics). Energy saving design is equally important for non-marketed 'renewable energy' sources. Data is usually available only for individual projects and initiatives; however, the International Energy Agency (IEA) Renewable Energy Working Party states:

> Through the next several decades, renewable energy technologies, thanks to their continually improving performance and cost, and growing recognition of their environmental, economic and social values, will grow increasingly competitive with traditional technologies, so that by the middle of the 21st century, renewable energy, in its various forms, should be supplying half of the world's energy needs. (IEA, 2001)

This suggests that the rate of growth in renewable energy sources is likely to increase markedly after 2025, coinciding with the predicted plateau of

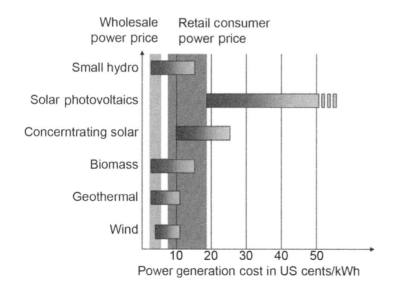

Source: IEA (2005)

2.3 *Cost comparison of renewable technologies*

conventional oil exploitation. Non-marketed passive renewal energy solutions currently face resistance in initial costs, particularly photovoltaics (PVs). The good news is that the measurable trend for initial costs will decrease with improvements in efficiency and volume of production. This is well demonstrated by the cost reduction per kilowatt (kW) peak installed that is reported by Japanese solar industry sources (see Figure 2.3).

COUNTRY TRENDS

Australia

Richard Hyde

Trends towards sustainable housing

The trends towards sustainable housing in Australia are comparable to those in many developed countries; the use of this form of housing is the exception, rather than the rule. Housing is designed primarily to meet social and economic needs. There is an increasing concern about the energy intensity of housing, particularly in warmer states where there is a shift from passive houses (without heating or cooling) to hybrid houses, which have active systems for climate control. A technical shift has occurred whereby the availability of low cost, easily installed 'split-system' air conditioning has replaced the window air conditioner as the system of choice. The marketing of such systems by energy companies has produced an alarming increase; for example, Brisbane, located in the subtropical zone of south-east Queensland, saw a 16 per cent increase in 2001, with the rate of increase growing further (Brisbane City Council, 2001). The trend towards solar sustainable housing is seen mostly in demonstration projects. Energy efficiency in housing is promoted by the introduction of energy codes.

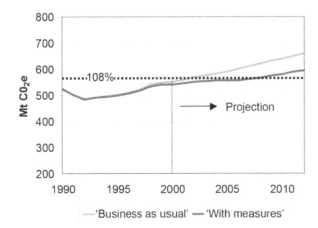

Source: AGO (2005b)

2.4 Greenhouse emissions as predicted for 2008–2012

Government policy

The Commonwealth government's funding of Aus$1 billion for climate change abatement, through initiatives such as Safeguarding the Future: Australia's Response to Climate Change and measures for a better environment, was announced as part of Australia's new tax system in 1999. Australia has agreed to a Kyoto Protocol target of limiting emissions to 108 per cent of 1990 levels over the period of 2008 to 2012. These measures are designed to arrest the probable causes (Howard, 1997).

The main agency involved with this work is the Australian Greenhouse Office (AGO). AGO suggests:

> *Climate change is an issue of major significance for all of us. Most of the world's leading scientists agree that global warming caused by human activity is occurring. New and stronger evidence that humans are having an influence on the global climate through greenhouse gas emissions [was] presented in a report prepared by the Intergovernmental Panel on Climate Change in 2001. The ultimate objective is to stabilize greenhouse gas concentrations in the atmosphere at a level that will prevent dangerous human-induced interference with the climate system.* (AGO, 2004)

The government's agenda therefore targets emissions abatement on many fronts, including the following initiatives:

- boosting renewable energy actions and pursuing greater energy efficiency;
- investing significant resources into greenhouse research and monitoring Australia's progress towards its Kyoto Protocol target through the National Greenhouse Gas Inventory;
- studying the landscape of Australia through the National Carbon Accounting System;
- investigating the possibility of a domestic emissions trading scheme;
- encouraging industry, business and the community to use less greenhouse-intensive transport;
- fostering sustainable land management practices; and
- providing impetus for greenhouse action through the Greenhouse Gas Abatement Programme (AGO, 2005a).

The building sector, in agreement with the Australian government, has resolved to eliminate the worst energy performance requirements in buildings. The Building Code of Australia, in conjunction with the AGO, has developed research into the impacts of these code changes. Working with the Australian

Building Codes Board, a plan for implementing energy codes for all building types has been established (AGO, 2005b). Minimum energy performance standards for housing were established in 2003. The method of assessing compliance follows either of the following:

- deemed to satisfy provision; or
- energy performance rating using an accredited software tool.

A compliance benchmark is nominated for each rating tool, which creates a standard, equated to four stars. This allows for variations in the standard between climates and locations.

The warm climate variations to the code include the provision of highly ventilated buildings and window shading.

The norm of energy use

The overriding concern within industry is twofold. First, there is the gap between 'predicted energy use' as determined by the rating tool and actual use. There are inherent contradictions with buildings that rate very poorly, yet are very comfortable to live in (that is, those that use no air conditioning but are designed for natural ventilation and shading). Inherent paradoxes that exist with regard to energy norms are anomalies that also occur in terms of the proportion of energy used and the influence of the size of house. Studies from Victoria show that the average house size has increased by 35 square metres during recent years, which has distorted the energy use figures. Furthermore, in warm climates the proportion of energy use for space heating/cooling is lower than that of other energy uses (transport and solar hot water heating).

Hence, establishing standards for energy use for space heating and cooling seems to represent only a small aspect of the energy demand equation. Eco-homes that take a more integrated approach to energy conservation address this problem.

In view of this kind of complexity, the Building Code of Australia is considering a range of frameworks to assess potential performance, from absolute standards of predicted energy, to more regulations deemed to satisfy provisions that reflect the number of energy-demand reduction strategies used in a house. Top-end demonstration houses use solar hot water, high efficiency appliances, photovoltaic systems and high performance envelopes, as well as hybrid solar gas cars. The inherent contradiction is that many of these buildings receive low ratings from current energy rating tools (Prelgauskas, 2003).

The emergence of a 'comfort' standard within the Code is needed to evaluate passive building design and to avoid problems of contradiction that are so often found when passive buildings are assessed as if they were using air conditioning. Consideration of occupancy is also crucial if overall energy consumption is to be predicted accurately. The move to complement the use of *design measures* prior to design approval with *performance in use* measures has been advocated. This has been achieved through the development of a new rating tool called the National Australian Built Environment Rating System (2007). It has the potential to provide a more robust method of performance assessment than is currently available.

Marketing

The AGO recognizes that global warming presents challenges for the way in which we live and work. Addressing this will require changes for industry, governments at all levels and the community at large. The government recognizes that meeting the challenge of climate change requires collaborative action. An important sector identified by AGO is the housing sector:

> *One fifth of Australia's greenhouse gas emissions come from households. There are 7 million households in Australia, each producing more than 15 tonnes of greenhouse gas emissions every year (1 tonne of emissions would fill a family home). Energy use, car use and waste are the largest sources of household emissions.* (AGO, 2004)

A key argument is that small but significant strategies can be used, which, when added together, contribute to an important reduction in greenhouse gas emissions. The AGO has a range of initiatives to assist communities and households in reducing their greenhouse gas emissions:

- Household Greenhouse Action is working to reduce emissions from households by focusing on home heating and cooling, refrigeration, lighting and water heating.
- Cool Communities is providing information, support and financial assistance to help communities take easy practical actions to reduce household greenhouse gas emissions.
- Cities for Climate Protection Australia is assisting over 160 councils throughout Australia to develop and implement greenhouse reduction action plans in both their corporate operations and within their communities (AGO, 2005c).

The improvement of the energy efficiency of homes and appliances is one of the most effective ways of reducing greenhouse gas emissions. In partnership with state and territory governments, the Commonwealth is working to improve the energy efficiency of many domestic appliances. Key publications for this work are *Global Warming Cool It: A Home Guide to Reducing Energy Costs and Greenhouse Gases* and *Your Home*, which has a suite of design guide materials for creating stylish, comfortable, energy efficient and environmentally friendly houses (AGO, 2005d). This literature contains advice on the use of a range of rating tools. Some aspects of building performance that can be rated include:

- energy performance of the building envelope;
- energy efficiency of appliances and services;
- performance of individual components (such as windows, insulation and wall construction);
- life-cycle environmental impact of the materials used in terms of emissions and depletions; and
- environmental impact of whole buildings.

With regard to the rating of appliances, it is currently mandatory for all of the following electrical products offered for sale in Australia to carry an approved energy label (Australian Government, 2005):

- refrigerators and freezers;
- clothes washers;
- clothes dryers;
- dishwashers; and
- room air conditioners (single phase).

In summary, the AGO has a range of information that provides designers with decision-making information regarding the environmental impacts of buildings. It has a clear policy concerning the reduction of energy use and greenhouse gas emissions; but it remains to be seen whether these incentives are sufficient to shift the market towards more sustainable housing to meet targets for greenhouse gas abatement.

Brazil

Márcia Agostini Ribeiro

The residential sector in Brazil accounts for 24 per cent of the total energy consumed, energy used for water heating alone amounting to 26 per cent of this total – more than 6 per cent of all energy produced. An energy crisis in Brazil in 2001 led to the introduction of other primary energy sources, especially natural gas, to complement the electricity produced today by hydropower. The use of solar energy for heating domestic hot water is becoming common in Brazil. There are 1.2 million square metres of installed collectors.

Trends towards solar sustainable housing

Brazil's power generation is predominantly hydroelectric, providing about 90 per cent of its installed capacity. In the long term, other renewable energy has the potential to supply 23,000MW of installed capacity to the national grid. During the next 20 years, as part of the PROINFA Programme, targets are 50MW peak of photovoltaic (PV) solar energy; 3 million square metres of solar collectors for water heating; and 3000MW peak of wind, biomass and micro-hydropower.

Brazil has an extremely large solar resource, with promising applications, including PV for rural electrification, hybrid systems with mini-grids in isolated areas, grid-connected systems in urban communities, water heating, bioclimatic architecture, and solar thermal energy for heating and cooling. Investments by the Brazilian electrical sector and the promise of incentives for energy efficiency policies and alternative energy sources show the potential of the future of solar energy in Brazil, even though there is a lack of specific legislation for this field.

Government policy, legislation and structure

In the last few years, the development of quality programmes for housing has been discussed all over the country. The Brazilian Programme of Quality and Productivity in Housing Construction (PBQP-H) is an example of what has been done in order to raise quality in the construction sector. It focuses on service and material quality, as well as on the efficiency of the construction process. This programme offers a 'quality stamp', which is a prerequisite for the acquisition of financing.

For the thermal/energy efficiency in housing, programmes are now under discussion in many states of the country. The National Electric Energy Agency (ANEEL) encourages programmes of energy efficiency and regulates the programmes that may be implemented by the energy companies of each state.

In 1985, the National Programme for Electricity Conservation (PROCEL) was instituted in order to coordinate measures to save electricity in the country. Resolution no 271, 19 July 2000, from ANEEL established criteria for the application of financial resources to procedures for saving energy, and to research and technological development of the Brazilian energy sector. At least 1 per cent of the annual income of each energy company in the country must be applied to such activities. There is also a specific programme for domestic hot water solar systems, aimed at replacement of conventional electric showers. In this case, the solar systems must have the National Institute of Metrology, Standardization and Industrial Quality (INMETRO)/PROCEL stamp, which guarantees the quality and efficiency of the system.

In 1991, the Technological Research Institute of São Paulo (IPT-SP) started a research study based on accepted environmental standards for public housing. The Minimum Performance Criteria for Public Housing aims to establish a methodology of evaluation for the housing sector and is under evaluation by the Brazilian Association of Technical Standards (ABNT). Five parameters were developed, focusing on thermal comfort:

1 human thermal comfort parameters;
2 typical climate conditions;
3 occupancy profile;
4 thermal performance of the construction; and
5 evaluation of thermal performance.

The thermal performance of constructions can be evaluated during development (auxiliary methodology) or after construction (evaluation methodology). Auxiliary methodology can be done by means of either computational simulations or following constructive parameters. Evaluation must be conducted using monitoring methods. This standard also establishes a bioclimatic zoning for the country and defines different construction recommendations for each bioclimatic zone (Givoni, 1992, pp11–23) considering the following issues:

- size of openings for ventilation;
- sun control of openings;
- opaque structure, thermal transmittance, thermal delay and solar factors; and
- passive thermal conditioning strategies.

Initiatives to promote solar housing

Four projects are under way to promote solar housing:

1 The project *Abordagem Integrada da Eficiência Energética* is under development in partnership with the solar laboratory (GREEN Solar) of the Catholic Pontificate University of Minas Gerais (PUC-MG), the Federal University of Minas Gerais (UFMG) and the Energy Company of Minas Gerais (CEMIG). It consists of four areas of research, of which sub-project 1: Sustainable Solar Buildings, aims to develop basic directions for architects, constructors and users of public housing in order to emphasize bioclimatic architecture and sustainable solar technologies.
2 The Sapucaias Project, started in 1999, is a partnership between the solar laboratory of PUC-MG and the state-controlled electricity company, ELETROBRAS, and was established to design and install solar water heaters for the Sapucaias housing project in the state of Minas Gerais. The

main goals were to evaluate the savings of energy and money provided by the solar water heaters and to familiarize users with the new technology. One hundred houses received solar water heaters in August 2000. Families also received basic training, and electricity consumption was monitored. After the first year of operation, the results showed an average reduction of 30 per cent in electricity consumption and 40 per cent in electricity costs. An important result of this project was the acceptance by the Caixa Econômica Federal (a federal bank) of the grant of subsidies for installing solar heating.

3   Demonstration and Applied Research Centres of Energy Efficiency (CDPAEEs) were built in 2000 at PUC-MG, the Federal Centre of Technological Education of Minas Gerais (CEFET/MG) and the Federal University of Minas Gerais in a partnership with CEMIG. These centres are used for qualification, research and demonstration on how to reduce electricity demand and energy waste. They have also been built as exemplary bioclimatic architecture buildings in accordance with the idea of energy conservation.

4   Since 1984, the Brazilian national certification board (INMETRO) has been managing a successful labelling programme for electrical appliances: the Brazilian Labelling Programme (PBE). GREEN Solar, INMETRO and all participating manufacturers jointly manage the voluntary programme. Although it is not mandatory to have the tests performed in order to sell these products in Brazil, manufacturers who do not participate face restricted access to government financing and tender processes.

Further work is focused on photovoltaic systems and solar hot-water heating systems.

Since the mid 1980s, PVs have been used in Brazil for several applications, including cathodic protection, telecommunications, traffic signals and control. Recently, the use of this solar technology in rural electrification has surpassed all other applications in scope and size. The current 10MW of installed capacity is distributed throughout the country, with more than 40,000 PV systems.

The federal government programme, PRODEEM, is the primary government-sponsored electrification programme focusing on off-grid technologies. PVs are now widely used for rural electrification of schools, with more than 5MW of PVs installed since 1996. To meet cost constraints, the systems and components were purchased through a centralized method in order to make use of the economy of scale in procurement. There are several other initiatives in various stages of implementation, such as the PRODUZIR programme, which has installed approximately 15,000 solar home systems in rural areas of Bahia.

One of the most important and current PV rural electrification programmes in Brazil is directed and sponsored by CEMIG, the Energy Company of Minas Gerais. The primary objective is to facilitate the access of lower-income people to education, lighting and communication.

Second, there has been progress in the development of solar hot water heating. Brazil is Latin America's largest manufacturer and user of solar water heating. By the end of 2003, the total installed area of solar domestic hot water systems was about 2 million square metres, with an annual growth rate of 30 per cent between 1997 and 2000. Despite this significant growth, the area of collector surface per inhabitant, a ratio of only 0.01:1, is still very small in comparison with countries such as Greece and Austria, at around 0.25:1.

Thermosyphon technology dominates in Brazilian domestic systems, and systems with forced circulation are used for larger applications. There are now more than 170 solar collectors and 140 hot water tank types made by Brazilian manufacturers. Since 1998, solar collectors and hot water tanks have had to be certified by the Brazilian national certification board, and *Programmea Brasileiro de Etiquetagem* has adopted national and international standards. It has already certified around 100 solar collectors and 120 hot water tanks. The tests for the certification procedure are conducted in the Brazilian National Solar Laboratory, located at PUC-MG in Belo Horizonte.

CEMIG has a large programme with PROCEL to implement solar collectors in large building blocks in the urban areas of the main cities of Minas Gerais. The goal is to replace electrical showers by solar collectors. CEMIG gives a 20 per cent tariff reduction to stimulate the replacement. As a result, there are more than 300 large upper- and middle-class residential building blocks using these systems.

In summary, the use of solar energy (PV and solar heating water) has had a long history in Brazil. There are numerous PV and solar thermal research projects throughout the country that have had an impact on new technologies and their use, reliability and deployment. The ANEEL regulation of decentralized renewable energy technology has been a major step forward and will solve many problems before the systems reach the field. A fraction of the potential future consumers of solar energy display characteristics that would indicate off-grid PV systems are competing to provide electricity to rural low-income consumers. Integrating this technology has called for a regulatory framework, certification of equipment and PV systems and, especially, training for maintenance and operation. These concepts are the basis for providing a viable sustainability model. Because many consumers are below the poverty level and do not have the ability to pay all of the costs associated with bringing in and using electricity, subsidies are frequently required to extend electrical services to rural and poor regions of the country. Consumers served by the electric grid are charged differential tariffs and are classified in accordance with certain socio-economic characteristics established by Resolution 456. In residential and rural sectors, if consumption falls below a certain threshold, social lump-sum tariffs apply. This differs between areas. Typically, the tariffs vary between US$1 and US$3 per month. Because this tariff does not cover the costs incurred, the concessionaries (or 'permissionaries') will recover these investments through the cross-subsidy tariffs, meaning that these low-income consumers will be covered by all of the concession area consumers. Most of the consumers already electrified by photovoltaic solar energy in Brazil are eligible for the low-income tariff.

Iran

Vahid Ghobadian

Iran is a rich country with regard to fossil fuel sources and a founding member of the Organization of the Petroleum Exporting Countries (OPEC). During the last decade, fossil fuel consumption has become an important issue for three main reasons:

1   The total annual increase rate of consumption during the last decade was 4.1 per cent, which is above the world rate.
2   Air pollution in large cities such as Tehran is at a dangerous level and is far above the standard health level for air quality.
3   Finding other sources of energy (especially renewable and clean energy) is essential for the sustainable development of the country.

Trends towards solar sustainable housing

Solar sustainable housing is rather a new subject in Iran, and although there have been some climatic houses built during the past few years, they were essentially designed and built with saving energy in mind. In Tehran and Azad universities, especially in the architecture and environmental design departments, solar sustainable housing is being taught and discussed, although work on other aspects of sustainability, such as economy, culture and society, have not yet been considered adequately.

Government policy, legislation and structure

Government policy in the third five-year plan, which started during the year 2000, is to decrease and optimize energy consumption in all of the economy's different sectors. With regard to housing, the Ministry of Housing and Urban Development, the municipalities and the Ministry of Petroleum are the responsible authorities.

The Building and Housing Research Centre, which is a research institute under the Ministry of Housing and Urban Development, is responsible for scientific research and laboratory tests for different building materials, products, construction methods and energy audits. This centre devises and proposes building codes and regulation for the ministry. The ministry, in cooperation with the National Engineering Organization and the universities, ratifies these matters.

Since the beginning of the new Iranian financial year, on 21 March 2004, the design and construction of all new buildings in the country with total floor areas of 800 square metres or more have had to meet national building regulations. The reduction of energy use in buildings is one of the 20 main issues of these regulations. The municipalities and national and local engineering organizations supervise and ensure that these regulations and codes are met during the construction of new buildings.

The Iranian Fuel Conservation Organization (IFCO), a subsidiary of the National Iranian Oil Company, has set a goal for conserving all energy carriers, defined in the sustainable energy programme of the country. The main projects of the Department of Conservation of Energy in its commercial and building sector, a part of IFCO, are as follows:

- Prepare national building codes and standards, and make them compulsory, focusing on insulation materials, double-glazed windows and the energy-saving architectural design of buildings. This part accounts for more than 50 per cent of departmental activity.
- Revise codes and standards for gas- and oil-burning appliances, with the cooperation of the Institute of Standard and Industrial Research of Iran (ISIRI).
- Implement energy labelling schemes for gas- and oil-burning home appliances.
- Execute an energy auditing and energy management system in public buildings, such as schools and hospitals.
- Provide direct assistance to major manufacturers of gas- and oil-burning space and water heaters to produce and distribute highly efficient products around the country and to support them with financial subsidies.
- Financial subsidies have been granted to four plants and factories for the renovation and/or construction of gas home-appliance production lines. Efficient water heaters and un-vented heaters from these programmes will show up in the market very soon.
- Ensure the maintenance and direct the tune-up of burners in public and large buildings in Tehran.
- Introduce the application of solar energy to replace oil products in some rural and isolated areas.

With positive outcomes from the pilot plan, a project has been proposed for the production of 215,100 solar water heaters and 1000 public solar baths in six provinces for US$48.8 million in a buy-back scheme.

There are other ministries and organizations which are indirectly involved in energy optimization and renewable energy use in buildings, such as the Ministry of Power, the Ministry of Industry and Mines, the Ministry of Planning and Budget, the Atomic Energy Agency and non-governmental organizations (NGOs) such as the Iranian Solar Energy Organization.

Norms of energy consumption for buildings

The commercial sector, houses and household appliances consume more energy than any other economic sectors in Iran. This breaks down to 35.2 per cent for oil products, 53 per cent for natural gas and 20.7 per cent for electricity. The value of energy consumed in the year 2001 amounted to US$5.5 billion and the cumulative forecast until the year 2020 is US$157.6 billion.

According to one survey covering the whole country, energy consumption per square metre of buildings is equivalent to 30 cubic metres of gas per year. This will slim down to 20 cubic metres in 2020 with the implementation of the IFCO programmes, still high compared with the European index of 5.5 cubic metres of gas per year. Currently, the average energy consumption for buildings in Iran is 310kWh/m^2. The goal is to decrease this to 160kWh/m^2 by the year 2010.

Initiatives to promote solar housing

As mentioned earlier, there are very few solar houses in Iran and they are mainly built by the private sector. However, since government policy is to optimize energy consumption, there are plans by IFCO, in cooperation with the Ministry of Housing and Urban Development and the universities, to design and build a few of these houses in order to promote their construction in the country.

Italy

Francesca Sartogo

Residential energy consumption in Italy during 2000 was over 25 per cent of total use, with a growth trend of 5.7 per cent in one year because of the growing use of domestic cooling equipment and electronic devices. Houses consume about 120kWh/m^2 each year: a very high value compared with the energy experts' and environmental associations' long-term goal of 15kWh/m^2 and also with mid-term goals of 60 and 30kWh/m^2 (Legambiente, Municipality of Bolzano, Faenza).

In 2000, final energy use was 60 per cent for methane, followed by 21 per cent for diesel, and 18.5 per cent for electric energy and biomass (wood). Most of the energy is used for heating and for hot water; but there is an increasing use of electricity for cooling.

Government policy

In 1991 an energy code was approved for the rational use of energy, energy saving and renewable energy development. This code covered local energy

Table 2.1 Special Islands Programme

Island	Project accepted				Contribution	
	Wind power	PV	Solar thermal	Energy saving systems	Thousand Euros	Percentage
Pantelleria	660kW	100kW	758m²	Urban Regional Energy (URE)	1,137	40%
Ventotene		126.29kW	494m²	Lights, pumps	1,063	60%
Gorgona	50kW		455m²	Thermal stations	407	35%
Giglio	535kW	13.65kW	346m²	Pumps	466	27%
Panarea		33kW	60m²		205	

Source: Francesca Sartogo

plans, economic incentives for renewable energies, control of energy consumption, energy certification of buildings, factories and farms, and obligations for renewable energy use in new or restored public buildings. Since then a lack of regulations has meant that there have been no more notable initiatives in energy saving policy. The aspect of building heating was regulated with DPR 412/1993 that uses a variable value called FEN (standardized energy needs), which set the maximum allowed limit of heating-energy losses through the building envelope (La Sapienza, 2007).

At present, only a few cities have an energy plan and little has been implemented with regard to establishing energy certificates and energy consumption regulation for building cooling. Nevertheless, there are some measures that have formed the basis of a better energy future:

- *Sustainable Development National Plan (1993)*: local execution of Agenda 21 programme;
- *Urban Restoring and Quality Recovering Programmes* (DMLLPP), 22 October 1997: a law that financed experimental projects regarding eco-efficiency and environmental quality of settlements;
- *Formal Agreement of Kyoto Greenhouse Gas Reduction Protocol*;
- *Inter-ministry Commission of Economic Programming (CIPE), 6 August 1999 Deliberation*: obligation for energy producers to sell at least 2 per cent of the excess over 100GW derived from renewable sources;
- *Solarized Municipality Project* (DMA, 4 February 2000): a programme that finances the installation of solar collectors on public buildings to an amount of 9 million Euros;
- Law no. *224/2000* (DMA, 6 December 2000): deliberations of the Energy and Gas Authority that allow power grid connection for small private energy producers; this opens the way for domestic PV plants;

- *10,000 Photovoltaic Roofs Programme (DMA, 19 March 2001)*: direct financing of small PV plants up to 70 per cent of total cost;
- *High Architectural Quality PV Programme (2002)*: 58 projects on new and historical buildings for a production of 2.9MW;
- *Special Islands Programme (2002)*: has received project proposals amounting to 1036kW, of which 268kW have been selected; and
- *Solarized municipality project*: has a provision of 15,000 square metres of installations in two years.

Local policy

Since the end of the 1990s, more and more local administrations (regions and city governments) have been adopting Agenda 21 goals as their own. Some building regulations and urban plans in the Toscana, Emilia-Romagna, Lombardia, Umbria and Marche regions have already been changed to performance-based ones to simplify architectural integration of renewable energies plants and passive systems. Regione Lombardia Code No 26 of 20 April 1995, for example, allows a wall to be built with a thickness of over 30cm without including the excess part, up to 25cm, in the calculation of legal volume. Other local codes provide for a 'volume incentive' (for instance, up to 15 per cent in the new Rome Urban Plan) for bio-energetic building efficiency (MBE).

Table 2.2 High Architectural Quality Programme

Project accepted	Power (kW)	Latitude (°N)
New buildings	411	8
Retrofit	2015	36
Historical buildings	561	14

Source: Francesca Sartogo

Trends

The trend for solar sustainable systems has been both in the area of active and passive approaches.

Active systems

Domestic renewable energy plants are not yet very prevalent; but there is an increasing interest in high-efficiency heat metres and micro-co-generation (heat plus electricity). Italy, in spite of its amount of sun radiation, does not have an acceptable solar direct gain hot water heater collector distribution: only 0.003 square metres per inhabitant, one fifth of the Austrian standard. Until a few years ago, domestic PV plants, except for some experimental projects, were practically unknown.

In the past there has been a deficiency in education and information for technicians and users. Recent new public energy measures have given hope for a photovoltaic building integration explosion. For years the Italian solar industry has been dormant; but now economic statistics show an important production growth during the last year, which, for major industries, has been about 50 per cent. At the end of 2001, the start of Italian PV programmes produced a substantial boost of PV installations. The first and the second phase of programmes will produce about 4200 new low-power (less than 20kW) plants over the entire Italian territory in two years. The last government objective was to reach 50,000 new PV plants in six years. Solar thermal promotion programmes, initiated in 2001, should produce new collector installations of 750,000 square metres in private buildings and 35,000 square metres in public buildings in three years. Unfortunately, the economic crisis and the lower interest in environmental matters shown by the new government will hamper this positive tendency.

Passive systems

Italy had a successful period of experimentation during the 1980s with active–passive integration in some residential and office complexes. Universities, municipalities and associations have increased their annual human and economic resources involved in bio-climatic research; but in spite of this, except for some inbuilt projects, there are no substantial bioclimatic buildings to serve as examples at meetings of energy experts. At present, only a few buildings can demonstrate the benefits of passive heating, and even fewer have passive cooling. The best examples are limited to public or commercial buildings, such as the new Guzzini Headquarters in Recanati, designed by Mario Cucinella. Nevertheless, there are a number of projects, some of which are being financed by specific sustainability-oriented codes, that represent new interest in the use of passive and active systems.

Japan

Motoya Hayashi

The Japan Solar House Promotion Conference was established in 1986 to support the spread of solar houses in Japan. Housing manufacturers and societies, PV developers and building parts manufacturers participated in the conference. They exchanged information on their solar technologies, and did all they could for their dissemination and diffusion – for example, arranging study bus tours, producing pamphlets and offering other information to the government.

A loan system for buyers of solar houses has been in place since 1986. A builder of solar houses can borrow money at low interest from a public banking agency; the maximum amount of loans is larger if a solar house system is used. Governmental assistance is available on condition that the ratio of heating and cooling energy to that in the standard house without solar systems is less than 70 per cent. The energy performance of solar houses and the thermal performances of houses, in general, are improving.

The conference tried to promote many kinds of solar houses: wooden (2 x 4 inch) stud houses, wooden prefabricated houses and steel prefabricated houses, and so on. In the case of large house-makers, solar houses were designed as model houses and built in the factory yard or in exhibition parks. Energy performance is calculated with simulation program and measured in the model houses. The data is used for public information with the use of pamphlets and television commercial messages.

Solar house builders, borrowing money at low interest, can produce a better indoor climate, prepare for their old age and contribute to a healthier environment.

Market situation and government policy

The population of Japan is projected to increase up to the year 2010 and to decrease afterwards. The number of households is also estimated to decrease after 2010; the number of people in an average family will be 2.5 in the future. The total number of houses amounted to 42 million in 1993: 109 per cent of the figure in 1988 and three times as large as that in 1958. The ratio of houses to households was 1.22:1 in 1993, and 10 per cent of houses are not used. However, the number of houses constructed in a year amount to almost 1.2 million, although this number is now decreasing, with the recession affecting all of Japan.

At the foundation of the conference in 1986, the government and well-educated people had a discerning attitude towards the problem of saving energy due to the oil shock of 1973. But most people are still ignorant about solar systems because there is not enough information available. Business conditions improved steadily from the 1970s to the 1980s, but then rapidly worsened. As a result, most buyers are not able to plan for the future with confidence. On the other hand, most buyers have become more interested in a better indoor climate. The trend has expanded throughout Japan with the spread of insulated houses and with improvements in energy-saving performance standards for houses, in general. The government introduced several policies, including defining standards for solar house systems in order to estimate their energy saving performance. A public banking agency for house purchasers created loan schemes for the builders of solar houses in 1986.

Some solar house models were produced and sold by companies and societies. These were acceptable to buyers; but their share of the housing market was still very low. On the other hand, solar hot water units, consisting of a solar collector and a tank, were produced for use on the roofs of existing houses. Sales of these units expanded to cover all of Japan because they were not expensive.

In one company, the sales volume of a solar house model increased rapidly to reach a few hundred houses during the first few years; but after that, sales decreased and the model was deleted. The reason was that in large companies, in the view of company management, the solar house was suitable only for a minority of buyers. In a society that supplies solar systems with roof collectors, the sale of solar houses increased only slowly.

Commissioners have tried to promote solar houses. During the 1990s, PV systems became practical. In a photovoltaic system, solar radiation is changed into electrical energy through the use of a solar collector set on the roof; this covers the energy use of a family, and the extra energy can be sold to the connected electric company. In 1993, the conference decided to include PV systems in the criteria for solar houses. In 1994, NEF began to subsidize such systems, with the result that PV houses have become a popular form of solar houses in Japan.

The numbers of solar houses supplied by the members of the conference reached 2000 to 3000 per year during the 1990s. During 2000, PV supply amounted to 36MW in Japan.

Malaysia

Peter Woods

The *Eighth Malaysia Plan* identified energy efficiency and a fifth fuel policy as central features of national development (Government of Malaysia, 2001). This applies to the identification of annual energy targets for commercial buildings, with tax concessions for energy efficient equipment and the encouragement of the use of renewable energy, particularly biomass (but also solar energy).

Norms of energy use

Energy targets for commercial buildings in Malaysia are actively under consideration by the Energy Commission and Ministry of Energy, Water and Communication for inclusion in building by-laws. At this stage, the suggestion is 135kWh/m^2 per annum. The Ministry of Energy's new headquarters at Putrajaya has a target of 100kWh/m^2 per annum and is described as a low energy office (LEO) building. There is no existing or proposed legislation at this stage relating to the domestic sector, either in building performance or domestic appliance ratings.

Trends

The Malaysian climate has a relatively constant ambient external temperature range of 21 to 34°C, and a mean monthly relative humidity range of 70 to 90 per cent throughout the year (Malaysian Meteorological Office, 2005). Wind speeds are low, with a preponderance of calm days. Minor variations occur within the monsoon periods, mainly in east Malaysia, and there are also some slight variations in the hours of sunlight; but it can be stated that apart from a few hours during the night, the ambient conditions are outside or at the upper limit of any recognized thermal comfort range in many houses without the assistance of mechanical air movement or cooling. At best, modern terraced housing with mechanically assisted ventilation, applied shading, roof insulation and lightweight wall construction (using Standardized Environmental Temperature (SET) to assess comfort; see Szokolay, 2004, p21) results in comfort conditions for 10 to 15 hours, but this is mainly during night time and in the early morning (Hanafi, 1991).

It is ironic that the traditional dwelling form in Malaysia, usually referred to as the Malay house, can achieve a degree of passive control to achieve comfort conditions (Rahman, 1994). The important factors are that these are stand-alone buildings, normally single storey, elevated above the ground, with high ceilings, few internal partitions and with openings on all sides. The construction is thermally lightweight (timber) with, traditionally, a thick thatched roof with high thermal insulation characteristics. Roof slopes are steep, with substantial overhangs shading the external walls. Peak indoor daytime temperatures range from 34 to 36°C, about 2 to 4°C higher than the external temperature. The low thermal mass allows the building to cool down rapidly; thus, the night temperature drops close to the external temperature of 23 to 26°C.

Development of housing in urban areas has been substantial over the past 20 years, driven by increasing population and urban migration. Consequent pressure on land availability has encouraged a particular house plan form for lower- and middle-class housing. This type is either single- or double-storey terraced or link housing. Plot sizes are generally small, with a built-up area of 50 to 60 per cent of the plot, with a floor plate of about 50 to 70 square metres. The frontages are narrow to maximize site density, with plan aspect ratios of 0.5 to 0.7. Orientation of the building façades is largely determined by site access and efficient road layout, rather than by solar protection. The typical construction is thermally heavyweight, consisting of a reinforced concrete frame, with cement-rendered brick infill and a cement tile roof. The interiors are heavily subdivided with cement rendered brick walls. Owners frequently convert the front of the property into a car porch, reducing both potential cross-ventilation and daylight penetration. Externally, hard surfaces are regarded as being most practical, increasing the potential for reflected gains. To maintain thermal comfort, the use of mechanical air movement is universal, with air conditioning, in bedrooms at least, fast becoming universal.

Passive cooling opportunities are thus very limited. The preferred window design with security grilles gives a limited free area. The traditional urban house form, the shop house, has one or several internal courtyards that can assist in air movement (Lim, 2000). Attempts to introduce this into modern housing are either confounded by the tight plan or is seen as a security risk and effectively closed up by the homeowner. Night flushing is seen as a possibility, particularly for single-storey development. This would require some mechanical assistance and a radical change in user behaviour. To maximize the usefulness of flushing between the hours of 4 am to 6 am, the building should be effectively sealed and shaded for the rest of the day.

The windows are normally flush to the outside wall surface, providing little opportunity for self-shading. External awnings or shades are not frequently found in modern domestic buildings.

The one surface where some passive measures could be adopted – the roof – has been extensively researched (Zakaria and Woods, 2002a, 2002b). Computer simulations (TAS software) varying the roof parameters of colour, insulation and ventilation, show that a change from a dark- to a light-coloured roof with absorptivity of 0.3 reduced the roof space temperature by 7°C and peak hour temperature in the living spaces by 1°C, while 200mm of insulation reduced the peak hour temperature by 1.1°C.

Altering the roof pitch (from 22° to 45°) has more effect on a darker-coloured roof, lowering the roof space temperature by 2.4°C and the space below by 0.3°C. Roof ventilation at the optimum rate of 10 air changes per hour (ACH) reduced the temperature in the roof space by 3.9°C and that of the indoor living spaces by 0.8°C for a low pitch and a dark roof. For low pitch and a light roof, the temperature was lowered by 2°C in the roof space and 0.6°C underneath.

Thus, small but worthwhile reductions in internal ambient temperature could be achieved by using light roof finishes. This, unfortunately, runs counter to preferred roof colouring; but some evidence suggests that owners, if not developers, may adopt this strategy.

The use of renewable energy has been studied as part of a strategy for lowering energy consumption. Photovoltaic installations can generate a reliable year-round contribution; but the impediment here is cost. A 3kW peak installation is capable of providing up to 30 per cent of daily energy needs, including some air conditioning, but at an installation cost of roughly 80 per cent of the construction cost of a semi-detached unit (NLCC Architects, 2002).

PERFORMANCE ASSESSMENT

Computer simulation thermodynamics

Veronica Soebarto

There are a number of questions that concern both the building designer or architect and the client during the design process of a building in warm climates. How thermally comfortable will the building be if it is not air conditioned? When will be the most uncomfortable times in the building and for how many days or hours in a year will this occur? How much shading is needed for the windows and what is the best design for the shading devices? How much energy will be used and how much will it cost to operate the building? Will the space receive enough natural light and can, therefore, the energy use for lighting be minimized? Some of these questions can be answered by using published design advice, rules of thumb or by applying first principles.

Simulation, however, can provide more accurate answers in a relatively shorter time as it takes into account the dynamic nature of the microclimate surrounding the building, as well as the building operations.

Rationale for using simulation

With the increasing use of computers in building design offices, the role of simulation in helping to design comfortable and low energy buildings is expected to rise. There are a number of reasons why simulation is essential in the design process. First, it can be used for design advice. Second, it can help in testing design alternatives or solution sets in the preliminary stage. Third, it can assist in fine-tuning the design, including selection of materials, by conducting parametric runs in the design development or at later stages. It can also be used as a verification tool in the assessment stage (for building rating or approval) of a new design, as a diagnostic tool for an existing building or as a testing tool during retrofit designing.

Simulation as design advice

During the design process, simulation can be used as a design instrument in considering various design strategies such as the building form, orientation, materials, shading strategies, daylighting, ventilation, heating and cooling systems. The benefit, compared to reference books, of using simulation is that it can provide dynamic advice for specific questions or problems that the designer is dealing with. For example, a designer may want to find out the impact of different solar orientations and a shading device on the heat gain and temperatures in the space. By using a simple rectangular form, or a sample building that might be supplied with the simulation program used, and placing it on the actual project location, one can simulate the building with different solar orientations and quickly see the impact of each orientation. Using the same building, the impact of a shading device can also be tested. Although the device may be different from what the actual building will have, the designer can have some idea of how much a shading device can reduce heat gain during the day. Simulation then becomes advice for 'live' design even before the actual design is formed.

Simulation as a testing tool

Once the preliminary design has been formed, simulation can be used to test the impact of design alternatives or solution sets either on the internal comfort (for example, the number of hours in a year when the building will be too hot) or on energy use and total costs. The designer can then rank the solution sets according to the magnitude of their impact. Figure 2.5 shows an example of ranking solution sets using the Energy-10 program from Sustainable Buildings Industry Councils and National Renewable Energy Laboratory in the US. This type of analysis can help the designer to make decisions on which solutions will be applied in the project.

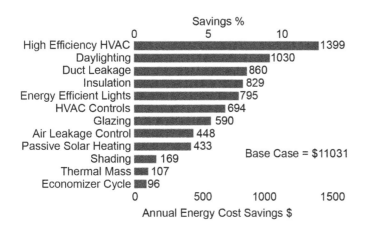

Source: Veronica Soebarto

2.5 Simulation with Energy-10 used to rank energy efficient strategies. Costs are calculated in US$

Simulation as a fine-tuning tool

Simulation can also be used to 'fine-tune' the design. From the preliminary runs the designer may find, for example, that applying horizontal overhangs and roof insulation would be the most effective solution. The designer can then find out how wide the overhangs should be, or how thick the insulation, by conducting 'parametric runs' – that is, by varying the width of the overhangs or the thickness of the insulation systematically for each run until the best results are achieved. One can also do parametric runs to find the most appropriate materials for the walls, roof, floor, etc. by varying their respective thermal properties and costs on each run.

Simulation as a verification/assessment tool

At the end of the design process, simulation can be used to assess the overall performance of the building to meet certain requirements or codes in order to

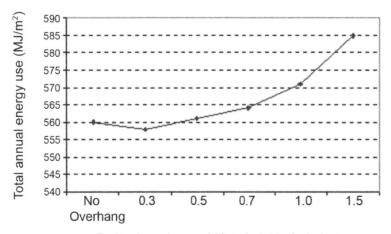

Source: Veronica Soebarto

2.6 The impact of varying overhang width on total energy use is shown in a house in Adelaide, South Australia. Increasing the width of the overhang will reduce energy for space cooling, but will increase the need for space heating, resulting in increasing total energy use, whereas using no overhang will increase space cooling energy more than it reduces space heating. In this case, having an overhang width of 0.3 times the height of the window is shown to be the best solution

obtain building approval. In Australia, for example, since 2002 new energy efficiency provisions have been introduced in the Building Code of Australia (BCA), requiring new residential buildings to achieve certain performance. One way to do this is by using a simulation tool recommended by the code to demonstrate that the building design achieves at least the minimum number of 'stars' required.

Simulation as a diagnostic tool

Simulation can also be used to diagnose the cause(s) of existing problems related to thermal comfort and energy use in an existing building. Most thermal simulation programs predict and show the user the energy use from elements of the building such as heating, cooling, ventilation, lighting, equipment and domestic hot water. In addition, the magnitude of the heat gains and losses through walls, roof, floor, windows, lights, equipment, ventilation, infiltration and people is given. By analysing these, one can diagnose the parts of the building that are making the major contribution to the problems. Without conducting a simulation, the retrofit strategies proposed may not be the right answer for the problem and therefore the retrofit project can become a wasteful exercise. For example, to make the space more comfortable one may decide just to add more roof insulation, only to find out later that it still does not improve the space comfort. By using simulation to analyse the heat gain distribution, one may find that the windows are, instead, the major source of

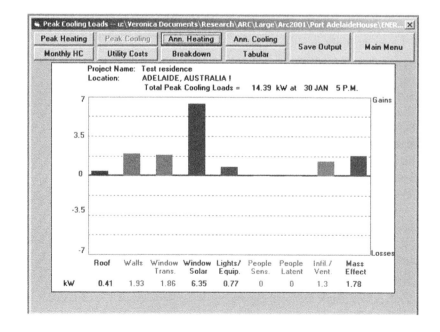

Source: Veronica Soebarto

2.7 Distribution of annual cooling load of an existing house in Adelaide, South Australia, simulated with ENER-WIN (from Texas A&M University, US): in this example, the window is the main source of heat gain

heat gain, followed by the roof. Therefore, the right step is to reduce the heat gain from the windows (for example, by installing external or internal shading devices or simply planting more trees); only if necessary will more roof insulation be added later. This way, the retrofit solutions are targeted to solve the actual problem rather than being decided arbitrarily.

Simulation programs for warm climates

Because each design stage has its own purposes and constraints, one may opt to use a variety of simulation programs for the different stages. Although this is possible, however, it is not particularly practical. It is also more economical if the program to be used is suitable for all stages of the design process. Some are intended for large buildings and are to be used by architects, as well as mechanical engineers, to accurately size mechanical heating and cooling equipment and to predict whole-building energy consumption; others are oriented to building designers to assist them in the design process. The former programs usually give very accurate and detailed results, but at the same time require very detailed input and therefore are not suitable for early design processes when not all the information about the building is available. The latter programs usually have more designer-friendly interfaces, require less complex input, present the results graphically (which makes it easier for building designers to understand), but may not be as powerful and accurate as the former models. The question is: how does one select which simulation program to be used for all stages of designing a building in warm climates?

The following section discusses the required capabilities of any simulation program if it is to be used in these climates and presents three software packages – DOE-2, Ecotect and ENER-WIN – to illustrate this issue. Some points about how to use simulation tools in the design process will also be discussed.

Program capabilities

Simulation programs for designing buildings in warm climates must have the capability to simulate, with an acceptable accuracy, the following design parameters: solar orientation; building forms; numbers of floors and spaces (more than one); roof pitch; roof, wall and floor materials (and their thermal properties); insulation; glazing (type, size and location on the wall); shading devices; shading by other structures; thermal mass; natural ventilation; and natural (day) light. The program should also allow the designer or user to specify a variety of use schedules or operations for the space, shading devices, windows, lighting and other electrical equipment that is likely to be applied in the building. A capability for simulating active systems (such as solar hot water systems) would also be advantageous, although not essential as these can be simulated separately.

The user must have access to several types of simulation outputs, which should include: hourly temperature of every zone or space in the building; some figures of when the building or certain spaces will be comfortable or uncomfortable; load breakdowns or some figures that show the distribution of heat gains (or losses) through the building elements (for example, roof, walls, windows and floor); and total energy use, as well as energy use breakdowns. Another factor to be considered is the ability of the program to perform a life-cycle cost analysis. This is about the impact of the design solutions to be tested on the initial, maintenance and operating costs. The cost factor is what clients are generally most interested in since as it gives a realistic measure of what the clients will have to experience (some clients may have difficulty in understanding what certain kilojoules or kilowatt hours of energy use actually mean).

To a certain extent, most simulation programs have these capabilities. Some programs, however, limit the number of spaces or zones that can be simulated (for example, Energy-10 from Sustainable Buildings Industry Council in the US). Some cannot simulate sloping roof and complex three-dimensional (3D) geometry, although they are intended to be used in early design processes (such as Building Design Advisor from Lawrence Berkeley National Laboratory in the US). Some programs are intended to specifically simulate airflow and natural ventilation (for example, LESOCOOL from Solar Energy and Building Physics Research Laboratory in Switzerland) but cannot perform whole-building simulation or test all other design parameters, as suggested above. Some programs have fixed or limited operation schedules and therefore are not able to simulate a naturally ventilated building where the windows may be shaded, opened or closed differently during various times of the day or year. Some programs use less complex algorithms or simulation models (instead of detailed or hourly simulation), and even though they are intended to analyse passive solar design and natural cooling strategies (for example, Builder Guide from National Renewable Energy Laboratory in the US), they are not suitable for detail design of shading devices since they do not take into account the dynamics of solar paths. Simplified methods were traditionally adopted in programs to be used in the early design process to reduce computing time. However, computer speeds are no longer an issue today and therefore one should opt to use programs with detailed hourly modelling methods.

Program accuracy

Besides the program capabilities, an important issue is program accuracy. In simulating a new design, the designer or user may not be totally sure whether or not the simulation results are accurate because the building has not yet been built and there is nothing to compare the results with. A way of testing the accuracy of the program is to use it to simulate an existing building and to gather measured or monitored data for the building to be used as a reference point for the simulation results. These measured data may include hourly indoor temperature and humidity, and hourly (if possible) or monthly energy use (for example, gas and electricity) or even energy use breakdowns (such as. electricity use for lighting and equipment). One can then 'trust' the simulation program if the simulation results match the measured data closely.

In a retrofit project of an existing building, the calibrated model can then be used to further analyse the design and to propose retrofit solutions, as previously discussed. A calibration procedure is essential in simulating existing buildings and will be discussed further in the section on 'Building monitoring'.

There are established methods to test the accuracy of a simulation program. These include IEA BESTEST (Building Energy Simulation Test), which evaluates the ability of detailed hourly simulation programs to adequately model the envelope dynamics of buildings, and HERS BESTEST, which evaluates detailed and simplified programs with an emphasis on modelling houses. HVAC BESTEST is a procedure to test the ability of whole-building simulation programs to model space conditioning equipment.

Six examples of software for thermal simulation

First, the high end programs such as DOE-2 (from Lawrence Berkeley National Laboratory in the US) is one of the oldest and most powerful simulation programs (others include BLAST from the University of Illinois, US, TRYNSYS from the University of Wisconsin, US, and ESP-r from the University of Strathclyde, UK). It is capable of simulating all of the design parameters mentioned above for buildings in warm climates. It has also been shown to provide results comparable with measured data.

The VisualDOE program (from Architectural Energy Corporation in California, the US), with its graphic-oriented user interface program, allows the user to simulate a building with DOE-2 in much shorter time than using DOE-2 (which is a text-based program) alone. VisualDOE, or any DOE-2-based program, however, requires very detailed input and is therefore not suitable for the preliminary design stages. It may be more appropriate to use this program in the verification or evaluation stage, when most of the information about the building has become available.

Second, the EnergyPlus program is a more recently developed building energy simulation program. The program is based on the most popular features and capabilities of DOE-2 and BLAST; however, it includes many innovative simulation capabilities that were not included in the two programs. Both DOE-2 and EnergyPlus are text-based simulation programs. They are stand-alone programs without a graphical interface. There are a number of user interface programs that can be used to create and run DOE-2 or EnergyPlus input files and display simulation results graphically. These include programs such as DesignBuilder, EPlusInterface, Green Building Studio, VisualDOE, DrawBDL, and others.

(a)

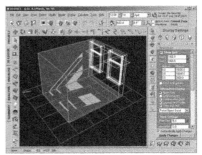

(b)

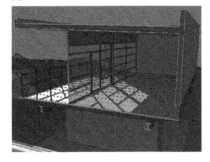

Source: Ecotect, www.squ1.com/site.html

2.8 Using Ecotect for (a) testing the shading device design and (b) analysing solar access

Third, the Ecotect program (from Cardiff University in the UK) is one of the few tools suitable for all stages of the design process that has its own 3D modelling interface. It is also equipped with extensive solar, thermal, lighting, acoustics and cost-analysis functions. The program uses the heat balance method to perform hourly energy calculations. It allows modelling of complex geometry and any types of building materials, insulation and glazing. Furthermore, it can be used to accurately design or test the design of shading devices. It also allows for predictions of the space temperature in a naturally ventilated building to see just how frequently, and for how long, an occupant is likely to be uncomfortable in any of its spaces.

As a minimum requirement, all the user needs to do to utilize this program is to draw the building geometry and assign materials (which can be picked from the program database) to the building envelope, and to specify the location of the building. The simulation results are presented in various forms, including the spatial distribution of both mean radiant temperatures within a model and predicted comfort levels, indoor hourly temperature for any day of the year, heating and cooling loads, load distribution, and discomfort times in a non-air-conditioned mode.

Fourth, the ENER-WIN program (from Texas A&M University in the US) performs hourly whole-building energy analysis for up to 98 zones in a one- or multiple-floor building by calculating transient heat flows, daylighting, energy consumption, demand charges, life-cycle costs and temperatures in unconditioned spaces. The program is supported with an hourly weather data generator for more than 1000 cities worldwide, although users can also enter their own hourly weather data. The initial inputs required are defining the building type (which will then retrieve numerous default values from the database), the building location and the building geometry, which is input through a sketching interface for drawing the building's floor plans. A number of outputs are produced, including hourly space temperature and humidity; hourly and monthly breakdown of energy use (space heating, cooling, ventilation, lighting and equipment); annual and peak loads; monthly energy costs; energy and cost savings by using daylight; and an estimation of heating and cooling equipment sizes. Finally, the life-cycle cost of the building design or elements is also reported. Similar to Ecotect, for naturally ventilated building simulation, this program will report the number of degree hours for which the spaces would be comfortable or uncomfortable, with the comfort range defined by the user.

Finally, the TRNSYS program (from the University of Wisconsin in the US) is a transient simulation program with a modular structure. It is able to simulate the hourly thermal conditions and energy use of multi-zone buildings and their equipment. TRNSYS is a user-friendly design tool and it allows the user to apply it during any stage of the design process. The TRNSYS library prepares many components required in a simulation, such as sophisticated building models, routines to handle input of weather data, time-dependent forcing functions and

(a)

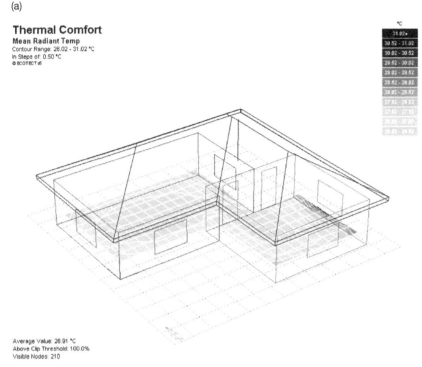

Thermal Comfort
Mean Radiant Temp
Contour Range: 26.02 - 31.02 °C
In Steps of: 0.50 °C
© ECOTECT v5

°C
31.02+
30.52 - 31.02
30.02 - 30.52
29.52 - 30.02
29.02 - 29.52
28.52 - 29.02
28.02 - 28.52
27.52 - 28.02
27.02 - 27.52
26.52 - 27.02
26.02 - 26.52

Average Value: 26.91 °C
Above Clip Threshold: 100.0%
Visible Nodes: 210

Source: Ecotect, www.squ1.com/site.html

2.9 (a) Thermal comfort analysis showing the hot and cold areas in a house and (b) output report on heating and cooling loads during a year

(b)

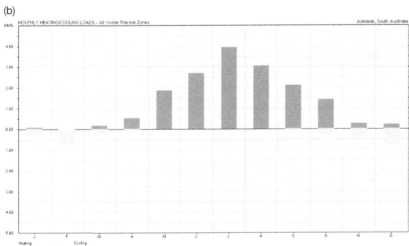

output modules of simulation results. In building design for warm climates, natural ventilation is an important factor for cooling solution sets. By coupling TRNSYS with COMIS, the effect of cross-ventilation with opening windows can be evaluated. In the thermal building model, TRNSYS, the airflow between the rooms and from outside are defined as input values. However, in natural ventilation, these values depend upon external wind condition and indoor and outdoor temperature conditions. Therefore, it is necessary to link the multi-zone airflow model COMIS with the thermal model TRNSYS (see Figure 6.5).

The COMIS program is a multi-zone airflow simulation program. It has been developed within the framework of international practice and verified by many experts. The COMIS program is able to simulate the hourly airflow conditions of each room and therefore allows the user to obtain effective information on cooling potential with cross-ventilation and the design of openings on envelopes and inner walls. This coupling model may require detailed input data related to airflow modelling; but it is desirable for use in the preliminary design stages.

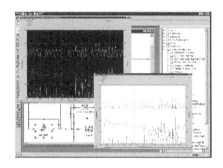

Source: The Transient Energy System Simulation Tool, www.trnsys.com

2.10 Online plotter on TRNSYS

The program has been used to analyse the thermal performance of a number of naturally ventilated houses in Australia, where the simulated indoor temperature has been calibrated to measured data, i.e. actual monitored data from 'real' conditions in the homes.

Notes about using simulation tools in design

Before using the tool, the user must be familiar with the program and understand the types of input required by the program. The programs Ecotect and ENER-WIN require minimal input when used in a preliminary design stage, but will require more detailed data when more detailed and accurate simulation is required. Often, the most time-consuming task is to gather all of the necessary input.

Regardless of the simulation program used, the user should be aware that any program should perform the simulation based on the input that is given by the user. The accuracy of the output depends upon the accuracy of the input. The user must therefore carefully check the input data. Frequently, a mistake made is as simple as omitting a decimal point in a value entered into the program, which can have a fatal result. Using common sense is therefore necessary when analysing the results presented. The user must also not be distracted by results presented in graphical or 3D formats. For example, in checking the results of solar access or shading analysis, one should remember that solar radiation always comes from a certain direction for

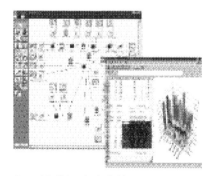

Source: http://international.cstb.fr/

2.11 IISibat assembly panel on COMIS from CSTB

a certain time and location and the shadow presented must therefore make sense.

It is also worth remembering that simulation can never replace reality. Because of the complexity in simulating reality, any simulation program employs some assumptions. Therefore, some knowledge of building physics or thermal performance is critical in understanding what to enter in the program and how to interpret the simulation results.

Information

A thorough listing of simulation programs with links to their individual websites is provided by the Building Technologies Programme of the US Department of Energy and can be found at www.eere.energy.gov/buildings/tools_directory/

Computer simulation – fluid dynamics

Valerio Calderaro

Models based on computational fluid dynamics (CFD) are used to estimate the effects of wind action in the external environment and stack effect in the internal environment. The term CFD generally refers to numerical solution of the partial differential equations governing a flow field, such that the velocities and temperatures at all points in the field are predicted. In ventilation studies, the term usually refers to numerical solution of the Reynolds-averaged Navier-Stokes equations, which means that for bioclimatic conditions, such as turbulent flow, average velocities and temperatures are predicted.

In essence, numerical analysis consists of using differential equations to describe the form for each element in a model, applying these equations to a large number of discrete elements and finding a solution that satisfies the boundary conditions by iterative procedures. In reality, the process is very difficult and a wide variety of mathematical procedures has been developed. Usually the equations are described in what is known as finite volume form; but finite element form is also used.

There is no fundamental difference between steady and unsteady flows; the finite difference form of the unsteady terms is obtained in a similar way as the other terms in the equations. To some extent, the time dimension can be viewed as an extra space dimension. However, the choice of time step for unsteady flows is an additional item of importance because of the need for stability. Moreover, the specification of initial boundary conditions requires that the complete velocity field and other relevant parameters are specified. The computation then proceeds in the designated time steps, producing a converged solution at each step before moving on to the next one.

An important preliminary to the computation is the choice of computational grid or mesh. In the early days simple Cartesian grids were employed; but these have been superseded by grids that increase the efficiency of the numerical analysis – that is, the grid is made finer where this is needed. Body-fitted coordinates (BFCs) offer such improvements, based on the geometry of the surfaces. The so-called adaptive grid takes into account other computational factors to optimize the grid more efficiently. Clearly, a necessary condition for a solution to be valid is that it should be reasonably insensitive to the form of the grid.

In some respects, the use of CFD for external flows is easier than for internal flows – for instance, because of the absence of buoyancy. On the other hand, the geometries of groups of buildings are inherently much more complex; furthermore, there is likely to be a need for a larger number of cells in the grid. This follows from the need to cover the large dimensions of buildings adequately while giving adequate coverage of the relatively thin boundary layers on the surfaces. For simulations, the Fluent model has the best characteristics of high precision, complete flexibility of the mesh and the ability to resolve complex geometries.

Fluent is a CFD model for predicting fluid flow, heat transfer, mass transfer, chemical reactions and related phenomena by solving mathematical equations that represent physical laws, using a numerical process. Such models are based on the finite volume method in which the space being studied is split into a finite set of control volumes or cells. The equations that are solved simultaneously to estimate the flow field take account of the conservation of mass, momentum, energy and so on.

Application of Fluent was used as part of a study for the EcoCity program (European Union, www.ecocity-project.eu/). The a new sustainable housing development was proposed for the city of Umbertide (see Case study 3.1 in Chapter 3), The development and the Umbertide city context was used to create a Fluent model. This model has been used to optimize ventilation characteristics in the proposed development. With this model, satisfactory ventilation conditions for the external environment at the urban level, and for the interior at the building level, have been achieved. The approach to energy use aims to reduce energy demand and environmental impact and to improve the indoor–outdoor comfort level.

The bioclimatic methodology for ventilation in the urban area includes several steps:

- analysis of the wind conditions at various seasons by a fluid dynamic simulation model;
- fluid dynamic simulation models of the wind conditions in the area for various possible building aggregations and obstruction/canalization of the wind (trees, ventilation corridors, etc.) in order to improve the external

Source: Valerio Calderaro

2.12 Fluent simulation of area A of the proposed sustainable housing development located in Umbertide

ventilation conditions over the seasons (protection from cold winds in the winter and increase of wind penetration for cooling in the summer); and

- comparative evaluation of the energy performance and environmental comfort in the solutions adopted and verification of their correspondence to dynamic simulation models.

In order to improve the microclimatic conditions in the two main seasonal conditions in the project area, the following solutions have been created:

- In the winter, the wind comes mostly from the north; a shelter belt of trees is placed on the north part of the site to protect areas to the south.
- In the summer, the wind comes mostly from the south; the orientation of the buildings is to the south, which promotes solar access in the winter. Courtyards and the provision of wind corridors are provided to allow for an adequate degree of penetration of the wind.
- Provision of a green zone is provided for evaporative cooling; this comprises a small lake and streams which, combined with transpiration from the plants to improve the conditions of the local microclimate, have a positive influence on reducing the energy requirements of the buildings.

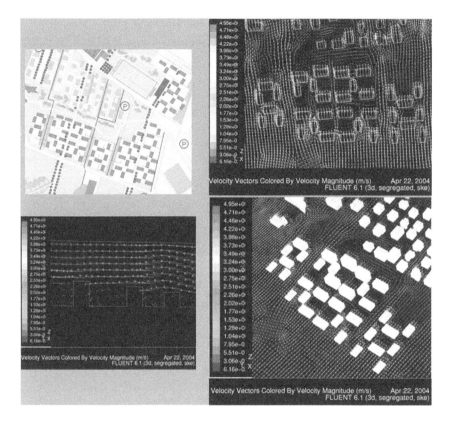

Source: Valerio Calderaro

2.13 Fluent simulation of area B of the proposed sustainable housing development located in Umbertide

Some simulations with Fluent have been done to estimate the effect of the wind in order to improve ventilation conditions during summertime. For the condition of a south wind with a speed of 5 metres per second, the main wind flows have been estimated for two zones, with respect to the values obtained for each zone. The images, taken from the Fluent simulation results, are relative to the main aggregations of the areas A and B and to the following reference conditions (see Figures 2.12 and 2.13):

- horizontal sections to a height of 1.2m from the ground (typical height for a person);
- vertical sections corresponding to the most significant aspects of the buildings;
- total volume; and
- a colour scale of speed values in metres per second, or pressure in Pascals (Pa), on the left of each image.

In this case, consideration is given to two main building dispositions that are typical for the design and for which speed and pressure profiles have been determined. These simulations have identified the areas that are not sufficiently

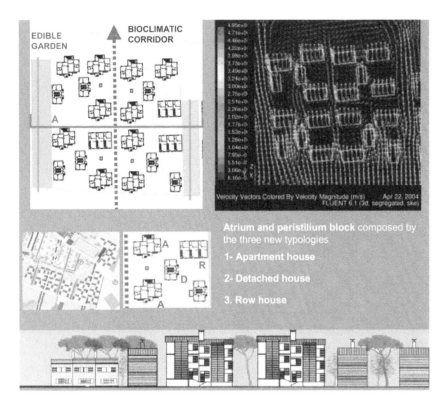

Source: Valerio Calderaro

2.14 Bioclimatic corridors, courtyards and building typology

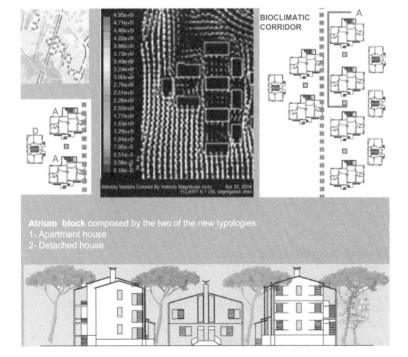

Source: Valerio Calderaro

2.15 Simulation of the typical areas of the building design

ventilated. As a result, modifications are made to the disposition of the buildings with the aim of improving the conditions of ventilation in the two areas. The results of these adjustments are shown in Figure 2.14. Axonometric views, as well as horizontal and vertical sections, are used to indicate air movement. Here, air speed is examined at several points and is represented by various coloured arrows. The air speed values for the various colours are shown on the vertical axis. Where the arrows are more numerous, this indicates a greater movement of the air, and where the arrows are longer this indicates a greater speed.

The use of both wind corridors and courtyards has considerably improved external conditions. Wind corridors funnel summer dominant winds because they are parallel to their direction. The presence of the corridor produces a turbulence effect in a lateral direction, as well as a depression effect, with consequent entrainment of the surrounding air. The open courtyard system, combined with interference between the wakes of the various buildings, produces particularly favourable summer comfort conditions (see Figure 2.15).

Interventions at the building level are:

- the use of a cooling system for ventilation through a stack effect, integrated with a convective loop system that is combined with a chimney to obtain an effective summertime system of cooling and ventilation; and
- a good microclimate and adequate interior conditions.

The system is based on the use of three ventilation chimneys: two for air admission and one for expulsion. It comprises the following elements:

- The first (outermost) chimney, for solar energy capture, is composed of a double, external, low emissivity glazing, a hollow space (chimney) and a wall that works as a thermal storage.
- The second chimney has a wall in common with the previous one and delimits a hollow space with another inner wall.
- The third chimney is inside the building and has the function of expelling the air to the outside.

In summer, during the night, because of the lower external air temperature, air cooling of the chimneys and building structures occurs through stack effect. During the day, external air enters from the cellars and passes through the central (second) chimney (both cooled during the night), cools the rooms and is expelled from the external (third) chimney (inside the building). This chimney has a solar collector system that increases the air temperature in this section, increasing the rate at which the air is expelled to the outside.

In winter, the transparent part of the chimneys of the south façade (protected with shielding elements from solar radiation in summertime) generates a convective loop, because of solar radiation, and transfers heat to the rooms. Heat is transferred from the thermal mass of the inner chimneys to the air. The heat accumulated in the mass of the chimneys serves to heat and ventilate rooms during the night. In this way, apart from controlling the air temperature, the system also controls air change in all seasonal conditions. Air movement is represented in the modelling by use of horizontal and vertical sections of the building. In particular, sections are chosen that correspond with the ventilation chimney, where a significant increase of air velocity is seen.

From the results of the simulations, it is possible to determine the courtyard shape and to design the optimum disposition of the single buildings in order to create sufficient and comfortable ventilation for both external and internal areas. The buildings' external environment is represented by axonometric views, as well as by horizontal and vertical sections, in order to demonstrate air movement.

Further information

See Fluent Computer Simulation Applications at www.fluent.com/ (accessed 26 August 2005).

Building monitoring

Veronica Soebarto

Monitoring building performance is the only way of finding out whether or not the building has performed as intended, expected or simulated. The monitoring results will comprise very useful information for designers to use in their next projects. Monitoring also helps to detect the source(s) of problems affecting the thermal and energy performance of a building and therefore should be the first step to be conducted before any retrofit design strategies can be proposed and implemented. Monitored data can also be used to 'calibrate' the simulation model to ensure that the model closely represents the actual building. Monitoring also serves several purposes in building science research, including helping to understand certain phenomena of building elements, materials or construction, and providing data to test and validate certain theories regarding those phenomena (see Figures 2.16 and 2.17).

It is, however, still not a common practice for many design offices to conduct monitoring of the building that they have just completed. One reason is that verifying the actual performance of the building (which may lead to fixing the problems) is not part of the commissioned work. In any country that requires energy rating of a new design, monitoring is not a mandatory process either. The rating is conducted at the design verification stage only, and many of the assumptions in the rating tools are based on generalized conditions that often do not reflect the actual circumstances of the project building. For the occupants, this can result in a surprise when a high-star rated house uses much more energy than predicted or when it was rated. It

Source: Veronica Soebarto

2.16 One month of ENER-WIN simulated results and measured temperatures of the living room in T-house (architect: Max Pritchard)

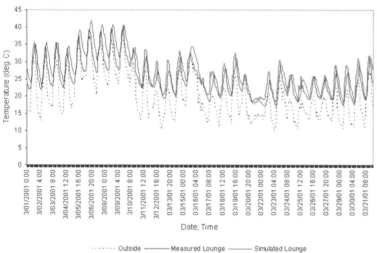

is also ironic that most of the so-called environmental design awards are usually given to buildings that have just been built and have only had a few months of occupancy. No actual performance is reported and the awards are given based on the designer's description of the building design through text, photographs and, sometimes, predictions made through simulation.

A common reason for not conducting building monitoring is the costs of the equipment and labour, or lack of information about them. This section attempts to provide basic information on the types of monitoring that can be conducted by any designers, design offices or clients, the monitoring equipment needed, suitable monitoring periods, and where to install the monitoring equipment. A few examples of the use of monitoring for design evaluation and calibration of a simulation model will also be presented.

Source: Veronica Soebarto

2.17 One month of ENER-WIN simulation results and measured temperatures of the bedroom in GM-house (architect: Glenn Murcutt)

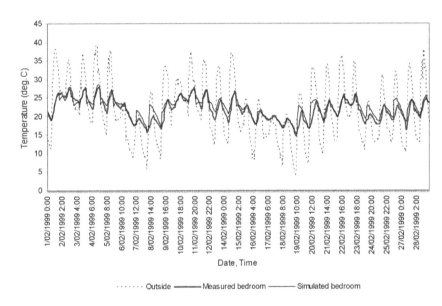

Type of monitoring and monitoring equipment

There are a number of monitoring types that can be conducted to assess the performance of a building. They are monitoring of:

- the indoor environment (temperature, humidity and light level);
- the outdoor environment (temperature, humidity, solar radiation, wind speed and direction);

- total energy use (electricity and gas); and
- breakdown of electricity use (that is, electricity use for lighting and individual electrical equipment).

Each of these can be monitored according to various time intervals (for example, every minute, five minutes, ten minutes, and so on), although the most common practice is to monitor either at 30-minute intervals or hourly. The data monitored are stored in the monitoring equipment (or data logger) for a period of time depending upon the capacity of the memory chip of the data logger. Some monitoring equipment allows the data to be downloaded remotely to a computer using a modem for access; other equipment requires the data to be downloaded directly from the data logger. Some equipment models can only record one or two measurements (for example, temperature only, temperature and humidity only, or electricity use only), while some have the ability to record several measurements at the same time. Obviously, the latter type costs more than the former. Electricity- or gas-use monitoring will also require the building electrician to install the measuring devices, such as current transducers (CTs) or power transducers (PTs), and therefore will incur higher labour and insurance costs. This type of monitoring is likely to be conducted in large buildings or for research only.

This is not to say that monitoring of small or residential buildings is not possible. Indoor monitoring can be done with simple data loggers that have recently become affordable and are easily obtained (information on these is

Source: Veronica Soebarto

2.18 A weather station installed on a roof, ensuring that it is not shaded by any trees or other structure

available on numerous websites). These portable data loggers are usually small and non-intrusive, making it possible to obtain the occupants' permission to monitor (and even their involvement in monitoring) the indoor performance. One disadvantage of this type of loggers, however, is they do not have the facility of downloading the data remotely.

Outdoor environmental monitoring can be conducted with individual data loggers to monitor the external temperature, humidity, solar radiation and wind speed, or with all the loggers assembled together to form a 'weather station'. More expensive weather stations will provide more accurate data; but there are options for affordable weather stations for small projects, which may not be as accurate but will still provide acceptable data.

The decision on which monitoring equipment should be used obviously depends upon the funds available. If the monitoring is expected to yield large energy-cost savings, then more expensive equipment is justified. If the purpose is only to assess indoor thermal comfort, then simple and portable data loggers should be sufficient.

Where to put the monitoring equipment?

The best location for the equipment or data loggers to monitor indoor air temperature and humidity (as well as light level) is in the middle of the room at the height of an average person or slightly above it. The problem is that this is usually not possible, given the fact that it can be intrusive for the occupants (and it may invite the curiosity of the occupants to touch or play with it, which will affect the accuracy of the data recorded). The next option is to attach it to an adjacent wall as long as the equipment will not receive direct solar radiation or heat from any other heat source, such as the heat from a cooking appliance, computer and the like.

The best location for external data loggers or a weather station is in an open space, not shaded by any trees or other structures. If this is not possible, then the next best option is to install the weather station on the roof, although accessibility may be a problem. One important point is that any device to monitor temperature should be shielded from direct solar radiation.

How long should the monitoring be conducted?

The answer to this question depends upon the purpose of the monitoring. If the purpose is to assess the thermal or comfort and energy performance, then the monitoring should cover at least different periods of the year (for example, dry and wet seasons, or winter and summer), although obviously the longer the monitoring, the better. If the purpose is to calibrate a simulation model, then the monitoring can be conducted over a shorter period. Two to four weeks in summer and two to four weeks in winter are generally considered to be sufficient.

(a)

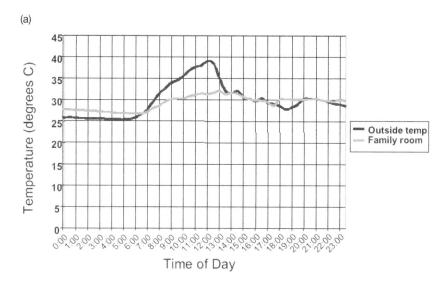

Source: Richard Hyde

2.19 Examples and results of monitoring conducted inside and outside a house in order to assess the performance of the design. The monitored data was also used to calibrate the simulation model, and the comparison of the measured data and simulated results for a month is presented

(b)

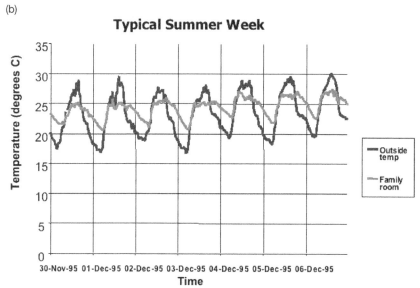

In summary, monitoring is used to understanding the thermal performance of buildings. It enables design professionals to carry out post-occupancy studies of buildings and can be linked to other tools, such as user feedback questionnaires, to help validate data gained from the monitoring process.

Monitoring information is normally used for testing prototype buildings in order to assess new technologies and design concepts. For example, in the Healthy Home project, data for average summer conditions were used to test performance, but also to test the house model in extreme conditions.

Case study 2.1 Bandaragama House, Colombo, Sri Lanka

Indrika Rajapaksha and Upendra Rajapaksha

2.20 Front elevation of Bandaragama House, Colombo, Sri Lanka

PROFILE

Table 2.3 Profile of Bandaragama House, Colombo, Sri Lanka

Country	Sri Lanka
City	Colombo
Building type	Detached
Year of construction	2001
Project name	Bandaragama House: a passive residence
Architect	Upendra Rajapaksha and Partner

Source: Indrika Rajapaksha and Upendra Rajapaksha

PORTRAIT

Bandaragama House sets an innovative design trend for environmentally sustainable architecture in Sri Lanka. The design explores the use of bioclimatic principles to create an architecture with a local identity. In this respect, the design theme is centred towards an exploration of architectural concepts and principles of the traditional courtyard built forms of the country as a source of deriving passive design strategies, which concentrate on reducing cooling loads. The aim of the study is to examine the adopted design strategies and to assess their effectiveness for thermal comfort in warm humid tropics. The methodology is based on the thermal performance results of a field investigation and simulated airflow patterns using a computational fluid dynamic analysis.

CLIMATE

Warm, humid tropical climates are found in the region extending 15° north and south of the equator. Sri Lanka (latitude 5°55' to 9°49'N and longitude 79°51' to 81°51'E) is an example of this climate and has a lack of seasonal variations in temperature (see Figures 2.21 and 2.22). The mean monthly temperatures range from 27°C in November to 30°C in April; relative humidity varies from 70 to 80 per cent during a typical year. The daily maximum temperatures are high as 25 to 33°C and the daily pattern in the dry season (September to November and March to May) has a diurnal temperature range of 7 to 8°C.

The work of Humphreys (1978), and its continuation by Auliciems (1997), gives the neutrality temperature as a function of monthly mean outdoor temperature. The neutrality temperatures for Colombo are within 27.3°C and

(a)

Source: Indrika Rajapaksha and Upendra Rajapaksha

2.21 (a) Location of Sri Lanka; (b) graph showing the neutrality temperature (Tn) in relation to minimum and maximum outside temperatures (To(max) and To(min))

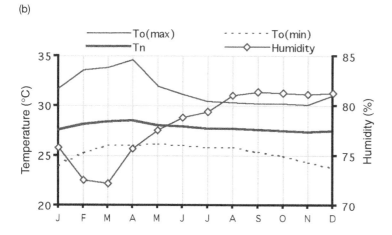

(b)

28.5°C. A closer look at the ambient climate conditions during February May in Colombo reveals that daytime temperatures remain above the extended upper limits of dry bulb temperature for most parts of the daytime hours (see Figures 2.21 and 2.22). Therefore, a strategy to reduce daytime maximum indoor air temperature, in addition to the provision of indoor airflow, is required.

PASSIVE CLIMATE MODIFICATION STRATEGIES

The main characteristics of warm humid climates, from the human comfort and building design viewpoint, is the combination of high temperature and high humidity, which, in turn, reduces the dissipation of body surplus heat. Givoni's bioclimatic chart, aimed at predicting the indoor thermal conditions of the buildings from outdoor weather conditions, recommends increased air movement to provide indoor thermal comfort. It indicates that high mass envelope, coupled with nocturnal ventilation, can restore indoor thermal comfort for February, March, April and May: the warmest period of the year (see Figure 2.22).

(a)

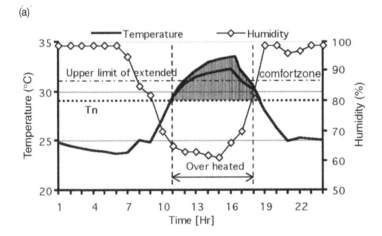

(b)

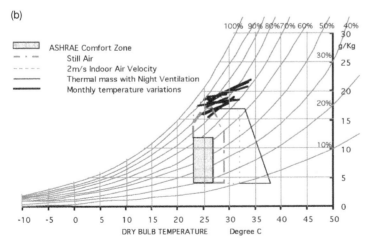

Source: Indrika Rajapaksha and Upendra Rajapaksha

2.22 (a) Daily pattern of relative humidity and dry bulb temperature in relation to Colombo, Sri Lanka (during the overheated month of April) (b) Givoni's bioclimatic chart applied to Colombo, Sri Lanka (6.5°N and 15m above sea)

PASSIVE DESIGN STRATEGIES OF THE TRADITIONAL COURTYARD BUILT FORM

A typical courtyard built form in Sri Lanka attempts to enhance wind-induced cross-ventilation and functions as an air funnel promoting maximum air circulation to the interiors. The guiding principle is to enhance cross-ventilation through a series of openings from the entrance door through the central courtyard and out of openings in the building fabric at the leeward side. The entrance veranda on the windward side acts as a wind tunnel, focusing the incident wind into the courtyard that lies on this air funnel, which, in turn, ventilates the indoor spaces surrounding the courtyard (see Chapter 9).

Furthermore, to avoid the heat of the wind-induced ventilation from reaching the interior, the perimeter openings are protected with heavy shade, in most cases achieved using wide eaves and other intermediate spaces such as verandas.

In tropical regions, the highest intensity of solar radiation falls on the roof and west walls. For ventilation efficiency reasons, the layout of tropical courtyard buildings is usually of a spread-out form, which then sets out a larger roof-to-volume ratio. A consequence of this is the exposure of the building, and particularly the internal courtyard, to a higher intensity of solar radiation for a greater part of the day.

The traditional courtyard built form presents some solutions to minimize the direct fall of solar radiation to the internal spaces around the courtyard. In most cases, a veranda – a place of transition from inside to the courtyard – is found along the courtyard perimeter. A spread-out plan form with internal courtyards enables better integration of transition spaces in the building composition. The floor of the internal courtyard, unbearably hot during most of the daytime owing to the high-altitude sun in the tropical region, is usually not used for any human activities. The height of the courtyard is in proportion to the human scale and is thus shallow, while demarcated by the boundaries of the roof that drain into the courtyard. In most cases, the plan form of the courtyard is either square or rectangular. The traditional courtyard has been adapted for this building (see Figure 2.23).

CFD simulation studies have been used to test the effectiveness of the adapted courtyard in this building (see Figure 2.24).

THERMAL PERFORMANCE

Courtyard thermal behaviour

The hourly air temperatures at the body level within the courtyard were monitored and compared with the ambient values (see Figure 2.25). The temperature pattern within the courtyard has followed the ambient pattern. Maximum air temperature within the courtyard was 2°C below the ambient maximum by 3 pm, two hours later than ambient maximum was recorded. The exposure of high-mass brick walls to the airflow has resulted in the time lag and flywheel effect. However, during the night, the temperature within the courtyard was 0.5°C to 1°C above the ambient. It was observed at early daytime, between 8 am and 9 am, that courtyard temperatures closely followed the ambient.

(a)

(b)

Source: Indrika Rajapaksha and Upendra Rajapaksha

2.23 (a) View of the central courtyard through the main entrance; (b) main entrance and the central axis through the still water pond; (c) plan and (d) section of the courtyard

(c)

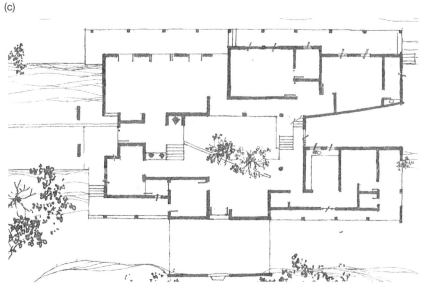

(d)

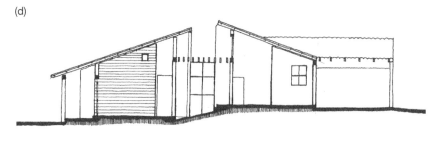

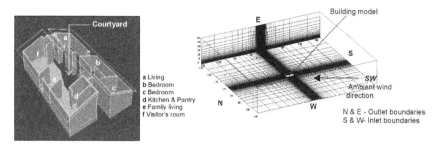

a Living
b Bedroom
c Bedroom
d Kitchen & Pantry
e Family living
f Visitor's room

N & E - Outlet boundaries
S & W- Inlet boundaries

Source: Indrika Rajapaksha and Upendra Rajapaksha

2.24 Simulated building model and surrounding flow field with boundaries of the orthogonal Cartesian grid

Thermal stratification pattern

Vertical thermal profile within the courtyard shows an ascending thermal stratification pattern with lower temperature values at the bottom level and increases through the height (see Figure 2.25). A similar stratification pattern was observed during the day and night. However, during the daytime maximum temperature within the full height of the courtyard remained below that of the maximum ambient.

A density difference due to air temperature is essential for stack effect to take place. It was observed that during the daytime, the temperatures within the courtyard were below the temperature levels at the openings found in the external envelope. Thus, the density levels at the openings were lower than in the courtyard, preventing a possible stack effect during the daytime.

Furthermore, the vertical temperature profile showed a heat gain from the courtyard's top. However, the presence of relatively lower temperatures within the courtyard indicates that the effect of heat gain from the sky opening has been moderated by the upwind forced ventilation promoted along the longitudinal axis (see Figure 2.28 for airflow pattern).

THERMAL MASS

Thermal 'flywheel' effect

The hourly wall surface temperatures of the axis, courtyard and internal walls were compared with external walls at the human body level (see Figure 2.26). All internal walls in each zone were monitored and averaged for a mean value.

Lower surface temperatures were observed for the interior walls than the external wall, as expected in high mass buildings. The temperatures were decreased by 3 to 4.2°C, well below the external values. Among them, the internal wall showed the higher temperature than the courtyard and axis wall. In comparison to the courtyard wall, the axis wall showed higher surface temperatures. This indicates heat absorption and, thus, the availability of incoming air through the axis.

The lowest wall surface temperatures were observed in the courtyard wall with a recorded maximum time lag of four hours. The maximum courtyard wall temperature was 28.8°C. These results indicate better interaction of high mass fabric and airflow in lowering the courtyard's surface temperature.

(a)

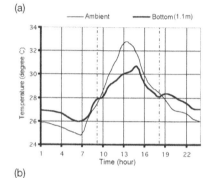

(b)

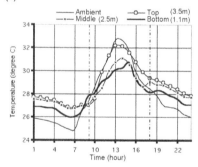

Source: Indrika Rajapaksha and Upendra Rajapaksha

2.25 (a) Air temperature differences for ambient and courtyard at 1.1m in height; (b) vertical temperature profile in the courtyard at 1.1m, 2.5m and 3.5m in height

INDOOR AIR TEMPERATURE

The courtyard directly influences the thermal environment of the internal zone to the west. Thus, the indoor air temperature of this zone was monitored and compared with the courtyard and ambient values to investigate the impact of the courtyard on the indoor environment (see Figure 2.27).

During the daytime, the internal zone temperature remained below that in the courtyard and both were lower than the ambient temperature. Variations in the internal zone temperature were similar to those in the courtyard temperatures. This behaviour correlates with the pattern of courtyard air temperatures, rather than the ambient temperature. The maximum internal zone air temperature was 29.8°C at 4 pm. This was a clear modification from ambient temperatures of 30 and 32.7°C. The indoor relative humidity during the daytime remained around 70 per cent, with an indoor air velocity of 0.4 metres per second. A comfort analysis shows that internal zone thermal conditions are comfortable during the daytime (ASHRAE, 1997).

However, the real problem occurs when the relative humidity moves to higher levels during the early evening, when indoor air temperatures remain around 28 to 29°C. This necessitates a slight breeze during the night for comfort.

AIRFLOW BEHAVIOUR

Stack effect at night

Typical ambient weather data for Colombo indicates still wind conditions during the night. Within this context, internal zones between the courtyard and axis openings were monitored for air movement. The measurements indicated the availability of ventilation between 0.2 to 0.4 metres per second in areas close to the openings and courtyard during night hours (see Figure 2.27).

Furthermore, a variation of pressure distribution due to a temperature difference of more than 1°C was visible at the axis openings and the top of the courtyard (see Figure 2.27). Relatively lower temperatures at the openings can enhance a slight breeze due to stack effect only in the night. This was evident with indoor air velocities recorded against almost still ambient wind conditions during the night. The benefit of such airflow is important for thermal comfort in the early night hours, along with relatively higher levels of humidity.

AIRFLOW PATTERN

Thermal investigation reveals that the indoor air temperature modification in a naturally ventilated high-mass courtyard building is dependent upon the airflow pattern through the building. Thus, the indoor airflow pattern was investigated using computational fluid dynamics. When the 3D building model was allowed to ventilate through openings 1 and 2, found in the building envelope and located within the courtyard on the longitudinal axis, it was found that the courtyard functions as a low pressure zone. In addition, other openings in the envelope functioned as suction zones. This pressure difference induces airflow through the building and thus optimizes the exposure of high-mass internal walls with incoming air. The air entering from

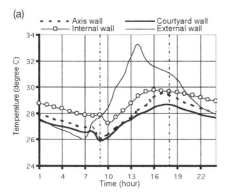

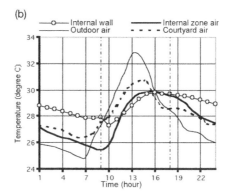

Source: Indrika Rajapaksha and Upendra Rajapaksha

2.26 (a) Temperature differences within the courtyard, with the axis through the internal and external wall surfaces (b) comparison of air temperature differences within the courtyard, internal zone and ambient with internal wall temperatures

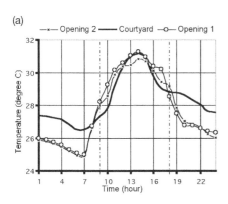

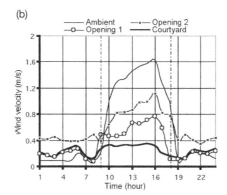

Source: Indrika Rajapaksha and Upendra Rajapaksha

2.27 (a) Air temperature differences at the axis openings and top opening of the courtyard axis; (b) comparison of air velocity differences at the axis openings and top opening of the courtyard with ambient wind at body level

openings 1 and 2 of the longitudinal axis travelled through the indoor spaces and finally discharged into the sky through the courtyard.

The results show that the courtyard does not admit any airflow from its sky opening but acts as an 'upwind air funnel' to discharge the indoor airflow into the sky.

INDOOR AIR VELOCITY

In upwind air funnel patterns, an increase of air velocity was seen with height in the courtyard (see Figure 2.29). The velocity varied from 0.3 to 0.6 metres per second, with lower velocities at the bottom level, increasing through the height of the void. A sudden increase in velocity was observed at 0.9m, where the bottom level of opening 2 is placed. This behaviour indicates the combined effect of both openings of the longitudinal axis. An increase of air velocities at

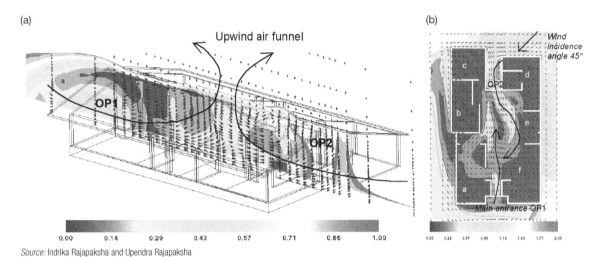

Source: Indrika Rajapaksha and Upendra Rajapaksha

2.28 (a) 'Upwind air funnel' through longitudinal axis, simulated vertical airflow distribution pattern; (b) simulated horizontal airflow distribution pattern

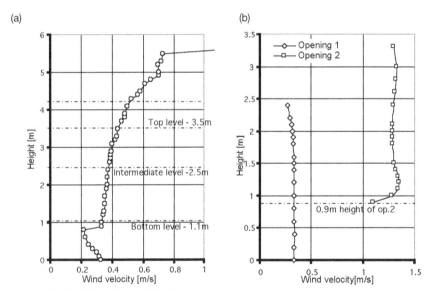

Source: Indrika Rajapaksha and Upendra Rajapaksha

2.29 (a) Simulated vertical air velocity profile of the 'upwind air funnel' within the courtyard; (b) simulated vertical air velocity profile at the envelope openings of the longitudinal axis

0.9m and above 2.5m reveals that the effect of opening 2 on the air velocities is greater than that of opening 1.

THERMAL COMFORT

In this study, thermal comfort was assessed using a thermal comfort index called the Predicted Mean Votes model (PMV), which has been proposed for warm humid climates (Kindangen, 1997). In this index, the operating temperature (To) of the Fanger's comfort equation is replaced with Standardized Environmental Temperature (SET). The parameters are activity levels of 1.1met, with tropical clothing of 0.4clo, mean radiant surface temperature and indoor air velocity at human body height in the internal zone E (Fanger et al, 1989).

Internal zones were assessed for pleasant comfortable thermal conditions with 6 per cent PPD (percentage of people dissatisfied) during day and night time. This corresponds to an airflow rate of 1.5 to 2 ACH (air changes per hour).

SUMMARY

The results obtained in this case study arose from thermal investigations onsite and computational analyses in the laboratory to explore the potential of a courtyard for passive cooling in warm humid tropics. The effect of courtyards on mass to air heat exchange and, thus, the lowering of indoor air temperatures below the corresponding levels of ambient shade temperature are correlated with indoor airflow pattern. Courtyards have a greater potential to act as passive cooling systems when they function as upwind air funnels, discharging indoor air into the atmosphere. The implication of this for architectural design is that useful guidelines are provided for designing naturally ventilated high-mass domestic built forms with internal courtyards in warm, humid tropical climates.

ACKNOWLEDGEMENTS

Facilities and expertise for this case study were provided by the Department of Architecture, Faculty of Engineering, Nagoya University in Japan and the Fujitsu Corporation of Japan. Assistance was also given by architect Nandana Karunasena. The help of staff members at Research-Architecture in Sri Lanka is highly appreciated.

REFERENCES

AGO (Australian Greenhouse Office) (2004) *Minimum Energy Performance Standards for Houses*,
www.greenhouse.gov.au/buildings/code.html#resdiential (accessed 1 September 2004)
AGO (2005a) *Australia's National Greenhouse Gas Inventory*,
www.greenhouse.gov.au, (accessed 22 August 2005)
AGO (2005b) *Tracking to the Kyoto Target*, www.greenhouse.gov.au,
(accessed 22 August 2005)
AGO (2005c) *Community and Household*,
www.greenhouse.gov.au/community (accessed 22 August 2005)
AGO (2005d) *Your Home*, www.greenhouse.gov.au (accessed 22 August 2005)

Australian Government (2005) *Energy All Stars*, www.energyrating.gov.au (accessed 22 August 2005)

ASHRAE (American Society of Heating, Refrigerating and Air-Conditioning Engineers) (1997) *ASHRAE Handbook Fundamentals,* SI edition*,* American Society of Heating, Refrigerating and Air-Conditioning Engineers, Atlanta, GA, Chapters 25–30

Auliciems A. (1997) 'Global differences in indoor thermal requirements', ANZAAS Conference, Brisbane, Australia

Brisbane City Council (2001) 'Discussions with council officers regarding energy code changes', *Building Australia,* October, pp25–26

Clini, C. (2004) *The Challenge of the Global Energy System*, www.iea.org/dbtw-wpd/index.asp (accessed 29 November 2005)

Fanger, P. O., Melikov, A. K. and Hanazawa, H. (1989) 'Turbulence and draft: The turbulence of airflow has a significant impact on the sensation of draft'*, ASHRAE Journal*, vol 31, no 7, pp1072–1077

Givoni, B. (1992) 'Comfort, climate analysis and building design guidelines', *Energy and Building*, vol 18, no 1, pp11–23

Government of Malaysia (2001) *Eighth Malaysia Plan*, Government of Malaysia

Hanafi, Z. (1991) *Environmental Design in Hot Humid Countries with Special Reference to Malaysia,* Welsh School of Architecture, University of Wales College of Cardiff, Cardiff

Howard, J. (The Honourable John Howard MP, Prime Minister of Australia) (1997) 'Safeguarding the future: Australia's response to climate change', available at www.pm.gov.au/news/media_releases/1997/GREEN.html

Humphreys, M. H. (1978) 'Outdoor temperatures and comfort indoors', *Building Research and Practice*, vol 6, no 2, pp92–105

IEA (International Energy Agency) (2001) *Renewable Energy Working Party*, www.iea.org (accessed 22 August 2005)

IEA (2004) *International Energy Outlook 2004*, www.iea.org (accessed 22 August 2005)

IEA (2005) *Catching Up: Priorities for Augmented Renewable Energy R&D Joint Seminar on Long-Term R&D Priorities*, IEA, 3 March 2005, Paris, France, www.iea.org/dbtw-wpd/index.asp (accessed 29 November 2005)

Kindangen, J. (1997) 'Window and roof configurations for comfort ventilation', *Building Research and Information*, vol 2, no 4, pp218–225

La Sapienza (2007) DPR 412/93, University of Rome, http://sae.amm.uniroma1.it/sae/Normative/dpr412.htm

Lim, T. N. (ed) (2000) *80 Years of Malaysian Architecture*, PAM, Malaysia

Malaysian Meteorological Office (2005) www.kjc.gov.my (accessed 25 August 2005)

National Australian Built Environment Rating System (2007) www.nabers.com.au/

NLCC Architects (2002) *Report of Solar House Project*, IGS Grant Committee, MOSTE, Malaysia

Prelgauskas, E., (2003), *Proving Energy Efficiency*, www.emilis.sa.on.net/emil_80.htm

Rahman, A. M. (1994) *Design for Natural Ventilation in Low-cost Housing in Tropical Climates*, Welsh School of Architecture, University of Wales College of Cardiff, UK

Szokolay, S., (2004) *Introduction to Architectural Science: The Basis for Sustainable Design*, Architectural Press, Oxford

Zakaria, N. Z. and Woods, P. (2002a) *Roof Ventilation for Houses in Malaysia*, Advances in Malaysian Energy Research, Malaysia

Zakaria, N. Z. and Woods, P. (2002b) 'Roof design and thermal performance of houses in equatorial climates', in Sayigh, A. M. M. (ed) *World Renewable Energy Congress VII*, Elsevier Science Ltd, Germany

Part II
Location, Climate Types and Building Response

Chapter 3
The Mediterranean:
A Cool Temperate Climate

Francesca Sartogo and Valerio Calderaro

INTRODUCTION

What is [the] Mediterranean? Thousands of things together. Not a landscape but innumerable landscapes. Not a sea but a sequence of seas. Not a civilization but a set of civilizations build side by side. Travelling in [the] Mediterranean means [meeting] the Roman world in Lebanon, prehistory in Sardinia, Greek cities in Sicily, the Arabian presence in Spain, Turkish Islam in Yugoslavia. (Braudel, 1987)

The orographic and climatic characteristics of the Mediterranean originate from an historical complex and geopolitical evolution. It has become, through the centuries, a defined geographical and geopolitical basin called Mediterranean Civil Ecumene, where first Egyptian and Greek, then Roman, through the city of Rome and its ten centuries of the Roman Empire, and finally Byzantine and Islamic dominations left us important common historical and cultural roots. 'The fundamental unifying basis of the Mediterranean zone is the climate, which brings landscapes and [societies] ... closer together' (Braudel, 1987).

To state it simply, hot dry summers and mild, cold wet winters with high daily thermal excursions characterize the Mediterranean climate. In these climatic conditions oranges grow from Spain to Lebanon and from Croatia to Algeria. Thermal comfort can be achieved without insulating clothing and people can live outside almost year round.

THE MEDITERRANEAN CLIMATE

The Mediterranean climate is experienced by those countries bordering the Mediterranean Sea. It is usually experienced between 30 to 50 degrees north and 30 to 40 degrees south of the Equator. Other areas of the world have Mediterranean-type climates; these climate types are characterized through their location and local weather patterns. For example, rainfall occurs almost entirely in winter from the westerly frontal storms. During the summer, the subtropical high-pressure zone dominates weather patterns, preventing rainfall. Mediterranean climates are generally located on the polar side of the great dry belt of subtropical deserts and the equator side of the zone of maritime temperate climates. Hence, the zones never occur on the east coasts of continents. The dry summer subtropical climate is also known as the 'Mediterranean' climate because the land that borders the Mediterranean Sea is a type locality for this climate. The wet winter/dry summer seasonality of precipitation is the defining characteristic of this climate. Summer drought places a great deal of stress on local vegetation; but plant structures have evolved to adapt to it. A number of climate zones can be distinguished in the European region.

Classification

The Köppen classification defines two different areas:

1   Mediterranean coastal zone < 22°C in temperature (Csa); and
2   Atlantic coastal zone: Portugal–Morocco > 22°C in temperature (Csb).

Recent Mediterranean climate trends could indicate a partial climate 'tropicalization', with rainfall concentrations during the transition periods (spring/summer and summer/autumn). The average standard Mediterranean climate, however, is determined both by south tropical air (high pressured) and North Pole cold currents.

The mean useful cooling energy for horizontal surfaces facing the sky factor is approximately $7Wh/m^2$. This has been determined for 28 Mediterranean cities and creates the conditions for passive cooling of buildings in this region (Santamouris and Askimopolous, 1996).

Irradiation

Daily global irradiation decreases with increasing latitude; but in January this trend is more pronounced than in June. This is not surprising in light of the fact that the daily horizontal extraterrestrial irradiation in winter depends largely upon latitude. Total diffused solar radiation during June to September changes from 217 to $308kJ/cm^2$ and comprises 41 to 47 per cent of the total annual value of direct solar radiation. The direct solar radiation's annual mean values are 35 to 44 per cent of the global radiation. The course of the mean air temperature in January and June has been mapped in Figure 3.2; the map indicates that there is a heating demand in the winter. In the Mediterranean area, cooling is necessary during the summer. Temperatures in Turin are about 5K lower than in Nice 150km away.

Humidity

Generally, the relative humidity of the Mediterranean area is uniform and its annual variability appears to be determined by local factors. During the spring and summer period, the minimum value can be observed in the western Mediterranean area, and the maximum in the central and eastern Mediterranean area. The medium value change is from 55 per cent (Nicosia) to 74 per cent (Gibralta). Normally, the area to the east has values lower than in the other basin areas.

In general, rainfall tends to be heavy and persistent on the windward side of mountain ranges lying across the westerly airflow. The depth of snowfall varies significantly from year to year and with location. In the coastal region, there is, on average, little snowfall. Altitude largely affects the amount of snowfall: relatively small differences in altitude produce very significant differences in depth of snowfall.

Wind

The effect of wind direction and wind speed on heating and cooling requirements depends upon the outdoor temperature and upon solar radiation. Cold northerly and easterly wind can, for instance, have more effect on heating requirements than the more frequently occurring gentle south-westerly wind. The annual average frequency of the wind direction illustrates that the wind can blow from any direction, with a dominant direction in some locations.

In the southern half of Europe, interactions between mountains, warmer seas and hot deserts to the south of the Mediterranean produce many well-identified local winds, such as the Mistral, a strong down-slope northerly wind, which penetrates the Rhone Valley and spreads along the French Riviera to the Mediterranean sea for long periods in winter. Another such wind is the Sirocco, which brings air masses containing large amounts of dust from the Sahara into Southern Europe. High wind values can be registered on the east side of the Mediterranean area: 6.6 metres per second in Skiros (Greece) and 2.2 metres per second in Palma de Majorca (Spain).

Italian climate zone

The Italian climate is generally a temperate zone. Normally, summer is hot, especially in the south. Spring and autumn are mild with fine, sunny weather. Winter in the south is much drier and warmer than in northern and central areas. Mountain regions are colder, with heavy winter snowfalls. Italy can be divided in three different climatic zones:

1   The north, between the Alps and Appennino Tosco Emiliano, is only mildly influenced by the temperate action of the sea and normally has a cold winter and a hot and humid summer.

2   The centre, between Liguria and Lazio, has a temperate climate with few differences between extreme seasons.

3   The south usually has a very hot and dry climate, while winters are mild.

According to the Köppen climate classification, the Italian climate, collectively, is humid and subtropical of medium latitude; but there are many climatic variations from north to south and from the coast to the far inland areas (see Figure 3.2). Köppen elaborated a climatic system by relating the presence of animals and vegetation to the real values of temperature, precipitation and wind movements. Köppen distinguishes five major types of climate and classifies them under capital letters.

The north regions are Cf type (without dry season), while the south regions are Cs type (Mediterranean climate with hot summer). The Adriatic side is colder than the Tirrenic side because the Adriatic Sea is not very deep and has only a minor mitigating effect on the coastal regions' climates (see Figure 3.2).

Rainfall tends to be higher in the mountains of the north, while summer droughts are not uncommon in the south, which is on the same latitude as Algiers. Hot, dry African winds such as the Sirocco and the Libeccio keep the humidity down (sometimes delivering Saharan sand in the process), while the cooler Tramontana and Mistral from the north will often bring storms and rain. Sea breezes have a warming effect on the coasts in winter and help to keep even the hottest days tolerable in summer, especially with the mountains near at hand. The average amount of precipitation in Italy is 982mm per year, which means a water input of 296km^3 per year; but the rain is not regularly distributed in time and space (weather data is available online at www.ucea.it and www.eurometeo.italian/home).

(a)

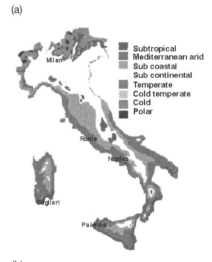

(b)

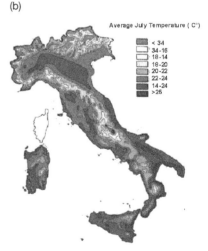

(c)

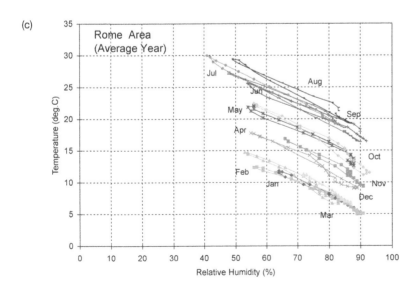

3.1   Main temperatures in July: (a) climate classification; (b) temperatures for July; (c) climograph for Rome showing a wide range of temperatures

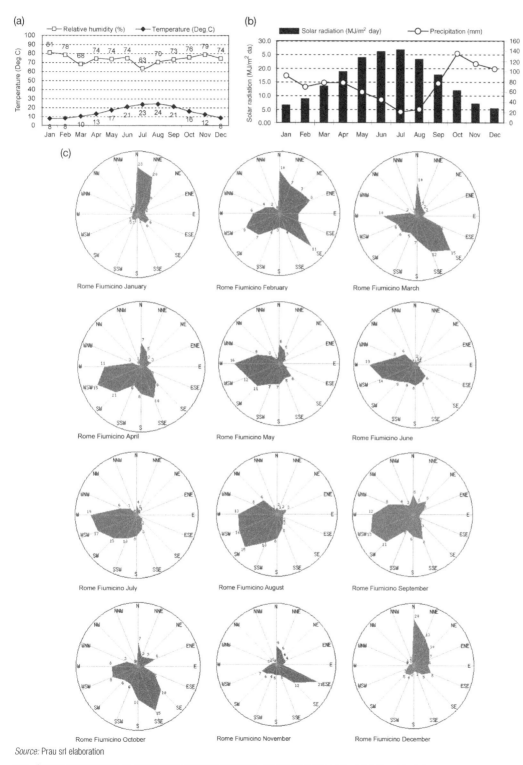

Source: Prau srl elaboration

3.2 Climate data for Rome: (a) temperature and relative humidity; (b) solar radiation and precipitation; (c) wind rose

(a)

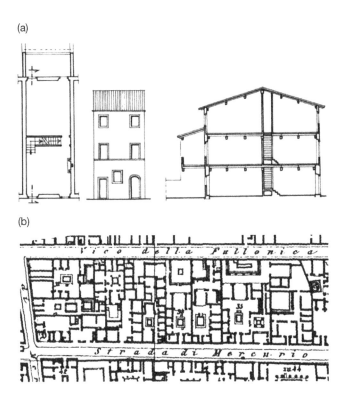

(b)

Source: Prau srl Archive (1700 cadastre map of Napoli and Piacenza)

3.3   Characteristics of urban form

The north regions receive rain during the hot season, and the south has a concentration of rain in cold seasons. The Adriatic receives less rain than the Tirrenic region. Evaporative transpiration is high in some areas of the country and during the summer period influences the water balance. Evaporative transpiration from the land and from vegetation cover has a level of 387mm per year, which corresponds to a water output of 116.6km^3 per year. Assuming zero net gain (or loss) of water from the country in a year, the runoff necessary to balance the precipitation inputs can be calculated to be 179.4km^3 per year.

MEDITERRANEAN SOLUTION SETS

Mediterranean urban and building characteristics

Traditional local Mediterranean cities and building textures are mostly organized according to hot temperate climatic conditions. The important focus on open spaces, such as streets, squares and courtyards, derives from the peculiar characteristics of Mediterranean cities' density and morphology. Ventilation and sun protection measures, together with appropriate materials and construction, represent the main issues of bioclimatic efficiency.

The morphological model of traditional urban design in Mediterranean areas is characterized by a framework of public open spaces – streets and squares – cut through a dense urban grain of low buildings (two to four storeys). From the viewpoint of climate changes, the density, direction and continuity of the street network and the shape of open spaces and density are the most important variables, apart from topography.

Density characterizes different kinds of urban structure and produces important effects on the microclimatic conditions in external spaces. In general, the greater the density, the lower the solar energy contribution and ventilation capacity, but the more stable the temperature.

Building form

In the Mediterranean tradition, the alignment of urban structures is often oriented to take advantage of the fresh breezes coming in from the sea and turns away from the hot continental winds. This strategy for climate control has been strongly influenced by cosmogonical legends about the foundation of cities. As for radiation, if the main alignment is predominantly east–west (with deviations up to 30°), this enables better use of solar radiation in wintertime, creating higher temperatures in the winter and lower temperatures in the summer.

Is not easy to analyse how these principles came to be adopted as a concrete Mediterranean urban tradition. Some examples must have derived from classical Greek cities, with their regular east–west axes, from Roman cities, with their north–south axes, and from Islamic cities, with their prevailing narrow, covered streets and systems of wind tower cross-ventilation. The composition of cities creates a form, therefore, which results in a homogeneous building fabric of cellular structure carved into public and private spaces.

Interrelation between Mediterranean public and private open spaces

The growth of the Mediterranean during the Middle Ages left a legacy of public and private open space. Public space is served by a matrix of connected streets, squares, courts and alleyways defined by building surfaces. The bioclimatic behaviour of each type of outdoor space depends upon the relationships that it has with surrounding elements. The lower the ratio between the size of buildings and the width of spaces, the lower, in general, will be the temperature in outdoor spaces and in buildings; this may be the reason that the dimensions of streets generally contract with latitude. The morphology of the Islamic compact urban texture is an example of this rule.

Another building strategy in the Mediterranean urban context is the extensive use of courtyards. The advantages of building in courtyards are as follows:

- There is minimal solar penetration during the daytime, which limits the heating of the courtyard's internal walls so that the temperature of the courtyard is lower than the external temperature.

- During the night, the courtyard retains a pool of cool air, which can be used for cooling adjacent buildings. The ancient house form in the extensive temperate areas of the Mediterranean culture, or in the valley cultures of China and India, gave rise to the 'courtyard house', setting a fundamental bioclimatic principle that still remains the basis of many buildings and urban structures today.

Gradually, through the centuries, this arrangement of building volumes within an outer perimeter wall has evolved into two fundamental systems:

- the *atrium*, an architecturally defined open space, often including an *impluvium* for water collection; and
- the *peristyle*, a larger and more open space, also useful for a vegetable garden.

Such systems not only present very suitable technological answers for compact urban aggregations in mostly high-density cities, but they also provide for better and more efficient bioclimatic equilibrium in places where surfaces and roof coverings are most exposed, and guarantee ventilation and natural daylighting despite limited open space. Even though, through the centuries, the typology has been transformed from simple courtyards to the mercantile courts of middle-class palaces and, ultimately, to those of high-class palaces, the open spaces of the atrium and the peristyle, as constant construction elements, have succeeded in retaining intact the system's original manifestation and all of its fundamental characteristics.

The ancient Roman city generally maintains along its principal solar axes a high proportion of courtyard houses with the original characteristics of the typology, a high occurrence of atria within the internal spaces, and the ancient, but still very efficient, bioclimatic system. On the other hand, the subsequent development phases of the texture of linear villages shows a greater fragmentation of the underlying typological organization, transformed to single houses and row houses, short form or adapted for multiple families, but always maintaining, even though they may be divided, the internal spaces of the original atrium and peristyles. This theoretical building typology follows the continuous urban development from the 12th century to the present, and is very common in Italian and other Mediterranean cities where, historically, urban growth has consisted of gradual development, with its highest increase from the 12th to the 15th centuries, a prolonged period of stasis between the 15th and 18th centuries, and finally a rapid expansion during the 19th and 20th centuries.

In the 19th and 20th centuries, city design for the new industrial expansion produced significant large-scale urban planning and more compact building aggregations. The concentrations of multi-family residence housing enlarged the ancient design of the building typology, running from single houses and row houses to linear buildings around a courtyard, and to linear blocks within a linear and vertical urban texture. Building structure typology, in addition, frequently lost some of its ancient bioclimatic characteristics.

Typological solution sets for the Mediterranean area

Hot dry summers and wet, moderately cold, winters characterize the Mediterranean area. The existing relationship between building typology and thermal behaviour of a building is well known. Even if it is relatively simple to study and find the appropriate building typology layout to use in cold or hot climates, the Mediterranean climate offers some problems for energy efficient building typologies because of the requirement for a combination of summer cooling and winter heating.

For instance, buildings can take advantage of winter sun radiation, while they must be protected from it in the summer. On the other hand, a good façade and internal ventilation is necessary in summer but not in winter. Fortunately, the wind direction in the hot season is usually different from that in winter. Whatever solutions can achieve a balance between these factors will satisfy our objectives, with the important consequence that plant design options can be chosen more freely.

For years, starting with the energy crisis of the 1970s, building energy consumption was calculated only for the winter period. Recently, in Europe, the explosion in the numbers of air-conditioning plants, and the consequent increased energy needs in summer, have aroused a new interest in natural passive cooling of buildings. Thus, examining the historical architectural character of the Mediterranean, one can recognize simple solutions that give efficient thermal control. Traditional building schemes have shown how important efficient climate regulation was for early house builders, and some of their solutions, with the addition of double-façade systems and solar roofing, can contribute to new solutions for housing in the Mediterranean climate.

With regards to the *single house typology*, the main examples follow the *courtyard* typology of the Roman *domus* (townhouse) and the *traditional single farmhouse* of the south-central area of Italy. The courtyard typology is usually associated with the compact urban schemes of Middle Eastern settlement and does not seem to be suitable for temperate climates. But if the cooling objective is important, this form can provide good lessons. The traditional single farmhouse was often a south-facing wide block (two basic cells wide) of two levels, built of solid baked-clay bricks. At ground level, usually on the north side, there were cowsheds or pigsties. North-facing windows were always smaller than south-facing ones and thick walls provided an efficient thermal inertia. In the case of the *terraced house*, we can analyse many structures in Italian historical centres.

For multi-family houses, most have followed the courtyard typology, directly derived from Roman *insulae* (apartment building) and successfully used in the largest Mediterranean cities from the 18th to the 20th centuries (Barcelona, Rome, Naples, Palermo). Recent studies warn us against uncritically considering the courtyard typology as invariably the more climatically appropriate solution. In any case, we suggest an aspect ratio (height/width) less than 0.3, corresponding to a wide courtyard, with dense deciduous trees and the use of water features. However, courtyard projects must always be verified using thermal simulation software. The linear block typology is a good alternative solution, is frequently used and offers a simple method of energy performance control.

In summary, a suggested typology set for the Mediterranean climate comprises a compact block shape for detached houses, stretched out along an east–west axis. House plan and orientation can be rather free. Terraced

Table 3.1 Solution set for the Mediterranean climate: single house

Characteristics
Form: compact block
Floors: three maximum
Dimensional ratio (length/width): 0.83–2.5 maximum
Orientation (0° = south): main façade variable from −45° to +45°
Roofing: pitched, ventilated
Solar protection: façade-shadowing systems
Active systems: photovoltaic (PV) and solar thermal collectors on roof
Passive systems: 'double-skin' bioclimatic system
Glazed/opaque surfaces ratio: south 30%; north 15%; east–west 20%
Thermal time lag: >8 hours
Ambient air exchange: 0.5ACH in winter, 10ACH in summer
Maximum yearly heating energy consumption: 30kWh/m^2
Reference U value: 0.3–0.4W/m^{2}K
Living room orientation: south

Source: Umbertide Case study – Prau Project

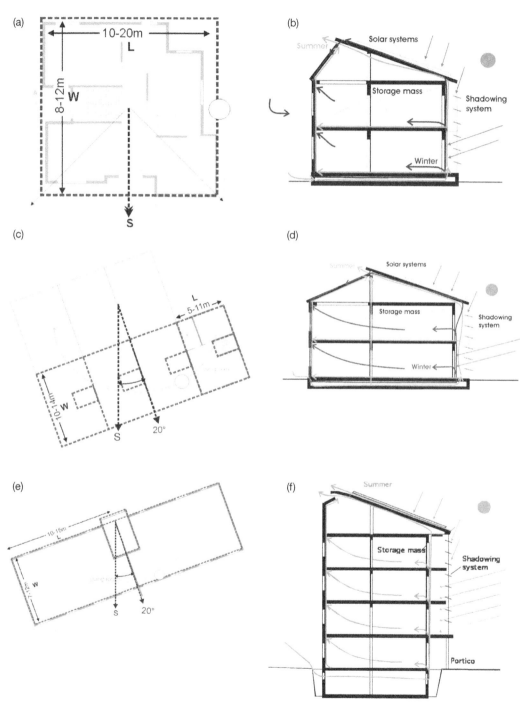

Source: Umbertide Case study – Prau Project

3.4 Solution sets for the following house types: (a) and (b) single detached house solution set;
      (c) and (d) row house solution set; (e) and (f) apartment solution set

houses should be at least two-storey dwellings, with the building rows arranged along an east–west axis. This typology has an excellent performance, particularly in the cold season, except for natural lighting. The simplest solution for multi-family typology is the linear block with two apartments per floor (see Figure 3.3a and 3.3b).

For all typologies, it is suggest that living rooms and main rooms should be located on the south sides of buildings; south-facing glazed surfaces should be protected from summer sun radiation with mobile screens; a pitched south-oriented roof should be chosen; walls should be insulated; there should be adequate thermal storage mass (perhaps a thick masonry structure) in the internal volume of the building; and there should be active solar roofing (photovoltaic and thermal) and a natural ventilation system with, for example, a solar chimney on the top, close to thermal storage masses for bioclimatic operation. Table 3.1 summarizes the main concepts regarding the most suitable typologies for the Mediterranean climate.

Typological solution sets for the Mediterranean climate

Solution sets are best understood as plans and sections. The solution set is a combination of the form and fabric of the building, designed to best meet the bioclimatic requirements of the climate and the needs of the building

Table 3.2 Solution set for the Mediterranean climate: row house

---

Characteristics

Form: row of two-storey houses

Floors: two maximum

Dimensional ratio (length/width): 0.35–1.1 maximum (single house cluster)

Orientation (0° = south): variable from 0° to 20° east

Roofing: pitched, ventilated

Solar protection: façade-shadowing systems

Active systems: photovoltaic (PV) and solar thermal collectors on roof

Passive systems: 'double-skin' bioclimatic system

Glazed/opaque surfaces ratio: south 40%; north 15–20%

Thermal time lag: >8 hours

Ambient air exchange: 0.5ACH in winter, 10ACH in summer

Maximum yearly heating energy consumption: 30kWh/m^2

Reference U value: 0.3–0.4W/m^{2}K

Living room orientation: south

Source: Umbertide Case study – Prau Project

programme. There are differences in the solution sets for the scale and configuration of the building type.

For the single detached dwelling, a compact form is used (see Figure 3.4a and 3.4b).

Table 3.3 Solution set for the Mediterranean climate: apartment

Characteristics
Form: block
Floors: four to five maximum
Dimensional ratio (length/width): 1.6–2.5
Orientation (0° = south): from 0° to 20° east
Roofing: pitched, ventilated
Solar protection: façade-shadowing systems
Active systems: photovoltaic (PV) and solar thermal collectors on roofs and façades
Passive systems: 'double-skin' bioclimatic system
Glazed/opaque surfaces ratio: south 40%; north 15–20%
Thermal time lag: >8 hours
Ambient air exchange: 0.5 (ACH) in winter, 10 in summer (ACH)
Maximum yearly heating energy consumption: 30kWh/m^2
Reference U value: 0.3–0.4W/m^{2}K
Living room orientation: south

Source: Umbertide Case study – Prau Project

Case study 3.1 Urban comfort as the new urban and building design culture: Umbertide new urban district

Francesca Sartogo, Valerio Calderaro, Giovanni Bianchi and Massimo Serafini with Carlo Brizioli, Valentina Chiodi, Isabella Calderaro and Pierpaolo Palladino

INTRODUCTION

This case study examines work on the development of bioclimatic housing for a housing development in Umbertide, Umbria, in Italy. The work involved participating with the local community on the design brief and master planning concept. Extensive climatic analysis and computer modelling of the proposed scheme has been carried out to test the design concepts.

Table 3.4 Profile of Umbertide, Umbria

Country	Italy
City	Umbertide, Umbria
Building type	Eco-city European Union Programme: Urban Development Towards Appropriate Structure for Sustainable Transport, Contract no EVRA-CT2001-00056
Year of construction	Urban Renewable Master Plan, June 2004
Project name	Urban comfort and sustainable transport as a new culture for city planning design

Source: Umbertide Case study – Prau Project

PORTRAIT

The purpose of the Umbertide Urban Renewal Project is to sketch an innovative development strategy for the city of Umbertide, its territory and its lively local economy by organizing a new territorial, ecological, energy-sensitive and bioclimatic residential development focused on an appropriate structure for sustainable transport. The availability of 63ha of dismantled industrial and railway areas provides an opportunity to build, in a brownfield area, a complex residential/mixed-use district for 1500 inhabitants in a holistic way, coherent with local demand for bioclimatic urban comfort and a local historical typological city-growing process.

Like almost all Mediterranean cities, which have a close correspondence between building form and urban grain, water system and climatic factors, the Umbertide case study project organizes the alternative mobility of the city through a set of bioclimatic north–south-oriented main axes, coincident with wind corridors and a hierarchy of bioclimatic urban open spaces. Unlike conventional urban planning organized along car routes, this project, almost completely car free, is organized according to bioclimatic wind axes, with an overlap of pedestrian and cycling paths. In this case, alternative mobility is the real core of the urban scheme; in fact, it provides the main structure of the entire urban design framework.

UMBRIAN CLIMATE

The orographic, hydrographic and geopolitical structure of Italy is mainly based on two important territorial systems. The first coincides transversally with the Alps and the Po River. The second coincides longitudinally with the Apennine Mountains and its main systems of rivers and valleys, the Arno-Chiana-Tevere, Tevere-Metauro and the Sacco-Garigliano-Volturno. This second system divides the central part of Italy into two longitudinal halves, one along the Tirrenian Sea and one along the Adriatic Sea, each with different historical and cultural roots. So, in the Umbria region, there are two of the eight climatic regions of the Italian classification: Versante Tirrenico Centrale and Versante Adriatico Centrale.

Source: Umbertide Case study – Prau Project

3.5 Umbria orographic region

The Umbria region is situated in the very heart of Italy, on the Arno-Chiana-Tevere River system, and is surrounded by important regions such as Toscana, Marche and Lazio. The region covers three valleys, the Nerina Valley on the River Nera with the city of Spoleto; the Umbria Valley on the Clitunno with the historical cities of Assisi, Todi and Perugia; and the Alta Valle del Tevere, north of the Umbria Valley, with its separate characteristics of a specialized local economy.

The Umbrian climate is a semi-continental and Mediterranean type, but is strongly influenced by the Appennine Mountains, which break the airflows coming from the Adriatic Sea. Airflow, normally coming from a north-east direction, does not pass easily over the Apennines. Furthermore, the mountains determine rainfalls in the west–east direction, with the exception of a deeper penetration in the Nerina Valley (see Figure 3.5).

Normally, the minimum temperatures are over 0°C, except Norcia, Città di Castello and Umbertide. In the Trasimeno area, there is little fluctuation between maximum and minimum monthly temperatures; the high differences in temperature are in the Alta Valle del Tevere and in the Terni area. According to the Walter and Lieth diagrams (www.usf.uniosnabrueck.de/projects/climate/cd/doc/index.htm#Diagrams) and the Umbrian observatories, the region is climatically subdivided into three fundamental weather types: dry period (Terni, Umbertide, Todi, Monte del Lago and Orvieto), without dry period (Norcia) and with humid period (Gualdo Tadino).

Umbrian wind data is limited to Perugia (direction and velocity) and Terni (direction). The general airflow of the region consists of one main wind direction with seasonal variations, from the north-east in the winter and from the south-east in the summer (see Figure 3.6). The wind velocity, registered by the Perugia Observatory, is 10 kilometres per hour. Winds are over 10 kilometres per hour for 20 per cent of the year, normally during the spring and winter seasons. In Terni, winds prevail from the south-east and the east, especially in the winter. There are few solar radiation observatories that give global radiation data (direct and diffused radiation) or as heliophany (direct lighting). The four stations of Perugia, Città di Castello, Todi and Terni notice a uniform solar radiation for the region. In winter, there are low values for daylighting due to the higher frequency of overcast skies (see Figure 3.6).

Umbertide climate

The city of Umbertide is situated in the middle of the Alta Valle del Tevere and has about 15,000 inhabitants. The city lies on the River Tevere, 30km away from Perugia to the south (the main town of the region, with 140,000 inhabitants) and 25km from Città di Castello to the north (40,000 inhabitants). Umbertide lies in a Mediterranean temperate climate with rainy winters and hot summers. The climate creates a significant requirement for energy for heating and cooling, which can be approximately 2192 degree days per year (degree day is a quantitative index used to reflect demand for energy to heat or cool houses). There is annual average sun radiation of 3.93kW/m² per day (winter 1.75kW/m² per day; summer 6.11kW/m² per day). During the summer months, light breezes (under 5 to 10 metres per second) from the south, east or west enter the city from

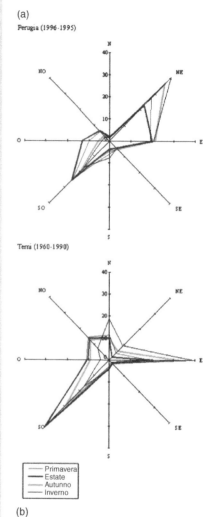

(a)
Perugia (1996-1995)

Terni (1960-1990)

Primavera
Estate
Autunno
Inverno

(b)

Source: Umbertide Case study – Prau Project

3.6 (a) Todi and Terni wind direction; (b) climate comfort profile the surrounding hills and are caught by the streets that separate the buildings, providing them with natural air conditioning. During the remaining seasonal periods, the prevalent direction of the wind is north-west.

The very comfortable months are in spring (May) and in autumn (September and October); the central months (from June to August) are hot and humid with light breezes. The maximum values of the solar radiation are in the central period of the year, with 22.9MJ/m² in June and 23.3MJ/m² in July; the minimum values are in December, with 5.1MJ/m², and in January, with 5.6MJ/m² (see Figure 3.7). Minimum temperature values occur in December and January, with 1.2 and − 0.9°C, and maximum temperatures occur in July and August, with 27.7 and 30.9°C (see Table 3.5).

CONTEXT

Following the new *Municipality General Master Plan* and the results of the *Citizens' Community Planning*, general guidelines for the *Urban Renewal Master Plan* for the Umbertide case study were established. At a meeting, 35 participants expressed their general agreement for the main eco-city concepts. The results were as follows. The maximum vote was for the City as a Renewable Energies Power Station, followed by, in order of preference, Car-Free City, Bioclimatic Comfort City, Cultural Identity and Social Diversity.

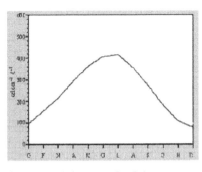

Source: Umbertide Case study – Prau Project

3.7   Monthly solar radiation

Table 3.5 Monthly temperature

Month	Minimum		Maximum		Mean
	Mean	Deviation from mean	Mean	Deviation from mean	
1	−0.9	−10.0	9.0	16.0	4.0
2	0.7	−8.0	11.7	20.5	6.2
3	3.0	−5.0	14.8	22.0	8.9
4	5.5	−1.0	17.9	25.0	11.7
5	8.4	2.5	23.4	29.5	15.9
6	12.2	5.0	27.7	34.0	20.0
7	13.6	8.0	30.9	37.5	22.2
8	13.5	8.0	30.3	37.5	21.9
9	10.1	0.4	24.0	34.0	17.1
10	7.8	1.0	20.0	28.0	13.9
11	4.7	−3.0	14.7	21.5	9.7
12	1.2	−6.0	9.9	16.0	5.5

Source: Umbertide Case study – Prau Project

The project promotes the use, reuse and reinvigoration of cultural heritage, and supports continuity of the urban cultural identity at different levels. History, microclimate and typology are the main responsive matrices of the urban design. The prevailing wind, with green areas and water, constitutes a solar-oriented bioclimatic spine that structures the urban and building texture.

The 'citizens' well-being' target aims to transform the area through a series of urban measures in order to achieve bioclimatic benefits, the mitigation of emissions and the control of air and noise. This has been called 'urban comfort' and has been assumed as the principal motivation for urban design and for the new sustainable transport structure.

A central objective was to utilize the 'skeleton' of the urban structure as the ventilation system. This links prevailing wind flow from the hillside over the river, in strict accordance with the existing city's oriented frame. The skeleton was transformed into the bioclimatic spine, which could funnel the wind to a number of areas. First, airflow from the river park is directed to the railway station and the ninth-century city. Second, the bioclimatic spine funnels the wind from the new urban area A to the ancient Borgo Minore. Third, the other two remaining 'wind corridors' run from the area B through the existing roads of the Workmen's Village – the Molino (an urban renewal area comprising commercial and artisan factory) – to the modern city.

The design of the urban texture consists of a system of buildings grouped around common external courts derived from the ancient Roman

Source: Umbertide Case study – Prau Project

3.8 Master planning to include bioclimatic principles: (area A) Borgo Minore Residential Development; (area B) residential and mixed-use development area; (area C) railway station and old industrial area; (area D) Central Bioclimatic Green Selotto, new residential development (1245 inhabitants, 64ha)

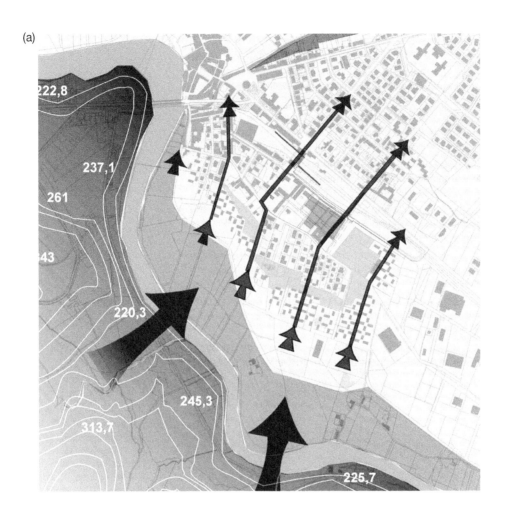

(a)

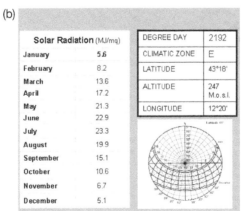

(b)

Solar Radiation (MJ/mq)	
January	5.6
February	8.2
March	13.6
April	17.2
May	21.3
June	22.9
July	23.3
August	19.9
September	15.1
October	10.6
November	6.7
December	5.1

DEGREE DAY	2192
CLIMATIC ZONE	E
LATITUDE	43°18'
ALTITUDE	247 M.o.s.l.
LONGITUDE	12°20'

Source: Umbertide Cast study – Prau Project

3.9　(a) Microclimate in the urban area; (b) microclimatic data showing solar radiation

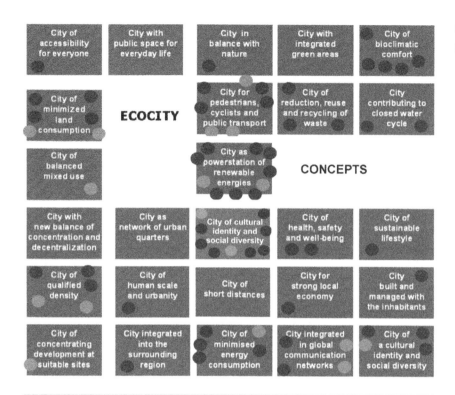

Source: Umbertide Case study – Prau Project

3.10 Umbertide: main intentions of the project

courtyard. The alternation of the form and function of this external space is organized according to the position of the site, in correspondence with the main pedestrian wind axis or with the secondary edible gardens lines: the first are organized in atrium blocks and the second in atrium and peristilium blocks. There are three different building typologies:

1   apartments;
2   detached houses; and
3   row houses.

Differences in construction systems, technological components, heights, density, and so on, of the buildings depend upon specific microclimatic needs due to the location. The resulting architectural language of the project naturally completes the spontaneous development process of the city through a new coherent urban design culture.

ENERGY USE

The solar sustainable housing *energy target* was assumed to correspond to the Italian Casa Clima certification concept. For the existing buildings' renewal, within the C area of the project, Category A, with a 30kWh/m^2 per year consumption target, was assumed; for the new settlements within the A and B areas, Category A Plus, with a 20 to 15kWh/m^2 per year consumption target and the use of ecological and natural materials, was assumed.

Both experts and citizens have been involved in a participation process, for the urban and building design, and also for the biomass district heating plant design. The energy saving reached is 75 per cent and the carbon dioxide (CO_2) emission reduction is 73 per cent, compared to the existing Italian standards.

ECONOMIC CONTEXT

Economic aims in the context of conventional buildings

According to the Renewable Energy Campaign, current agricultural and industrial structures could support an increase in cultivation for energy production. Land suitable for the production of bio-fuel for transport systems already represents 11.7 per cent of cultivated land. Another sustainable agriculture opportunity is represented by coppice woods, which could be included in the Probio national and European Union (EU) programme for maximizing the use and development of forest and agricultural wood for local biomass.

Funding

Activation of Agenda 21 forums and territorial agreements (Umbria Region and National Institution) represents the most common way of creating local partnerships that are able to finance interventions for urban development. There are other funding programmes, either for agricultural and territorial development, or for urban and building renewal – for instance, PRUSST, PRU and Contratti di Quartiere, periodically made available at regional and national level.

(a)

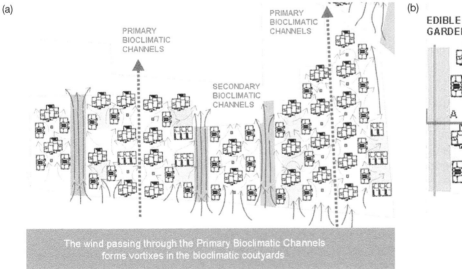

(b)

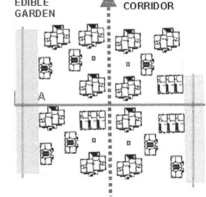

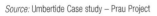

Source: Umbertide Case study – Prau Project

3.11 (a) Bioclimatic corridors; (b) details of housing cluster

Local Wind System   S-SSO summer wind   5m/s

Source: Umbertide Case study – Prau Project

3.12 Macro airflow effects are used to assist with setting up the computer simulation

SITE DESCRIPTION

Urban context

The area for development is situated south-west of the railway station; it has rail tracks and the Tiberina road to the north, and the Tevere River to the south, and existing buildings are industrial plants. The area is therefore of a

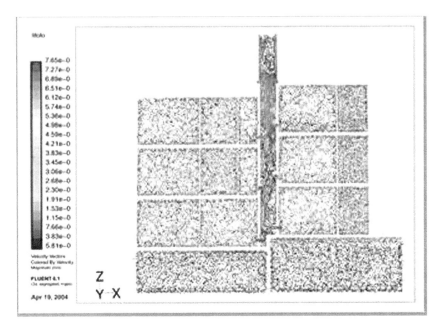

Source: Umbertide Case study – Prau Project

3.13 Fluent simulation for internal building ventilation

brownfield typology – that is, an area of land with a pre-use, which is derelict. The aim is to renovate the area including the railway, the industrial buildings, creating an agriculture park around the river and provide new residential housing. The city and landscape morphology derives from the geometry of the ancient Roman *centuriatio*, still visible in the flat area around the Tiberina Road, with its layout oriented according to sun and wind directions. The urban structure was planned in accordance with the water system, which functions to collect, distribute and drain water. The water, green public space and landscape network, all oriented towards the sun and the wind, respond to both the ancient and the modern city.

Site selection

The relocation of the existing industrial structure to a more suitable area provides the possibility of creating a new centre for the city, with a new residential, commercial and workshop development connected with the old city and through a new light-rail network to the entire Umbrian territory. This is the last available area for development in the city; but it does not present the most suitable microclimatic conditions, so the project has striven to gain maximum urban comfort from the existing microclimate by funnelling the airflow into appropriate wind corridors and into new architectural forms responsive to the climate.

BUILDING FORM AND STRUCTURE

The buildings range from two to five floors, depending upon different typologies, but most comprise four rooms (bedrooms, kitchen and living room) and two bathrooms. The form of the design acts as a reference for the continuity of the city's development. Key factors that were considered include the following:

- heights of buildings and the number of apartments;
- typology of existing buildings and district compactness;
- interrelation and spaces between buildings, together with the geometrical form and size of open spaces (these have been optimized through the 'fluid dynamic calculation simulation' in order to control urban comfort efficiency and relative density);
- dwelling size and configuration; row house (64.71 square metres) and detached house (114.55 square metres) configurations; ground floor: bathroom, living room and kitchen; first floor: bathroom, two bedrooms; apartment house (166.39 square metres): living room, kitchen, two bedrooms, bathroom; basement: there is a basement space, unheated, which is used for summer ventilation.

BUILDING CONSTRUCTION

The construction systems of the buildings are as follows:

- a structural system for row houses and detached houses consisting of load-bearing walls in traditional brick masonry; apartments are made from reinforced concrete frames with traditional brick cladding;
- external walls comprising ventilated cavities, with traditional brick masonry;
- roofs comprising a concrete structural system, sealer, air space and traditional roof tiles;
- services include the natural ventilation system and the integration of the auxiliary district heating system.

VENTILATION SYSTEM

At the urban level, the urban form has been designed to improve ventilation in local neighbourhoods and housing clusters. This has been achieved by harnessing the prevailing direction of the wind during the various seasons. Further techniques have been used to obtain protection from winter wind (rows of trees) and to promote summer wind penetration, with consequent reduction of energy demand for buildings. Urban natural ventilation system flows are tunnelled through the main wind-corridor pedestrian spines into the private 'corte open spaces' and the different buildings structures.

At the building level, the system is based on the use of three ventilation chimneys: two for air admission and one for expulsion. During the night in summertime, because of the lower external air temperature, such chimneys produce air cooling of the chimneys and of the building structures through stack effect. In wintertime, solar radiation on the transparent part of the chimneys of the south façade generates a convective loop, which transfers the heat picked up from the masses of the inner chimneys to the rooms.

The reason for choosing this method is that this simple and integrated bioclimatic system is able to obtain a high energy performance that improves

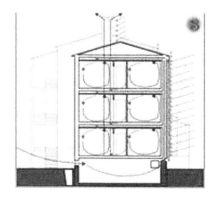

Source: Umbertide Case study – Prau Project

3.14 Ventilation system predicted using the computer simulation process the quality of the indoor and outdoor environment – that is, temperature, air and acoustic conditions. This system provides completely natural ventilation and is an innovative choice for the project; it is also important since the whole of the urban development is oriented to provide the best outdoor and indoor microclimate. The consequences for the design and the structure of the building are as follows:

- *Urban consequence*: the shape and configuration of both the buildings and the public and private open spaces derives directly from ventilation requirements; in fact, the disposition of each courtyard and of the surrounding blocks, and the selection of their heights, depend upon wind corridors. In this way, it is possible to guarantee a satisfactory degree of wind penetration in summertime when the wind comes from the south. On the other hand, a barrier against winter winds (rows of trees) has been planned for the northern part of the site along the railway track.
- *Building effects*: the southern façade of each building requires specific treatment in order to provide places for air channels and *brise soleil*; a similar problem has been faced in positioning air vents and chimneys appropriately with respect to the distribution of rooms. The heat accumulated in the mass of the chimneys will be used during the night to heat rooms and for ventilation. In this way, the system controls air changes, as well as temperatures, in all seasonal conditions.

APPLIANCES

The main energy saving appliance is the passive bioclimatic system of the building itself. Other appliances include:

- hot water supply for appliances;
- hot water heating and cooling (local district supply will be covered by a district heating network using the existing agricultural biomass, produced and converted to 'wood pellets' by local enterprises);
- energy efficient lighting;
- low energy-consumption appliances.

ENERGY SUPPLY SYSTEM

Energy supply for purchased energy is provided through a biomass local district heating and electric network. Energy storage systems are provided through the building itself, which features a thermal storage system. The heating system and hot water supply is provided by district heating. The heating distribution network will consist of buried insulated steel pipes. The biomass material is fed by a double-cycle system and also features a fire safety and security system, with water sprays and a firewall that operate in the event of a fire.

SOLAR ENERGY UTILIZATION

Daylighting is used to reduce energy; daylighting optimization is achieved with a *brise-soleil* system on the south façade of the building. Furthermore, direct heat-gain heating is used. Part of the south façade is completely glazed

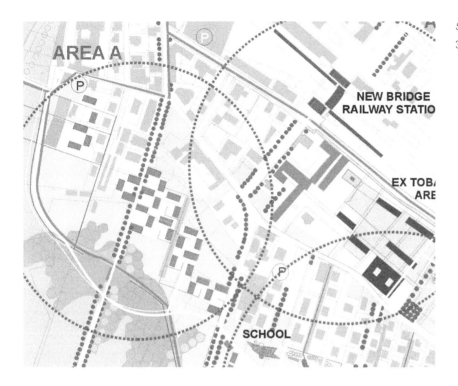

AREA A

P

P

NEW BRIDGE
RAILWAY STATIO

EX TOB,
ARE

SCHOOL

Source: Umbertide Case study – Prau Project

3.15 Urban form: details of the housing areas

in order to increase direct heat gain. Sun protection in summer is provided through selective shading, which allows winter direct heat gain from the sun, but screens out summer sun.

Surfaces are protected by a *brise-soleil* system controlled by a mechanical orienting system: the installation of obstruction/reflection elements in the external part of the windows allows solar radiation to be screened in hot periods, and direct solar radiation to be reflected onto the ceiling in winter. Therefore, the ceiling is used for thermal storage, as well as to reflect the remaining solar radiation, and thus achieves uniform natural lighting without glare.

Solar thermal systems that are provided have a separate heating convective-loop system. The system is based on the use of three ventilation chimneys: two for air admission and one for expulsion. The first (outermost), for solar energy capture, is composed of external low-emissivity double glazing, a hollow space (chimney) and a wall that works as a thermal storage; the second chimney has a wall in common with the previous one and delimits a hollow space between this and the inner wall; the third chimney is inside the building and has the function of expelling the air to the outside.

BUILDING HEALTH AND WELL-BEING

With regard to the internal environment, materials have been selected to minimize off-gassing of materials. In the external environment, lifestyle spaces are provided through the use of wind corridors, which improve microclimatic conditions. They operate for the two main seasons:

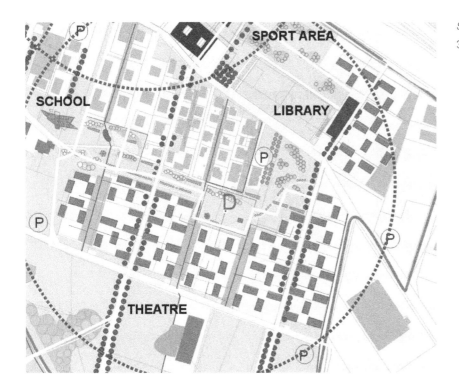

Source: Umbertide Case study – Prau Project

3.16 Urban form: details of sports area, school, library and theatre

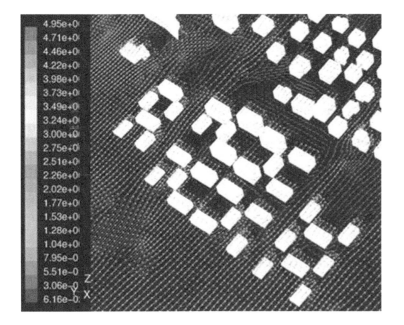

Source: Umbertide Case study – Prau Project

3.17 Vector plans using Fluent software

Source: Umbertide Case study – Prau Project

3.18 Shading on the external façade

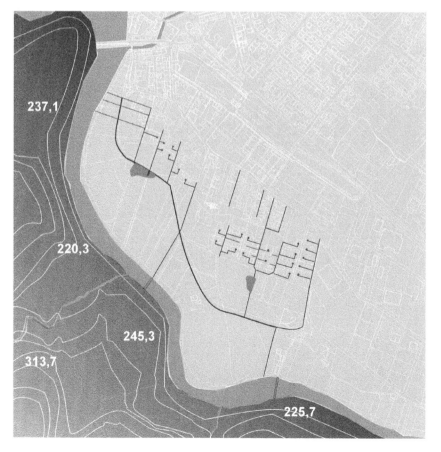

Source: Umbertide Case study – Prau Project

3.19 Courtyard spaces provide open space design for bioclimatic principles

1   Winter: rows of trees in the upper part of the area have been planted to form a barrier in wintertime when the wind comes mainly from the north-east; in summertime, when the wind comes from the south, the disposition of the courtyards (for winter solar capture) allows a satisfactory degree of wind penetration.

2   Summer: the use of wind corridors allows a satisfactory degree of wind penetration; the use of green zones, ponds, brooks and canals combines with the evaporative transpiration effect of the vegetation to improve the conditions of the local microclimate and to reduce the energy demand of the buildings.

WATER RECYCLING AND CONSERVATION

Water storage is provided through a central rainwater storage tank in each courtyard, linked to the building and to the water canal network of the vegetable gardens. There are two rainwater storage ponds along the street water canal and near the theatre for grey water filtering and depuration. Consumption control, distribution and recycling management have been initiated effectively, including advanced waste and water treatment and utilization of ultraviolet (UV) filters in the rainwater storage pond.

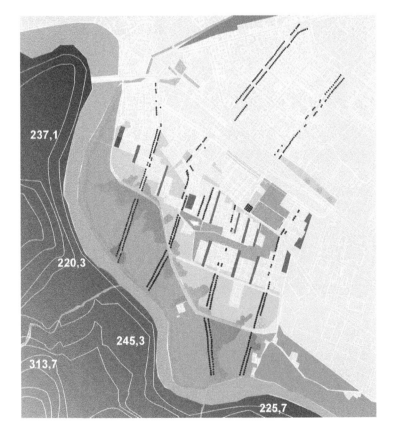

Source: Umbertide Case study – Prau Project

3.20 Water reticulation

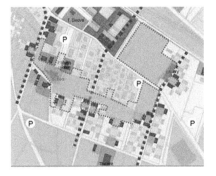

Source: Umbertide Case study – Prau Project

3.21 Water tanks

CONTROLS

In the first phase, energy management will occur by manual control; as a result, proper instructions need to be distributed to the users. In successive stages, this management will be automated (home automation and building automation) in order to optimize the regulation of daylighting, natural ventilation, thermal comfort, etc.

PLANNING TOOLS APPLIED

Considering the various indicators, such as processes, context, urban structure, transport flows and socio-economic factors, there is good performance information about the urban structure, especially about the urban comfort provided by the new urban renewal design.

The use of computer simulation tools has concentrated on the local microclimate, which is influenced by the river and by the nearby hill ridge, with summer breezes mainly from the south. Computer simulation provides the opportunity to create appropriate urban ventilation into four bioclimatic wind corridors in a series of urban open spaces, producing a good external and internal microclimate. The whole project has been tested and improved through continuous fluid dynamic software simulations (Fluent), and this has been implemented for the three different typical Umbrian typologies.

At the urban level, several simulations with Fluent (fluid dynamic model simulations) have been carried out to estimate the effects of the wind in order to improve summer ventilation conditions. Reference data for the simulation comprised a wind speed of 5 metres per second; simulation was completed for the whole intervention area, also calculating speed and pressure. We then carried out two further relevant building aggregations derived from the local microclimate. Simulations showed that some areas were not sufficiently ventilated. We then carried out the necessary modifications in the disposition of the buildings to improve ventilation conditions. Due to the results of such simulations, we were able to determine the courtyard shape and the bioclimatic corridors to obtain an optimized disposition of the single buildings in order to create sufficient and comfortable ventilation for both external and internal areas.

At the building level, the use of the stack effect (integrated with a convective loop system and with a chimney) to realize a system of cooling and ventilation resulted in a good microclimate and adequate indoor conditions during summertime. The development of this system has grown little by little and has been tested and improved by continuous Fluent simulations. Incoming air is transported through the building by different chimneys, either for admission or for expulsion functions, which work by stack effect. In the admission chimney, during winter, the air flows into the glazed section, becomes warmer as a result of the passive greenhouse effect and is then distributed to each room. In summer, the air flows into an insulated section and provides cooling due to its speed and continuous circulation; moreover, shading systems at the windows and at the glazed chimney prevent overheating.

ACKNOWLEDGEMENTS
Acknowledgements are due to Carlo Brizioli, Isabella Calderaro, Valentina Chiodi and Pierpaolo Palladino.

INFORMATION
For further information, contact the architect Francesca Sartogo at Prau@fastwebnet.it and Professor Valerio Calderaro at Università La Sapienza Roma, Laboratorio sperimentale di progettazione ambientale, Rome, Italy

Source: Umbertide Case study – Prau Project

3.22 Apartment house

(a)

(b)

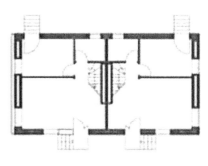

Source: Umbertide Case study – Prau Project

3.23 Detached house

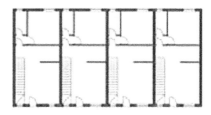

Source: Umbertide Case study – Prau Project

3.24 Row house

Source: Umbertide Case study – Prau Project

3.25 Housing courtyard

SUMMARY

At the urban level, different site characteristics (natural river and agricultural park; ex-industrial renewal; existing workmen's village and ancient 13th-century Borgo Minore; new railway connections; and new residential development) were coordinated within a unique, organic, energy-ecological holistic urban planning design. The urban comfort, microclimate, and site qualities are responsive to the urban design. Improving ventilation within the public open space has been carried out according to the direction of the prevailing wind during the various seasons (wind corridors); protection from winter wind (through rows of trees) and promotion of summer wind penetration have also been achieved, with a consequent reduction in building energy demand.

At the building level, microclimates have been created that are responsive to urban texture and building typologies. Optimization of the architectural form and functions of exterior and interior open and closed building spaces has occurred through Fluent simulations (fluid dynamic software simulations).

The building typology interventions utilize a cooling system for ventilation through a stack effect, integrated with the convective loop system and combined with the use of a chimney to achieve a cooling and ventilation system that is able to produce, in summertime, a good microclimate and adequate indoor conditions. The installation of obstruction/reflection elements outside the passive systems of the windows permits sun radiation to be screened during the hot period and reflects the direct solar radiation on the ceiling in winter time, thus using the ceiling as a thermal storage element and reflecting the remaining solar radiation in order to obtain uniform natural lighting system without glare.

REFERENCES

Braudel, F. (1987) *Il Mediterraneo, lo spazio, la storia, la tradizione*, Bompiani, Milano, Italy

Pollione, M. V. (1999) *De Architectura,* Bari Editore

Santamouris, M. and Askimopolous, D. (1996) *Passive Cooling of Buildings*, James & James, London

Chapter 4
Adelaide: A Warm Continental Climate

Veronica Soebarto

INTRODUCTION

This chapter discusses passive cooling strategies that are appropriate for a warm continental climate, with case studies located in the Adelaide metropolitan area in South Australia. Although the focus is on passive cooling, it is important to note that passive heating must also be taken into account because Adelaide experiences cool winters: for eight months of the year, the monthly average temperature is below 20°C.

The case studies presented were chosen to represent two major settings – urban (in the centre of the city of Adelaide) and countryside (the Adelaide Hills region). Three types of dwellings in the urban setting are presented – a townhouse unit (three-storey attached dwelling), an apartment unit (one-storey attached dwelling) and a detached cottage (there are three of these), while the house in the countryside is a single detached dwelling.

The indoor temperature and humidity of all dwellings were monitored hourly for a full year and a weather station was installed on the site.

CLIMATE

Adelaide, the capital city of South Australia, is located at 34.9° S latitude and 138.5° E longitude, 42.7m above sea level. It has hot and dry summers, cool winters, and is pleasantly dry during the other seasons. Adelaide is a particularly dry place, with an average monthly rainfall of 66mm during winter and only 23mm during summer. The hottest months are January and February, but the heat often continues until March. There can be several very hot days with very warm nights (days with maximum temperature of over 38°C) in a year and, in general, the diurnal range is quite significant (more than 10°C). In winter, several very cold days (days with maximum temperatures of less than 13°C) may occur for two or more days. Summer average relative humidity is around 42 per cent, while in winter the average relative humidity is 68 per cent.

The Adelaide Hills tend to be cooler than the city during summer days and nights, when they bring gully breezes to the Adelaide plains. The beachside suburbs also tend to be one to two degrees cooler during summer, due to sea breezes. On summer afternoons, the wind blows from the ocean (south-west); but sometimes a hot wind blows from the north, bringing hot and dry air from the desert. Not only is this wind unpleasant, it can also be dangerous as it may aggravate bushfires. On winter mornings, the wind blows from the north and north-east; but in the afternoons a cold wind blows inland from the ocean.

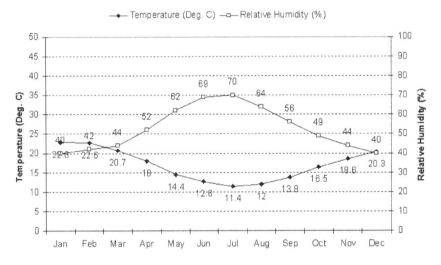

Source: Veronica Soebarto

4.1 Monthly average temperature and relative humidity for Adelaide, South Australia

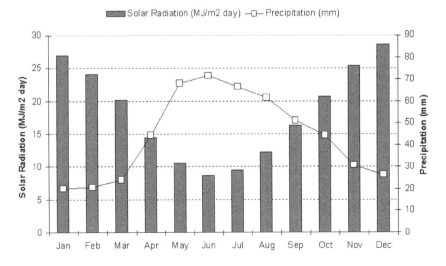

Source: Veronica Soebarto

4.2 Monthly solar radiation and precipitation in Adelaide, South Australia

SOLUTION SETS

Typological solution set for warm continental areas

The success in this climate of passively designed buildings (that is, buildings with no mechanical heating and cooling systems) depends upon a number of strategies that have to work together. They include proper orientation of the main rooms and windows, adequate size of the glazing, and the use of insulation, thermal mass and summer shading. As a general rule of thumb, whenever possible the longer side of the building should be oriented to the north. This will allow as much solar penetration as possible in winter, which is a requirement for passive heating. In summer, shading of the north-facing windows can be done with relative ease. West- and east-facing elevations should be kept to a minimum as the summer morning and afternoon sun can be quite intense. To minimize heat loss in winter, the roof and walls must be insulated and the use of double glazing is advisable. Thermal mass, such as uncovered concrete floors and internal brick walls, is required to help in moderating internal temperatures by averaging out the diurnal extremes. During summer days, this mass absorbs the heat, keeping the internal space cooler than outside. In winter, the mass stores the heat from either solar radiation or heaters, and releases it at night to help warm the internal space.

Table 4.1 Typological solution sets for a warm continental climate: detached house

Detached housing

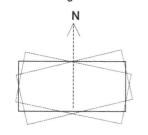

Roof can be extended to form a 'skillion' and enlarge the area for solar collectors

Summer sun

Winter sun

Smaller windows on south walls

Slab on ground

Form: compact unit

Floors: one, on ground

Solar Orientation (0° = north): 15°–20° when possible

Roofing: insulated pitched, minimum 20°

Solar protection: adjustable shadowing system to the north, east and west

Active system: solar hot water system

Passive system: passive ventilation, night ventilation/day closure, solar heating, thermal mass (floor and internal walls)

Glazed/opaque surfaces ratio: 30% for north-facing walls

Glazed surface/total floor area ratio: 20%

Layout: living rooms to the north, sleeping rooms to the south, services to the west

Daylighting/skylight orientation: 180° (south) if solar heat is unwanted; 0° (north) if winter solar gain is wanted, but must be shaded in summer

Insulation: required for roof and all walls

Source: Veronica Soebarto

Table 4.2 Typological solution sets for a warm continental climate: row house

Row housing

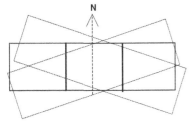

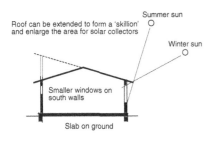

Roof can be extended to form a 'skillion' and enlarge the area for solar collectors

Summer sun

Winter sun

Smaller windows on south walls

Slab on ground

Form: row houses

Floors: three storey maximum

Solar orientation (0° = north): 15°–20° when possible

Roofing: top floor, insulated pitched, minimum 20°

Solar protection: adjustable shadowing system to the north, east and west

Active system: solar hot water system

Passive system: passive ventilation (stack effect), night ventilation/day closure, solar heating, thermal mass (floor and internal walls)

Glazed/opaque surfaces ratio: 30 per cent for north-facing walls

Glazed surface/total floor area ratio: 20%

Layout: living rooms to the north, sleeping rooms to the south, services to the west

Daylighting/skylight orientation: 180° (south) if solar heat is unwanted; 0° (north) if winter solar gain is wanted, but must be shaded in summer

Insulation: required for roof and all walls

Source: Veronica Soebarto

Except during hot days in summer and cold days in winter, comfort in non-air-conditioned houses can be achieved by simply opening the windows since most of the time the outdoor temperature and humidity are within the acceptable range. During hot summer days, most openings should be closed to keep the internal space cooler than outside. They can be opened later in the afternoon when the outside temperature has dropped. There are, however, a number of days in summer when the night-time temperature stays above 35°C. When this occurs, few passive cooling strategies can be used to achieve a thermally comfortable indoor space.

Summary

Suggested typology set for a warm continental climate:

- compact block shape, stretched along the east–west axis, if possible, with the main glazed area on the northern walls for passive solar heating;
- use of roof and wall thermal insulation;
- use of thermal mass (particularly in the floors), which should not be exposed to solar radiation;
- shading to exclude summer solar radiation, but allowing winter sun penetration;
- operable windows.

Four case studies are presented in this chapter. The first three (an attached townhouse unit, an attached apartment unit and a detached house, all found in Case study 4.1) are located in the same housing development in the city of Adelaide. The fourth (Case study 4.2) is a detached house in Norton Summit in the Adelaide Hills region, about 15km from the Adelaide central business district (CBD).

Case study 4.1 Christie Walk, Adelaide, South Australia

Veronica Soebarto

PROFILE

Table 4.3 Profile of Christie Walk, Adelaide, South Australia

Country	Australia
City	Adelaide, South Australia
Building type	Three-storey attached townhouse, one-storey attached apartment unit, three-storey detached cottage
Year of construction	2000–2002
Project name	Christie Walk
Architect	Paul Downton, Ecopolis Pty Ltd

PORTRAIT

The Christie Walk development is a mixed-density community housing project located in the heart of the Adelaide CBD, developed by Wirranendi Inc and created by Urban Ecology Australia Inc (UEA), a non-profit organization whose aim is to promote and create ecologically integrated human settlements in the city. The development is intended to be a pilot project demonstrating how communities can address the core issues for sustainable living in cities:

- water and energy conservation;
- material reuse and recycling; and
- healthy, people-friendly public spaces.

The project features onsite sewage and grey-water treatment; onsite storage of storm water; a solar hot water system; power from photovoltaics (when the project is completed); passive solar/climate-responsive design; use of recycled, non-toxic materials with low embodied energy; pedestrian-friendly spaces; local food production; a shared community garden and roof garden;

and reduced car dependency due to the inner city context. In the first two stages of the project three housing types were built: three-storey townhouses, apartments and detached cottages. The third stage will include a five-storey building with 13 apartments and a ground-level community area with a kitchen, dining or meeting room, library, toilets and laundry.

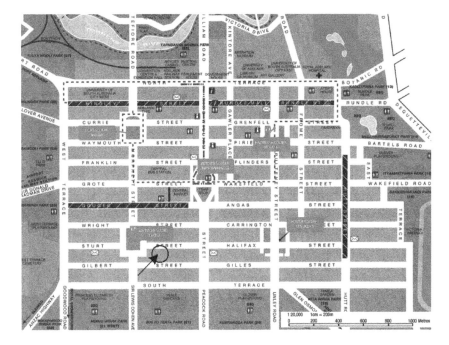

Source: Veronica Soebarto

4.3 The site of the first three case study buildings in the Adelaide central business district (CBD), as indicated by the arrow

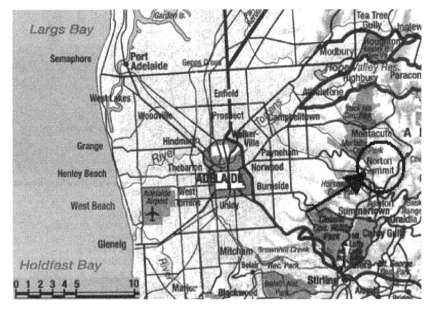

Source: Veronica Soebarto

4.4 Location of the fourth case study building, indicated by the arrow, in relation to the Adelaide CBD

Source: Ecopolis Architects Pty Ltd, Adelaide

4.5   Bird's eye view from the north–east direction of the Christie Walk development

CONTEXT

The idea for the project started in 1988 when a number of people in Urban Ecology Australia wanted to create and live in sustainable housing in the city. Part of UEA's focus was to remind people that our urban systems are part of what we call 'the environment' and, because they are where most of the people live and work, they are where the solutions for ecological living must be focused. The main objective of the design brief is to create healthy living with a high overall environmental performance.

The land for the development was owned by Wirranendi during construction, and individual properties were then sold on a community title. Each property owner also shares ownership and responsibility for the communal areas, including the community garden, a roof garden and a community house.

ECONOMIC CONTEXT

The project was initially funded by Wirranendi, whose members were drawn from the people who intended to live in the project. The non-profit structure of the development co-operative and building company played an essential role in keeping house prices within a range comparable to conventional inner-city properties in Adelaide. The buildings were mostly built with volunteer labour, which helped to reduce the start-up cost. House prices include all the community areas and facilities and range from Aus$120,000 to $350,000.

SITE DESCRIPTION

This development is located on the western side of the city of Adelaide. It is only two blocks away from the Adelaide Central Market, minutes from the main area of the CBD, and close to some of the parklands surrounding the city. The project is located on a relatively small (approximately the size of two quarter-acre blocks) and awkward T-shaped site, which is severely constrained, with buildings on or close to most of the boundaries. This site, however, was selected by UEA since the project was meant to demonstrate to the general public that it is not always possible to have an ideal site for passive solar orientation (as found in many individual passive solar houses in the suburbs or in remote areas), yet it is still possible to achieve a healthy, comfortable and environmentally sustainable dwelling, while maintaining the occupant's privacy, through careful and clever planning, design, use of building materials and landscaping.

BUILDING STRUCTURE

The townhouse building is located on the south part of the site and is elongated in an east–west direction. This location makes this building the only one in the site that has an ideal north orientation. Each unit has three floors; the ground floor consists of a small entry space, a sunspace, and an open living, dining and cooking space. There is a small courtyard at the back (south) side of each unit. The first and second floors consist of a north-facing balcony, a bedroom and a bathroom. Each level is connected by a spiral staircase, which also functions as a 'thermal flue' (see the section on 'Ventilation system').

The apartment block is located on the long part of the site, running on a north–south axis. This means that most of the openings are located on the east side of the building, and each unit has a very limited number of north-facing windows (or none at all). Each unit is a single storey with one bedroom, a bathroom, a living–dining space, a kitchen and an open space that can be made into a second bedroom. The unit on the ground floor also has a small courtyard on the west side.

There are three detached straw bale cottages on the site: two with two storeys and one with three storeys. Due to its location on the site, each of these houses also has a limited number of north-facing windows. The layout of each of these cottages is similar; the ground floor consists of an open space for living, dining and kitchen, while the first floor consists of a bedroom (or two bedrooms in the three-storey house) and a bathroom. The top floor of the three-storey cottage is an open space, used as a studio/office.

BUILDING CONSTRUCTION

The external walls of the townhouse and apartment buildings are constructed of 300mm thick load-bearing autoclaved aerated concrete (thermalite). Between the townhouse units a 400mm thickness of load-bearing, low-strength concrete ('earthcrete') is used for the internal mass. There is also some steel framing used in the apartment building. Load-bearing timber frames are used for the cottages and the external walls are made of rendered 500mm straw bales.

All of the buildings have reinforced concrete slabs on the ground floor. The townhouse building use *Pinus radiata* proprietary trussed joists for the first and second floor construction, while the cottages' upper floors use either plantation *Pinus* or recycled timber joists. A compressed straw panel (equivalent to particle board, but containing no woodchips or formaldehyde) is used for the floor decking of the apartment building and cottages. Reinforced concrete slabs are used for the apartment building. All concrete in the slabs and mass walls contains the maximum percentage of fly-ash.

All window frames are made of recycled timber. All fixed windows are double glazed, whereas the operable windows are single glazed and are equipped with aluminium flyscreens. Floor finishes are either *Marmoleum*, a modern variant of linoleum selected for its aesthetic merit and environmental credentials, or bamboo flooring. Locally produced ceramic tiles are used for all wet areas.

The roofs of the top floor of the apartment and the cottages are insulated with reflective foil and 200mm bulk (batt) insulation. The community roof garden also helps to cool the upper floor of the apartment building. Wall insulation of the townhouse and apartment buildings is provided by the 300mm thermalite walls, while the straw bales of the cottage walls also function as excellent thermal insulation.

All construction, finishes and paints are non-toxic. Either plantation or recycled timber is used for all woodwork, such as doors and cabinets.

VENTILATION SYSTEM

All of the buildings/units employ comfort ventilation most of the time; night-time ventilative cooling is used during summer. Vegetation surrounding the building filters and cools the outside air before it enters the building through small top-hung windows. These windows are set low on the wall to draw in cooler air. The outlets are high-level louvres, vents or ventable skylights. The latter are employed in the townhouse and cottage; the air is drawn to the skylights by convection through the stairwells. During hot summer days, all openings are closed, and external bamboo blinds are rolled down to shade the windows. The thermal mass helps to maintain the indoor temperature at a comfortable level. The small windows, vents or skylights are then opened at night to release warm air and bring in cooler outside air. This system helps to maintain the indoor temperature at a comfortable level.

(a)

(b)

(c)

Source: Toga Pandjaitan

4.6   (a) Townhouse building facing north; (b) apartment building facing east; (c) straw bale cottage facing west

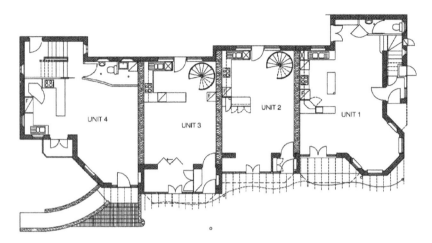

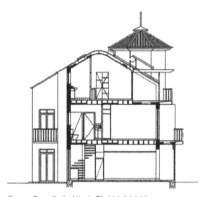

Source: Ecopolis Architects Pty Ltd, Adelaide

4.8   Cross-section of the townhouse

Source: Ecopolis Architects Pty Ltd, Adelaide

4.7   Ground floor of the townhouse building; the monitoring was conducted in unit 2

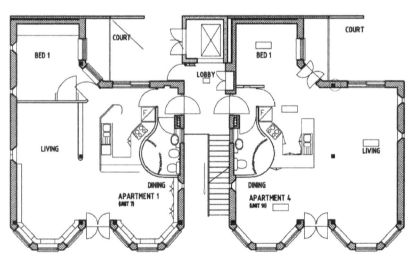

Source: Ecopolis Architects Pty Ltd, Adelaide

4.9   Floor plan of a ground-floor apartment unit

APPLIANCES

Most of the rooms in each dwelling are well lit, minimizing the need for electrical lighting during the day. A mix of compact fluorescent and incandescent globes is used for night-time lighting. None of the buildings is mechanically heated or cooled, although in some dwellings a small portable electric heater is sometimes used in the bedroom during cold winter nights.

Source: Ecopolis Architects Pty Ltd, Adelaide

4.10 Section of the apartment building

Ceiling fans are installed in some rooms to assist airflow during still days. Domestic hot water heating for all dwellings is provided by solar hot water heaters (with an electric backup system). Low-flow shower heads are used in every dwelling. In the earlier stage of the development, gas cookers were favoured due to their high energy efficiency; however, as the efficiency of electric cookers has improved, and since they give a cleaner indoor air quality, they have become the preferred option in the later stages of the project.

ENERGY SUPPLY SYSTEM

The mains electricity for the whole development is drawn from the grid; however, photovoltaic panels will be installed on the pergolas of the roof garden on top of the apartment building. It is expected that the development will generate sufficient excess electricity to be sold to the local energy utility since the dwellings use very little energy for space heating and cooling, water heating and lighting, having been carefully designed to be as self-sufficient as possible.

During 2003, the energy consumption of some of the dwellings, occupied by one or two persons, was monitored in a separate project. The results show that the energy usage for the one-person dwellings in this development is reduced by 60 per cent of the state average for similar homes, and 50 per cent for the two-person dwellings. The average electricity use of the one-person dwellings is 2293kWh per year and for the two-person dwellings is 3795kWh per year, compared with the state averages of 5469kWh per year and 7359kWh per year, respectively.

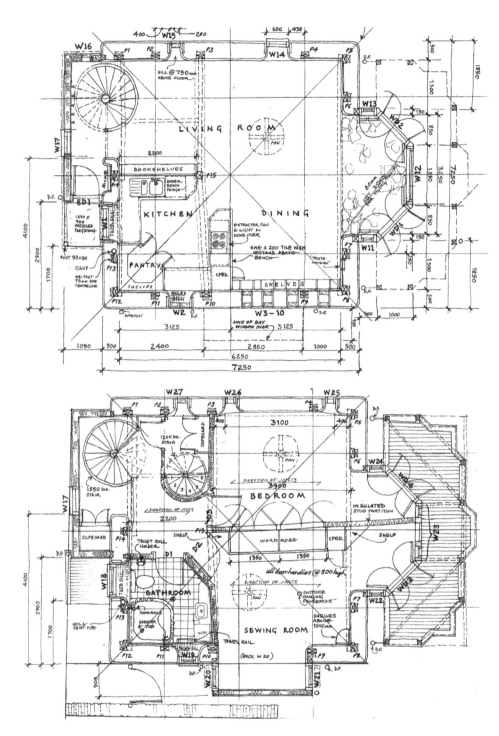

Source: Ecopolis Architects Pty Ltd, Adelaide

4.11   Ground and first floor of the straw bale cottage

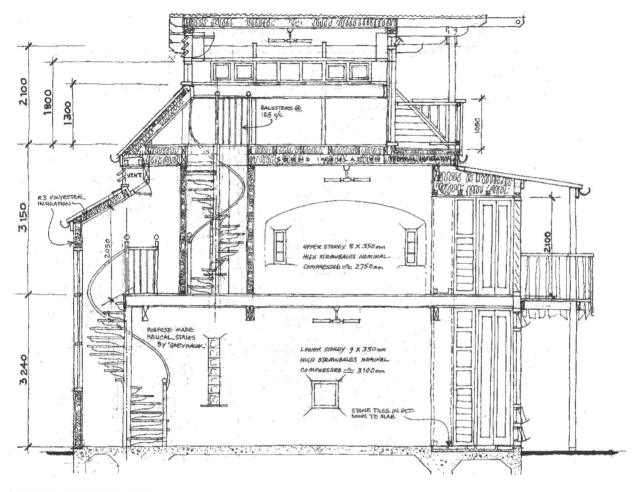

Source: Ecopolis Architects Pty Ltd, Adelaide

4.12 Cross-section of the straw bale cottage

Carbon dioxide emissions show similar reductions: the one-person dwelling produces CO_2 emissions of 2.54 tonnes per year and the two-person dwelling produces emissions of 4.21 tonnes per year, compared with the state average of 6.07 tonnes per year and 8.16 tonnes per year, respectively.

SOLAR ENERGY UTILIZATION

As described earlier, solar energy is used in active systems (solar water heating and, later, photovoltaic panels), as well as passively. Passive solar heating is avoided in summer and used in winter for all dwellings; however, due to the location and orientation of each dwelling, the amounts of winter solar gain are different. It is interesting to note, however, that although only the townhouse building has the ideal north-facing orientation, the winter

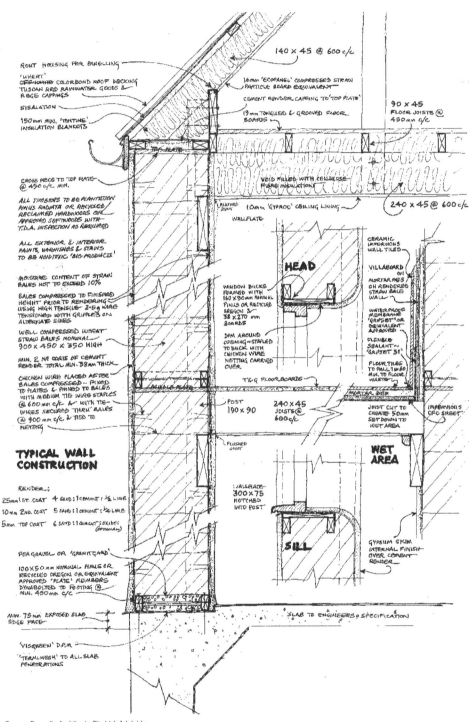

140 × 45 @ 600 c/c

ROUT HOUSING FOR PANELLING

'WHEAT'
OFF-WHITE COLORBOND ROOF DECKING
TUSCAN RED RAINWATER GOODS &
RIDGE CAPPINGS

16mm 'ECOPANEL' COMPRESSED STRAW
PARTICLE BOARD EQUIVALENT

CEMENT RENDER CAPPING 'TO TOP PLATE'

SISALATION

19mm TONGUED & GROOVED FLOOR
BOARDS

150mm MIN. 'TONTINE'
INSULATION BLANKETS

90 × 45
FLOOR JOISTS @
450mm c/c

TOP PLATE

CROSS PIECE TO TOP PLATE
@ 450 c/c MIN.

VOID FILLED WITH CELLULOSE
FIBRE INSULATION

ALL TIMBERS TO BE PLANTATION
PINUS RADIATA OR RECYCLED
RECLAIMED HARDWOODS OR
APPROVED SOFTWOODS WITH
T.D.A. INSPECTION AS REQUIRED

FLUSHED
JOINT

10mm 'GYPROC' CEILING LINING

240 × 45 @ 600 c/c

WALLPLATE

ALL EXTERIOR & INTERIOR
PAINTS, VARNISHES & STAINS
TO BE NON-TOXIC 'BIO-PRODUCTS'

CERAMIC
IMPERVIOUS
WALL TILES

MOISTURE CONTENT OF STRAW
BALES NOT TO EXCEED 10%

HEAD

VILLABOARD
ON
MORTAR PADS
ON RENDERED
STRAW BALE
WALL

BALES COMPRESSED TO FINISHED
HEIGHT PRIOR TO RENDERING
USING HIGH TENSILE 2·5⌀ WIRE
TENSIONED WITH GRIPPLES ON
ALTERNATE SIDES

WINDOW BRICKS
FORMED WITH
160 × 90mm NOMINAL
PINUS OR RECYCLED
OREGON &
38 × 270 mm
BOARDS

WATER-PROOF
MEMBRANE
'GRIPSET' OR
EQUIVALENT
APPROVED

WELL COMPRESSED WHEAT
STRAW BALES NOMINAL
900 × 450 × 350 HIGH

DPM AROUND
OPENING STAPLED
TO BRICK WITH
CHICKEN WIRE
NETTING CARRIED
OVER

FLEXIBLE
SEALANT
'GRIPSET 51'

MIN. 2 No COATS OF CEMENT
RENDER TOTAL MIN 38mm THICK

FLOOR TILES
TO FALL 1 IN 60
MIN TO FLOOR
WASTE

CHICKEN WIRE PLACED AFTER
BALES COMPRESSED – FIXED
TO PLATES & PINNED TO BALES
WITH MEDIUM TIE WIRE STAPLES
@ 600mm c/c A – WITH TIE-
WIRES SECURED THRU BALES
@ 900mm c/c & TIED TO
NETTING

T & G FLOOR BOARDS

MORTAR DPM

DOUBLE PLATE

POST
190 × 90

240 × 45
JOISTS @
600 c/c

JOIST CUT TO
CREATE 50mm
SET DOWN TO
WET AREA

IMPERVIOUS
CFC SHEET

FLUSHED
JOINT

**TYPICAL WALL
CONSTRUCTION**

WET
AREA

RENDER :

WALLPLATE
300 × 75
NOTCHED
INTO POST

25mm 1ST. COAT   4 SAND : 1 CEMENT : ½ LIME

10mm 2ND. COAT   5 SAND : 1 CEMENT : ¼ LIME

5mm TOP COAT   6 SAND : 1 CEMENT : OXIDES
(OPTIONAL)

GYPSUM SKIM
INTERNAL FINISH
OVER CEMENT
RENDER

SILL

PEA GRAVEL OR 'GRANITGARD'

100 × 50 mm NOMINAL PINUS OR
RECYCLED OREGON OR EQUIVALENT
APPROVED 'PLATE' MEMBERS
DYNABOLTED TO FOOTING @
MIN. 450mm c/c

MIN 75mm EXPOSED SLAB
EDGE FACE

SLAB TO ENGINEER'S SPECIFICATION

'VISQUEEN' D.P.M.

'TERMIMESH' TO ALL SLAB
PENETRATIONS

Source: Ecopolis Architects Pty Ltd, Adelaide

4.13 Construction details of the straw bale cottage

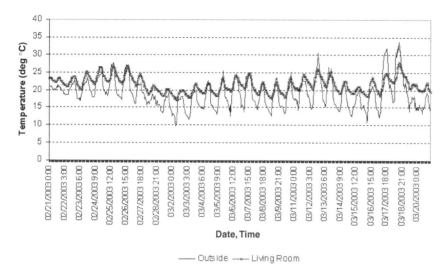

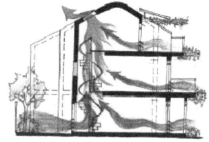

Source: Ecopolis Architects Pty Ltd, Adelaide

4.14 Stack ventilation system employed in the townhouse and cottage buildings

Source: Veronica Soebarto

4.15 One month's monitoring during summer 2003 of the townhouse unit, showing the indoor and outdoor temperature

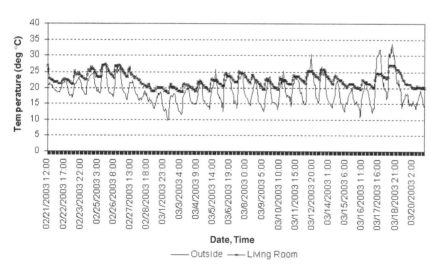

Source: Veronica Soebarto

4.16 One month's monitoring during summer 2003 of the apartment unit, showing the indoor and outdoor temperature

performance of the other dwellings is not significantly different. This is because solar gain is still made possible through the openings, even though they do not face north, and also because of the high insulation value of the window glazing and walls. Figures 4.17 to 4.19 show the relationships between indoor and outdoor temperatures of the three dwellings during the winter of 2003.

In summer, direct solar gain is avoided through the use of external shades (bamboo blinds), balconies, pergolas and overhangs, as well as vegetation. During the day, the external shades are usually drawn and all openings are shut. Reflective foil (and bulk insulation) in the roofs of the apartment building and cottage also help to reduce solar gain.

BUILDING HEALTH AND WELL-BEING

Providing a healthy living environment is the main goal of this development. Throughout the project, materials are selected very carefully. As described in the section on 'Building construction', only non-toxic construction and finishes are used and formaldehyde and PVC are avoided. All concrete used has a maximum content of fly-ash, a waste product of power stations, to reduce the amount of cement used. Cement production is one of the largest contributors of greenhouse gas emissions.

Noise control in each dwelling is provided by the massive walls and double-glazed windows, which also provide thermal advantages, as previously discussed. Air pollution is reduced due to the small number of cars within the site. Being located in the city, very close to good public transport, there is little need for car parks.

The landscaping of the site is intended to create an interesting and productive community space. This includes the addition of vegetable and roof gardens, which also demonstrates that it is possible for a tight urban site to produce food. The paving of the site was done by the residents themselves, resulting in a creative and attractive outdoor environment.

WATER RECYCLING AND CONSERVATION

Storm water from the roof, balconies and other surfaces is collected in two 20,000 litre underground tanks. After filtering, this water is used for irrigation on the site and for toilet flushing. Grey water and black water are treated in a chlorine-free sewage treatment plant. It is planned that the composted solids will be taken to rural sites for use as fertilizer, while the filtered effluent will be returned to the site as a second-class water supply through the onsite storm water system. Most of the vegetation consists of native plants with lower water needs; some exotic plants are also used, where appropriate, as part of the passive design strategies.

PLANNING TOOLS APPLIED

No specific planning/design tools were applied during the design process. The design was mainly done by the architect through the application of

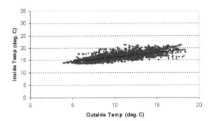

Source: Veronica Soebarto

4.17 Indoor and outdoor temperature during winter 2003 of the townhouse unit

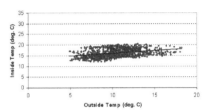

Source: Veronica Soebarto

4.18 Indoor and outdoor temperature during winter 2003 of the apartment unit

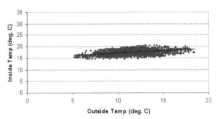

Source: Veronica Soebarto

4.19 Indoor and outdoor temperature during winter 2003 of the straw bale cottage

passive design principles for this climate. Thermal simulation was conducted by the author after the project was completed to analyse the effects of some of the building elements on the building performance. A calculation of the embodied energy of the construction was also performed after the project was completed.

The three dwellings described above were monitored by the University of Adelaide from February 2003 to February 2004. A weather station was installed on the roof of the apartment building. The energy use (electricity and gas) of six dwellings was monitored by Urban Ecology Australia, led by Monica Oliphant from the University of South Australia.

The occupants of the three dwellings were interviewed by the author and all of them expressed their satisfaction with the performance of their dwellings and of the whole development, in general.

INFORMATION AND OTHER REFERENCES

Commonwealth of Australia (2003) *Your Home: Design for the Lifestyle and the Future*, Section 7.3 'Medium density – Christie Walk', On CD and downloadable from www.yourhome.gov.au/index.htm (accessed 12 January 2004)

Downton, P. (2003a) Interview by Veronica I. Soebarto, 4 March 2003, tape recording; transcript available at the School of Architecture, Landscape Architecture and Urban Design, University of Adelaide, South Australia

Downton, P. (2003b) *FACT SHEET CW01 Urban Ecology Australia Whitmore Square EcoCity Project Christie Walk*, www.adelaide.sa.gov.au/SOC/pdf/ Christies_Walk.pdf (accessed 12 January 2004)

Ecopolis (2003) *Christie Walk*, www.ecopolis.com.au/projects/CWGBC (accessed 12 January 2004)

Occupants of the townhouse and apartment units (2003) Interview by Veronica I. Soebarto, 15 and 16 March 2003, tape recording; transcripts available at the School of Architecture, Landscape Architecture and Urban Design, University of Adelaide, South Australia

Oliphant, M. (2004) *Inner City Residential Energy Performance: Final Report*, Submitted to the State Energy Research Advisory Committee, Urban Ecology Australia, Adelaide, June

Sustainable Housing and Urban Setting (2003) *Proceedings of the 37th Australian and New Zealand Architectural Science Association (ANZAScA) Conference*, Faculty of Architecture, University of Sydney, pp575–587

(a)

(b)

Source: Veronica Soebarto

4.20 Landscaping done by the residents

Case study 4.2 Kawanda Muna, Norton Summit, South Australia

Veronica Soebarto

PROFILE

Table 4.4 Profile of Kawanda Muna, Norton Summit, South Australia

Country	Australia
City	Norton Summit, South Australia
Building type	Detached
Year of construction	1999
Project name	Kawanda Muna/Smith-Miller House
Architect	John Maitland, Energy Architecture

PORTRAIT

Set in the hills above Adelaide, this 123 square metre two-bedroom-and-study home is a realization of the owners' desires to build an ecologically sustainable family home for three at a price most people can afford. The house is self-sufficient for water and wastes, and uses minimal energy. The building construction uses recycled materials and minimizes embodied energy. The design of the house ensures that its ecological footprint is as small as possible.

The site location for the building was selected for maximum northern exposure (the house is called *Kawanda Muna*, which means 'north in front' in Kaurna, the local language) and to catch the breeze for natural ventilation, with minimum destruction to the site. Energy for space heating is reduced by allowing maximum solar heat through the predominantly northern glazing and storing the heat in the rammed earth walls and exposed concrete floor. No mechanical cooling equipment is installed and summer cooling is provided by natural ventilation and by reducing the heat gain with existing trees, overhang and internal blinds. All walls and the roof are insulated, and the building uses a significant amount of recycled materials. In 2000, the house received a commendation in the category of 'Environment' from the South Australian Chapter of the Royal Australian Institute of Architects.

CONTEXT

The main goal of building this new, privately built detached house was to create an ecologically sustainable house, which:

- is self-sufficient for water, wastes and electricity;
- uses as many recycled materials as possible without spoiling the aesthetics or function; and
- makes best use of the site, with minimal damage to it.

In 1998 the owners approached the architect, John Maitland, whose work has always reflected his concern for the environment. The design process started with several site visits by the owners and architect to establish the location of the house. The design was created by the architect in consultation with the owner; the architect was involved until the building process began.

ECONOMIC CONTEXT

The building has a floor area of $123m^2$ with an additional $50m^2$ of garage and workshop. This owner-funded house cost around Aus$125,000 in the year 2000 for the house construction alone; with all the fit-outs (kitchen, cabinets, rainwater tanks, etc.), it cost around Aus$171,000. In addition, a Aus$15,000 photovoltaic panel was installed, with a subsidy of Aus$5000 from the local utility company. The land value is Aus$90,000.

SITE DESCRIPTION

The house is located on Norton Summit Ridge, in the Adelaide Hills (35° south latitude, 360m above sea level). The 0.7ha site slopes down toward the north. There are a number of *Eucalyptus* trees that have been maintained as much as possible to provide a natural setting, as well as shading for the house. Native plants of local provenance are being directly seeded to restore the condition of the severely degraded understorey, which had been grazed by cattle and sheep and abandoned during the 1970s to introduced plants such as broom, blackberry and gorse. The site is 300m from a main road, has mains electricity and is some 15 to 20 minutes from the CBD.

BUILDING STRUCTURE

The house consists of two separated, rectangular, north-facing structures joined by a level-changing component, which is the main entry space. The north structure (on the lower level) is an open space comprising living, dining and kitchen areas. The south structure (on a higher level), which is turned 15° toward the east, consists of two bedrooms (one at each end), a study room and a wet area (bathroom and laundry). From the south wall of this part, which is partially underground (earth bermed to 1m), there is access to the garage. The roofs slope up 15° to the north to allow maximum penetration of sunlight, except for the roof over the entry, which slopes up 35° to the south to provide the optimal slope for the solar collector.

BUILDING CONSTRUCTION

The building has a hybrid system of load-bearing walls and timber frames holding the timber roof structure. External walls are constructed of stabilized rammed earth (400mm thick for the east and west walls, 200mm thick for the north and south walls), with 90mm concrete blocks for the south wall of the sleeping block and the east wall of the kitchen, all of which are insulated with R 2.5 fibreglass insulation on the outside and clad with corrugated metal (colourbond).

The 100mm polished concrete floors provide thermal mass and do not need tiles or other finishes. The north wall and roof structures are of used Oregon timber (except for the roof purlins), which had been collected by the owner for a number of years. The door and windows frames, however, are of new western red cedar. R 3.5 roof insulation was installed above the plasterboard ceiling.

VENTILATION SYSTEM

The building uses cross-ventilation through the openings, the siting of the building enabling breezes to enter easily. High window openings allow for a stack ventilation effect. A small pond containing rainwater in the ventilation path provides evaporative cooling. Ceiling fans are also used to create air movement when needed. The house has no mechanical cooling. Space heating is provided in the living space only, by a slow combustion heater, and occasionally a radiant heater is used in the study room.

APPLIANCES

Energy efficient appliances used include a 378W dishwasher, a 640W refrigerator, a 237W washing machine (no electric clothes dryer) and a gas stove/oven. Rainwater is the sole source of water, which is heated with a 6 square metre solar collector with a 380 litre water tank, connected to the slow combustion heater, and an instant gas booster for the kitchen. Natural light is used in all rooms during the day, while compact fluorescent lights (16W, 24W, 26W, and 36W) are used at night. The average lighting load is 2W/m^2 in the living/dining block and 2.6W/m^2 in the sleeping/bathroom block.

ENERGY SUPPLY SYSTEM

The house is connected to the mains electricity grid; however, it also generates power from grid-interactive photovoltaic (PV) arrays of twelve 86W modules installed on the property. The power generated by the photovoltaic panel is distributed back to the mains electricity grid, so the owners receive a rebate on their utility bills for the electricity generated.

The electricity is used for lights and appliances, while gas (LPG) is used for cooking and backup for water heating, and wood (from the site) is used for space heating.

SOLAR ENERGY UTILIZATION

The site enjoys a daily average of five sunshine hours in winter and ten hours in summer. Daylight is utilized throughout the day in all rooms in the building. Solar heat gain through the north-facing glass is the main source of space heating in the winter. During summer, the solar heat gain is reduced by the roof eaves and internal blinds, while the windows on the west walls are shaded by trees. Due to a beautiful view to the east, a larger window was installed on that side, shaded by an external awning.

BUILDING HEALTH AND WELL-BEING

The house employs four main principles: refusing to waste resources; reusing others' cast-offs; recycling as much as possible; and reducing overall size to conserve resources. Ninety per cent of the timber used in the building is second-hand. A number of materials, including electrical fittings, wire, and plumbing, were collected over the years by the owners.

Source: Energy Architects, Adelaide

4.21 Floor plan of Kawanda Muna house

Source: Energy Architects, Adelaide

4.22 West elevation of Kawanda Muna house

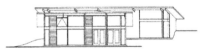

Source: Energy Architects, Adelaide

4.23 North elevation of Kawanda Muna house

Source: Evyatar Erell

4.24 North-east corner of Kawanda Muna house

Source: Evyatar Erell

4.25 Living room with rammed earth wall in Kawanda Muna house: the high window can be opened for ventilation

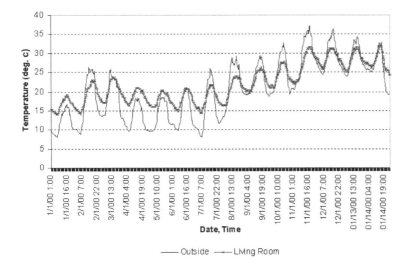

Source: Veronica Soebarto

4.26 Two weeks' monitoring during the hottest period in summer 2001

The building's annual energy use is much lower than that of a standard house in South Australia, as projected by the Australian Greenhouse Office, resulting in a lower greenhouse gas emission. A study conducted by the author in 2000 showed that the house's total annual energy use (electricity, gas and wood) was about 31.2GJ, whereas the projected energy consumption of a standard house in South Australia for the same period was 41.5GJ. The estimated CO_2 emission of all fuel energy used in this house in 2000 was 5.7 tonnes per year, whereas the state average for similar houses was 7.1 tonnes per year.

Despite the fact that the house uses less energy than a standard house in the same area, its performance can still be improved. Currently, the amount of window glazing is rather excessive, resulting in notably high levels of winter heat loss and heating energy. This large glazed window area also caused the indoor temperature to reach about 32°C late in the afternoon during hot summer days (when the outside temperature reached 37°C). If reducing the size of the window is not an option, due to the desire to maintain the view, double glazing should at least be used, and removable external shades may have to be fitted during hot summer days.

WATER RECYCLING AND CONSERVATION

The main source of water for this house is rainwater, which is collected in two 22,700 litre tanks and pumped up with a solar-powered pump to a 1000 litre header tank. Water for bushfire protection (since the house is located in a bushfire area) is obtained from a neighbouring storage tank of dam water.

Grey water and black water are treated with a Dowmus wet composting system with a biolytic filter. This system uses earthworms to process wastes, including kitchen scraps and cardboard, but has a normal flushing arrangement. The composted effluent will be removed every five to eight years.

PLANNING TOOLS APPLIED

No specific planning/design tools were applied during the design process; however, the roof eaves were designed carefully by considering sun angles. The building's indoor thermal performance was monitored for more than one year by the University of Adelaide. A weather station was installed on the north-east side of the house. Thermal simulation was conducted to analyse the design and predict its performance. Energy used was recorded and the embodied energy of the materials was calculated. The occupants were interviewed and they expressed satisfaction with the performance of the house. They found the house's openness, ease of communication and sense of 'outdoors inside' pleasant and relaxing. The house is comfortable and stimulating to live in.

OTHER INFORMATION AND WEBSITES

Commonwealth of Australia (2003) *Your Home: Design for the Lifestyle and the Future*, Section 7.6b 'Cool temperate – Kawanda Muna', on CD and downloadable online, www.yourhome.gov.au/index.htm (accessed 12 January 2004)

Soebarto, V.I. (2000). 'A low energy house and low rating: What is the problem?', in Soebarto, V. and Williamson, T. (eds) *Proceedings of the 34th Conference of Australia and New Zealand Architectural Science Association*, Adelaide, 1–3 December, pp111–118

Smith, J. (2004) kawanda@senet.com.au (accessed 12 January 2004)

Chapter 5
Tehran: A Hot Arid Climate

Vahid Ghobadian, Neda Taghi and Mehrnoush Ghodsi

INTRODUCTION

The Iranian plateau is situated in a dry geographic region. The dry climate of Northern Africa and the Middle East continues into Iran and Central Asia. Average precipitation in Iran is less than the world average. It should be noted that even though Iran is classified generally as a dry country, it includes many different climatic conditions. Iran can be divided into four basic climatic regions:

1   *Northern shores* (temperate climate): this region, which is to the south of the Caspian Sea, has the highest precipitation in the country and there are very dense forests in the highlands and intensive agriculture in the lowlands. It is cold in winter and hot and humid during the summer season. Average annual precipitation is about 1.5m. Relative humidity is above 70 per cent throughout the year.
2   *Mountain and high plateau region* (cold climate): the high mountain range of Alborz is situated to the north and Zagros is situated to the west of the country. This region is cold and dry in winter and mild and dry in summer. Average annual precipitation is about 30cm.
3   *Southern shores* (hot and humid climate): the hottest area of Iran is along the Persian Gulf and Oman seashores. It is mild in winter and hot and humid during the long summertime. In this region the average annual rainfall is less than 20cm and relative humidity is above 50 per cent throughout the year.
4   *Central plateau* (hot and dry climate): this region is larger than the other regions and covers most of the central Iranian plateau. It is cold and dry in winter and hot and dry in summer. Tehran, the capital of Iran is situated in this region.

CLIMATE

Tehran is at 35° 41' N latitude and 51° 19' E longitude and is situated 1191m above sea level.

The heating period is from mid November to early March. Most of the precipitation is from the middle of autumn to the middle of the spring season; of course, in winter, this comes as snow. Average annual precipitation is 218mm. The predominant wind direction in winter is from the west.

The cooling period is from early June to the middle of August. The predominant wind direction is from the south and south-east. There is very little rain in summertime. Because of the dry conditions, temperature fluctuation between day and night is rather high: on average, about 8 degrees in summer and 10 degrees in winter.

SOLUTION SETS

Typological solution sets for a hot arid climate

The hot arid region in Iran is hot and dry in summer and cold and dry in winter. Because of the lack of humidity in the air and the distance from the sea and ocean, temperature fluctuation between day and night is rather high in all seasons. The third element that is important is the prevalence of sandstorms from desert areas, which may blow during all months of the year in places near the central desert part of the country.

For human comfort in this area, there is a need for cooling in summer and heating in wintertime. Traditionally, inwardly oriented houses – with central courtyards – would protect the interior of the buildings from the frequent sandstorms. These houses are called 'four seasons' houses because the north side of the house, which receives direct sunlight, is used as the family residence during the cold months of the year. The south side of the house, which is always in shade, is used during the summer months. Wind catchers are usually built on top of the summer section (see Figure 5.4). These houses are built with heavy masonry walls and vaults, which act as thermal masses and therefore reduce temperature fluctuations between day and night.

The urban texture of traditional cities in this area is compact. In this way, the whole city is protected against sand storms, and winter heat loss through outside walls is also minimized.

An old Iranian proverb states: 'There is logic in proven practices.' Therefore, for the design of new sustainable solar houses with regard to environmental, social, cultural and economical factors and updated traditional practices, we suggest the following typologies for apartments, and detached and row houses.

Suggested typology set for a hot arid climate

A compact block shape for detached houses is recommended, extended along the east–west axis. House plan and orientation should be directed towards the

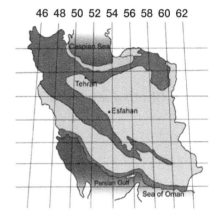

46 48 50 52 54 56 58 60 62

Northern shores -temperate climate
Mountain & high Plateau -cold climate
Southern shores -hot & humid climate
Central Plateau -hot arid climate

Source: Behrooz Pakdaman, Mehrnoush Ghodsi

5.1 Climatic regions of Iran

south. Row houses should be two-storey dwellings, including a basement, with building rows arranged along the east–west axis. For apartments, a building block should extend along the east–west axis, all apartments having direct access to sunlight from the south side.

For all typologies, the south side should be transparent with shading on top of it in order to receive maximum winter solar heat and to block summer solar radiation. The east and west sides of the buildings should be protected from summer sun by trees, adjacent buildings or anything that would cast shade on these two sides. The north side should have minimum openings to provide cross-ventilation in summertime and minimum heat loss in winter.

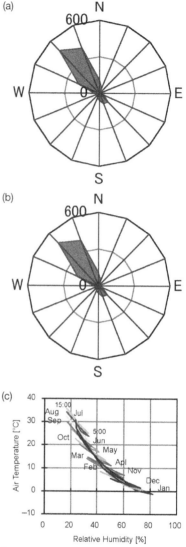

5.2 (a) Winter wind rose; (b) summer wind rose; (c) humidity

Source: Tamaki Fukazawa

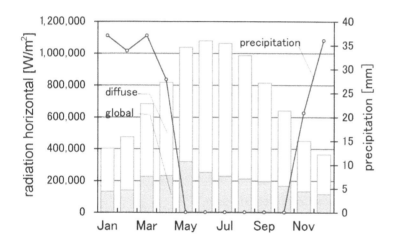

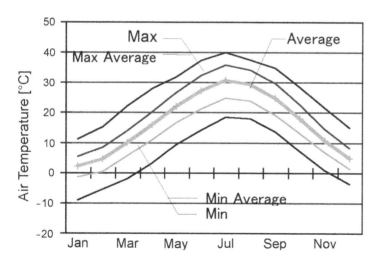

Source: Tamaki Fukazawa

5.3 Yearly temperatures for Tehran

Detached houses

The staircase can be used as a wind catcher for cross-ventilation or for its chimney effect when there is no wind. The main living spaces – such as living and dining rooms and bedrooms – are best located on the south side. The secondary spaces – such as staircase, bathroom, toilet and hallway – are best located on the north side to act as a buffer space.

The outer skin of the building should be completely insulated with adequate thermal storage mass (masonry structure) in the internal volume of the building.

Row house

There are five designs for row houses for the north-west of Tehran. The row houses are joined together on the west and east sides to minimize conduction heat loss through the walls in winter and solar heat gain in summer.

For the best winter solar heat gain, the main elevations of the houses are oriented towards the south. Most of the major spaces, such as living rooms, dining rooms and one of the bedrooms, are situated on the south side.

The south walls of the houses are transparent, with 110cm wide shades above the windows of each floor. These shades block the high altitude sun radiation from penetrating the rooms during summertime, while the winter sun can penetrate deep into the major spaces of the houses and bring warmth and light inside. Since the building material is masonry, the excess heat will be stored in the building fabric for night time.

Small swimming pools in the basements are placed right behind south-facing windows. Water in these pools is used as thermal mass. The heat gained from direct solar radiation during daytime will be stored in water for night-time heating. The sprinklers above the pools will keep the place cool – through evaporative cooling – in the summer. The windows on the north side are much smaller than on the south side because of the colder conditions on the north side of the building during the winter season. These windows provide views and daylight on the north side and facilitate cross-ventilation during the hot months.

A staircase is located in the middle of each house. The room at the top of the staircase on the roof is used as a wind catcher; since the summer wind is from the south and south-east, opening the door of this room allows the summer wind to come in. It should be noted that the spaces between stairs are open so that air can circulate through the staircase easily. During the summer, if the door is kept open even when there is no wind, the house will be cooled by the chimney effect.

The buildings are ground coupled, and the ground acts as a substantial thermal mass and insulator for the basements, so temperature fluctuations between day and night will be minimized and less energy will be consumed.

Except for the west and east walls, the outer skin of each house is insulated. The north wall is made of two layers of bricks (20cm thick for the inside layer and 10cm for the outside layer) and one layer of glass wool insulation (5cm thick) in between. There is one sheet of vapour barrier on each side of the insulation. On the roof, there is a 10cm layer of hard polystyrene.

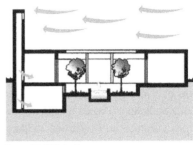

Source: Vahid Ghobadian

5.4   Section of a traditional four-season house

Source: Neda Taghi

5.5   Views of the south (equator-facing) elevation of row houses

All of the windows are double glazed. The type of glass chosen for the windows blocks ultraviolet radiation from entering the buildings.

Thick cloth curtains are used as night insulation for windows during winter nights. There are 4 square metres of flat-plate hot-water solar collectors on the rooftop of each house. This is adequate for a family of four in Tehran. Rainwater from the roofs and yards is directed towards flowerbeds in the courtyards.

Since the summer wind is mainly from the south, the yards with pools, sprinklers and plants are situated on the south side of the buildings. In this way, the wind will become cooler and more humid before entering the buildings. Deciduous trees are planted in the yard in order to provide shade for the

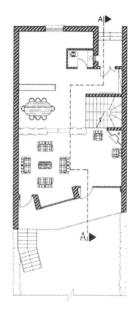

Ground Floor Plan

0 1 2 3 4 5

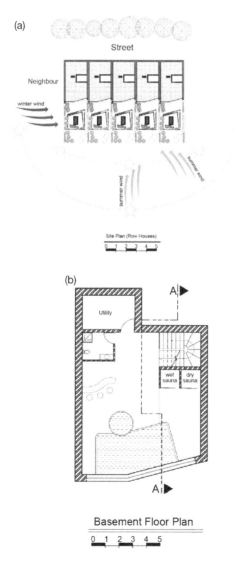

Site Plan (Row Houses)

0 1 2 3 4 5

Basement Floor Plan

0 1 2 3 4 5

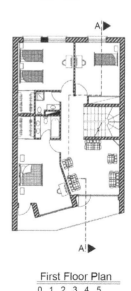

First Floor Plan

0 1 2 3 4 5

Source: Neda Taghi

5.6 Row house: (a) site plans; (b) basement floor plan; (c) ground-floor plan; (d) first-floor plan of a row house

buildings in the summertime. These trees also humidify the air and direct the summer wind towards the buildings.

Apartments

There are two four-storey blocks of apartments. In the first block, there are four apartments on each floor, totalling 16 apartments. Eight of them are 106 square metres in extent, with two bedrooms, and eight are 132 square metres, with three bedrooms. In the second block, there are only four-bedroom apartments, each 250 square metres in extent. On each floor there are two apartments: a total of eight apartments in the block. These blocks are extended along the east–west axis. All of the major spaces such as living, dining and bedrooms areas are situated on the south side in order to have the best view of Tehran and

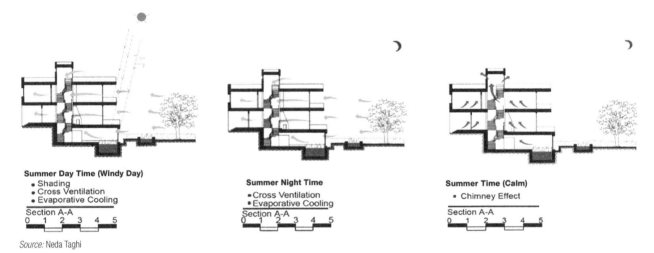

Summer Day Time (Windy Day)
- Shading
- Cross Ventilation
- Evaporative Cooling

Section A-A
0 1 2 3 4 5

Summer Night Time
- Cross Ventilation
- Evaporative Cooling

Section A-A
0 1 2 3 4 5

Summer Time (Calm)
- Chimney Effect

Section A-A
0 1 2 3 4 5

Source: Neda Taghi

5.7 Summer thermal strategies for the row house in Figure 5.6, as seen through section A–A

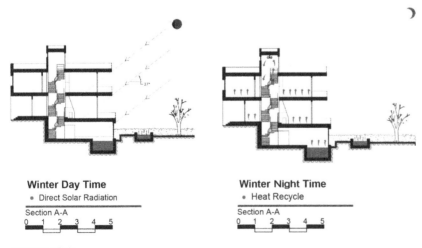

Winter Day Time
- Direct Solar Radiation

Section A-A
0 1 2 3 4 5

Winter Night Time
- Heat Recycle

Section A-A
0 1 2 3 4 5

Source: Neda Taghi

5.8 Winter thermal strategies for the row house in Figure 5.6, as seen through section A–A

to receive direct sunlight and heat in the winter. There are balconies in front of the south-facing openings that block direct solar radiation during the summer.

The southern orientation of the blocks has another advantage. The blocks are rather long and narrow and since the predominant summer wind in Tehran is from the south and south-east, this means that the summer wind can penetrate the buildings, so cross-ventilation can be easily achieved. There are no openings on the east or west side of the buildings. Pine trees on the west side of the site block the winter wind.

(a)

Source: Mehrnoush Ghodsi

5.9 (a) South elevation of the apartment complex; (b–c) typical floor plans of three-bedroom apartments

(b)

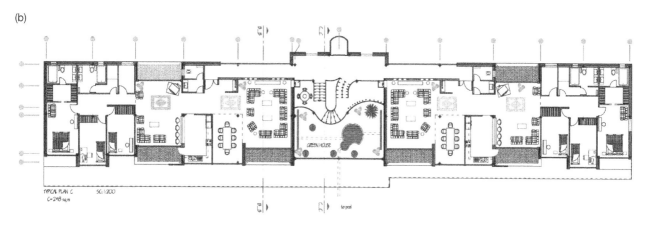

(c)

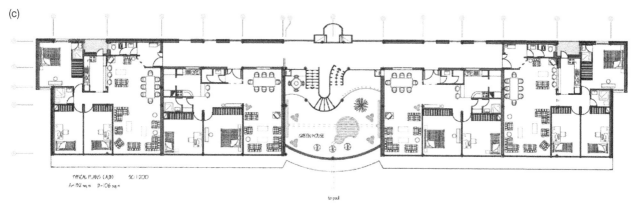

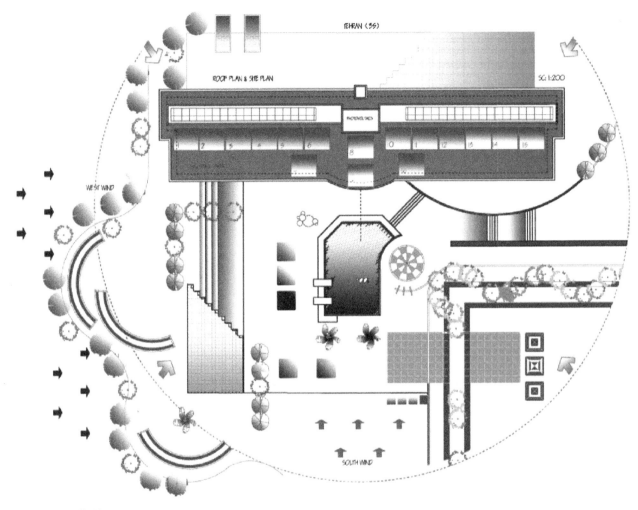

Source: Mehrnoush Ghodsi

5.10 Site plan: photovoltaic (PV) and solar collectors on top of the roof produce electricity and hot water for residents

There is a large greenhouse in the middle of each block. In wintertime, solar heat collected in the greenhouse during the daytime is transferred to a stone bin in the basement. At night, the heat from the stone bin is transferred to the upper storeys. In summertime, the sprinklers on the two side walls of the greenhouse will make the place cool and humid. This air from the greenhouse is transferred to the adjacent apartments.

There is another greenhouse on top of each block. These greenhouses collect solar heat in winter and transfer it to the lower storeys. In summertime, they act as wind catchers and direct the southerly wind to the residential floors.

(a)
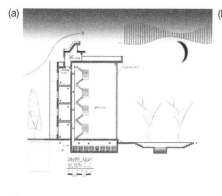

(b)
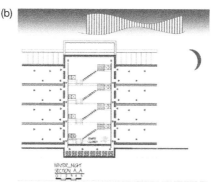

(c)
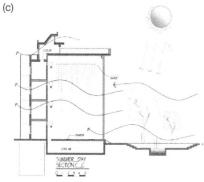

(d)
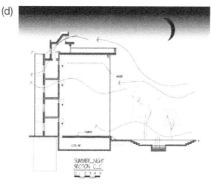

Source: Mehrnoush Ghodsi

5.11 The greenhouse acts as a passive heat collector in winter and direct evaporative cooler in summer. A roof pond is used to store the water. Rock bins in a basement below the basement are used to store heat. (a) During the day in winter, heat is collected in the greenhouse and stored in the rock bins below. At night insulation is placed over the glass façade to prevent heat loss; (b) during winter nights, a thermal siphon transmits heat collected in the rock bins to the upper floors during a winter night; (c) in summer, water from the roof pond is used to supply water to misters (water sprays) suspended from the ceiling of the greenhouse. The water evaporates and cools the dry air which is driven into the greenhouse by cross ventilation; (d) a similar process is used at night

(a)
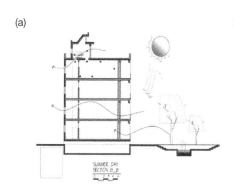

(b)
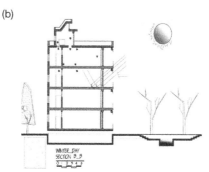

Source: Mehrnoush Ghodsi

5.12 Apartments: (a) Cross-ventilation during a summer day; (b) winter heating

(a)
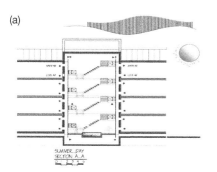

(b)
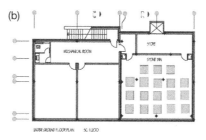

Source: Mehrnoush Ghodsi

5.13 (a) Typical section through the greenhouse. Air is moved from the basement through a plenum to adjacent rooms; (b) basement floor plan of apartments showing the rock storage bins

When there is no wind, the wind catchers act like chimneys and funnel the hot air from the inside to the outside.

SUMMARY

In Iran, buildings account for about 40 per cent of total energy use and in large metropolitan cities like Tehran, 30 per cent of pollution derives from buildings. By designing sustainable solar housing, which is a socially, economically and environmentally viable alternative to typical houses in Tehran, we can look for cleaner, healthier and more sustainable development in the future.

Case study 5.1 Boroujerdi House, Kashan, Iran

Vahid Ghobadian, Islamic Azad University

PROFILE

Table 5.1 Profile of Boroujerdi House, Kashan, Iran

Country	Iran
City	Kashan
Building type	Semi-detached, courtyard
Year of construction	1875–1876
Project name	Boroujerdi House
Architect	Ali Maryam

Source: Vahid Ghobadian

PORTRAIT

This building is a typical traditional four-season house in the hot arid climate of the central plateau of Iran. It is located in the city of Kashan, which is 258km south of Tehran. It was built approximately 130 years ago as a house; but now it is being used as the cultural heritage office in Kashan. The building is a good example of a sustainable traditional house in Iran. Economically and socially, the building was, and still is, appropriate for this place. The building materials, mainly brick and adobe, are recyclable. The energy used for construction, occupation and destruction of these buildings is at a minimum.

CONTEXT

The building design, materials and construction methods are all traditional, but may be modified for present-day requirements; the building was

(a)

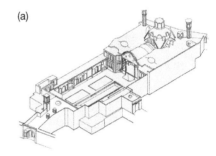

(b)

Source: Iran National Heritage Organization

5.14 Boroujerdi House, Kashan, Iran: (a) isometric; (b) front view

climatically designed in accordance with the economical and social context of the region. There are many lessons to learn from it for designing, building and using modern houses.

ECONOMIC CONTEXT

This house was built for a wealthy local merchant. It was renovated about 30 years ago, to be used as an office complex. At present, compared to a modern building of the same size, it costs less than half as much to maintain a comfortable environment inside the building throughout the year.

SITE DESCRIPTION

The lot is 1700 square metres in the city of Kashan, 34° N latitude, 955m above sea level. The city has cold winters and hot and dry summers, with frequent dusty winds from the deserts on the eastern side of the city. The house is located in the old neighbourhood of the city. All of the old houses in this area are detached or semi-detached, with central courtyards. The new ones are row houses.

BUILDING STRUCTURE

The building has a large central courtyard. The only opening to the outside is the entrance door. Except for this door, all of the openings are towards the central courtyard. In order to enter the house, one has to pass through the entrance door, a circular vestibule and a long corridor, only then reaching the central courtyard. Circulation, daylight, ventilation and views are all realized by way of the courtyard. The building is basically divided into two parts: the

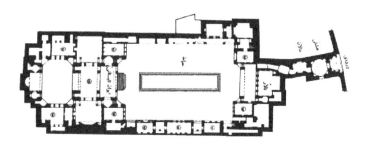

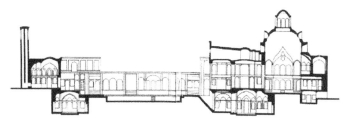

Source: Iran National Heritage Organization

5.15 Boroujerdi House: (a) floor plan; (b) section

Source: Farid Moalem

5.16 A view from under the dome of the summer residence

winter residence on the north side and the summer residence on the south side. The building has a ground floor and a basement. The basement is usually used during summertime, especially in the afternoons, since during this time it is cooler compared to the ground floor.

BUILDING CONSTRUCTION

The building is made with load-bearing walls, vaults and domes. The building material is adobe and brick, which can be easily reused for new constructions. The outside surface of the building is exposed brick and inside it is stuccoed with plaster.

The walls are approximately 60cm thick and act as a thermal mass, minimizing the house's temperature fluctuations between day and night.

VENTILATION SYSTEM

Since the building has only one opening (the entrance door) to the outside, summer cross-ventilation is provided through the three wind catchers, as well as through the openings on the dome on top of the summer residence and the openings around the courtyard. When there is wind, it is directed inside through the wind catchers. When there is no wind, the wind catchers act as chimneys, providing vertical ventilation through the chimney effect. The building is thus kept cool during the summertime.

ENERGY SUPPLY SYSTEM

There was no energy supply in the past; only wood or charcoal and lanterns were used for heating and light. At present, electricity is supplied directly from the national grid. Natural gas is used for cooking and heating.

SOLAR ENERGY UTILIZATION

This building is called a four-season house because, in winter, the residents live in the northern part of the house (winter residence), which receives direct sunlight and heat at that time of the year and is warmer than other sides of the house. At the beginning of summer, they move to the southern part (summer residence); since this is always in shade, it is permanently cool. During daytime, all of the rooms receive natural daylight.

BUILDING HEALTH AND WELL-BEING

The building envelope surrounds the central courtyard and protects it against the harsh outside climate, especially dust storms. Inside the courtyard, there is a pool, with sprinklers above it, and plants and trees. During the summer, the microclimate inside the courtyard is relatively cooler and more humid than the hot and dry macroclimate outside. During the hot months of the year, it is warm in the afternoon and evening inside the building; thus, all of the residents' activities are transferred to the courtyard from dusk to dawn. In the evening, residents turn on the sprinklers on top of the pool and water the plants; they also visit the courtyard. Here they put out carpets on large wooden platforms, where they sit, rest, socialize, have dinner and sleep. They go inside in the morning when the building fabric has become much cooler. In wintertime, they live in the winter residence, which is warmer and well protected against the cold winter winds.

Source: Farid Moalem

5.17 Rooftop of the summer residence showing the wind catchers

Source: Farid Moalem

5.18 Boroujerdi house: (a) view of the courtyard and the winter residence; (b) view of the courtyard and the summer residence

WATER RECYCLING AND CONSERVATION

In the past, residents used to collect rainwater and direct it towards the cistern in the basement. This water and water from a well was used for drinking water. These days, the cistern is no longer in use.

OTHER INFORMATION

Further information is available from Iran Meteorological Organization/Iran National Heritage Organization/Iran Fuel Conservation Organization at www.chn.ir.

Case study 5.2 Akbari House, Tehran, Iran

Vahid Ghobadian with contributions from Neda Taghi, Mehrnoush Ghodsi, Amir Javanbakht, Farid Moalem, Morteza Kasmaee and Behrooz Pakdaman

PROFILE

Table 5.2 Profile of Akbari House, Tehran, Iran

Country	Iran
City	Tehran
Building type	Semi-detached
Year of construction	To be built
Project name	Akbari House
Architect	Vahid Ghobadian and Neda Taghi

Source: Vahid Ghobadian

PORTRAIT

This case study comprises two houses, which were designed for the north-western part of Tehran with regard to the following criteria:

- Construction methods in Iran are still, to some extent, onsite labour intensive.
- Building technology is not as advanced as in Western countries.
- In Iran, there still exist many traditional sustainable houses, especially in the hot and dry region of the country.

Therefore, the concept for the sustainable design of these two houses is to learn from the past, upgrade and update the design and consider it fully, with respect to the social, cultural, economic, environmental and technological context of the place.

Source: Neda Taghi

5.19 Three-dimensional view of Akbari House, Tehran, Iran

The two houses are attached to each other at the side to form a semi-detached building. The dimensions of the site are 53 x 58m. The street is on the north side of the site. The buildings are located on the north part of the site and the main yard is on the south. Each house design is in the luxury range, with three bedrooms, swimming pools and many modern conveniences. The site is located in a relatively new neighbourhood.

CONTEXT

The houses are designed to provide comfort for the inhabitants, together with low pollution, noise and energy use during the 12 months of the year.

The building is privately owned. For the architectural, mechanical and environmental design of the building and landscaping, experts from universities and engineering consultants were involved.

ECONOMIC CONTEXT

The construction cost of this building is no more than that of a conventional building in the same range in Tehran; but the long-term expenses of upkeep, maintenance and utility bills are considerably reduced. Since the building is privately owned, the owner provides all of the funding.

SITE DESCRIPTION

The site is in a newly developed part of the city in the foothills of the Alborz Mountain.

It has a 15 per cent north-to-south slope and has a very good view of the city from the south and of the Alborz Mountain from the north. Most of the buildings here are four- to six-storey apartments, with some detached houses. The site is barren and there are no trees or provision of shading. The adjacent sites are also detached houses.

BUILDING STRUCTURE

To obtain the best winter solar heat gain, the main elevation of the houses is oriented towards the south. All of the major spaces, such as bedrooms, living rooms and dining rooms, are situated on the south side. The secondary spaces, such as garages, staircases, bathrooms and toilets, are on the north side, and act as a buffer space between the outside and inside of the building.

The south sides of the houses are largely transparent, with 120cm of shade depth above the windows of each floor. These shades block the more vertical sun radiation from penetrating the rooms during summertime; but winter sun can penetrate deep into the major spaces of the houses and bring warmth and light inside. Since the building material is masonry, the excess heat would be stored in the building fabric for night time.

Small swimming pools in the basements are placed right behind south-facing windows. Water in these pools is used as thermal mass. The heat of direct solar radiation during daytime is stored in water for night time. The sprinklers on top of the pools keep the place cool – through evaporative cooling – in the summertime.

The windows on the north side are much smaller than those on the south side because of the colder condition on the north side of the building during winter. The purpose of these windows is to provide views and daylight on the north side and to facilitate cross-ventilation during the hot months of the year. The buildings are ground coupled and the ground acts as a great thermal mass and insulator for the basements, so temperature fluctuation between day and night would be at a minimum and less energy would be consumed.

BUILDING CONSTRUCTION

Clay brick is a very common traditional building material in the hot and arid region of Iran. The raw material for constructing bricks is soil, which is plentiful and readily available, making bricks a very practical and economical material to use.

Another advantage of bricks is that they have a very good thermal capacity and can be used as thermal mass. By using bricks in the building fabric, and placing insulation on the outside of the building fabric, a large heat mass and thermal capacity will exist inside the houses, which would even out day–night temperature fluctuations. This fluctuation is about 8°C in summer and 10°C in winter.

Another advantage of bricks is that the construction skills for this material are well developed; brick construction has been used in Iran for more than 6000 years.

The final advantage of using bricks is that it is a very durable material and lasts for millennia. It is easily recycled and can be used over and over again. The bricks from an old building can be used for a new one with no difficulty.

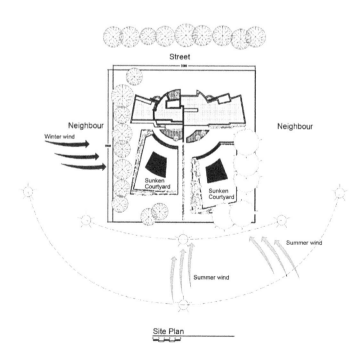

Source: Neda Taghi

5.20 Akbari House: site plan

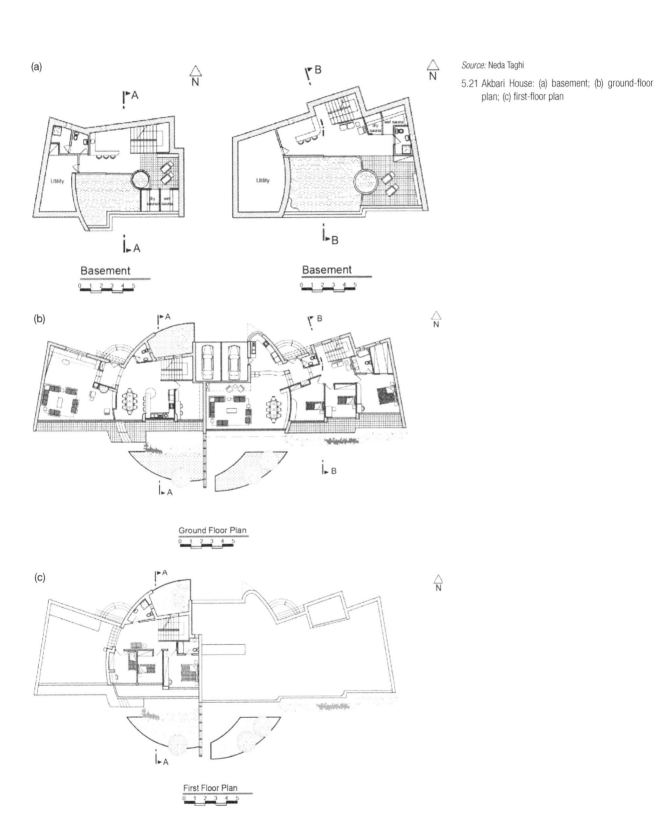

(a)

N

Basement
0 1 2 3 4 5

Utility

dry sauna | wet sauna

B

N

Basement
0 1 2 3 4 5

Utility

dry sauna | wet sauna

Source: Neda Taghi

5.21 Akbari House: (a) basement; (b) ground-floor plan; (c) first-floor plan

(b)

A

B

N

Ground Floor Plan
0 1 2 3 4 5

(c)

A

N

First Floor Plan
0 1 2 3 4 5

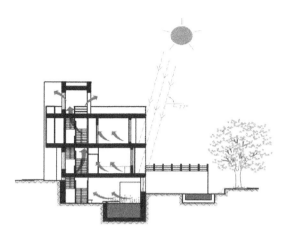

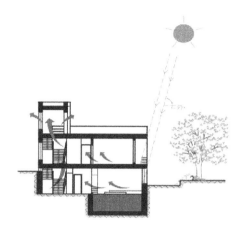

Summer Day Time (Calm Day)
- Shading
- Chimney Effect

Section A-A
0 1 2 3 4 5

Summer Day Time (Calm Day)
- Shading
- Chimney Effect

Section B-B
0 1 2 3 4 5

Source: Neda Taghi

5.22 Summer daytime (calm day); sections A–A and B–B of Akbari House

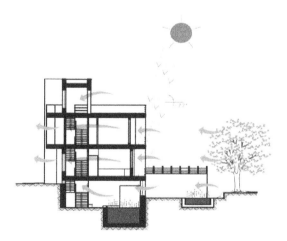

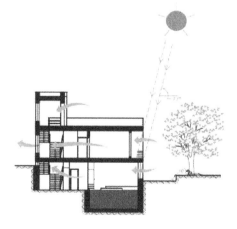

Summer Day Time (Windy Day)
- Shading
- Cross Ventilation
- Evaporative Cooling

Section A-A
0 1 2 3 4 5

Summer Day Time (Windy Day)
- Shading
- Cross Ventilation
- Evaporative Cooling

Section B-B
0 1 2 3 4 5

Source: Neda Taghi

5.23 Summer daytime (windy day); sections A–A and B–B of Akbari House

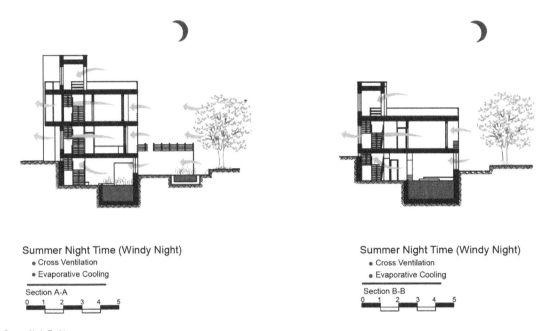

Summer Night Time (Windy Night)
- Cross Ventilation
- Evaporative Cooling

Section A-A

0   1   2   3   4   5

Summer Night Time (Windy Night)
- Cross Ventilation
- Evaporative Cooling

Section B-B

0   1   2   3   4   5

Source: Neda Taghi

5.24 Summer night (windy night); sections A–A and B–B of Akbari House

In any case, if old bricks are not reused and are dumped, they have no harmful effect on the environment since they are just baked clay.

The building has a conventional reinforced concrete frame. In order to make the building fabric lighter, especially above ground level, perforated bricks are used. Regular bricks are used for the basement. Reinforcing is also placed between the brick walls and structure in order to prevent walls from collapsing during earthquakes.

The roof is flat. Only the doors, windows, and mechanical and electrical equipment are prefabricated.

VENTILATION SYSTEM

The main concept for cross- and stack ventilation for these houses is the use of natural forces. The prevailing wind in summer is from the south and south-east.

Since the houses are oriented towards the south, the building ratio of width to length is 1:4, and all of the building openings are on the south or north sides, natural cross-ventilation in summer is easily achieved.

The room on top of the staircase on the roof is used as a wind catcher. Opening the doors on the south side of this room allows summer winds to enter. It should be noted that the spaces between the stairs are open, so air can circulate through the staircase easily. When there is no wind, by keeping the door open, the house is cooled during the summer through the chimney effect.

Summer Night Time (Calm Night)

● Chimney Effect

Section A-A

0 1 2 3 4 5

Summer Night Time (Calm Night)

● Chimney Effect

Section B-B

0 1 2 3 4 5

Source: Neda Taghi

5.25 Summer night (calm night); sections A–A and B–B of Akbari House

APPLIANCES

There are 5 square metres of flat-plate hot-water sun collectors with a tank on the roof for each house. This is adequate for a family of five in Tehran. High efficiency appliances, refrigerators and compact fluorescent lighting are used.

ENERGY SUPPLY SYSTEM

A mechanical system is only used as a backup system for these houses. In winter, the combination of solar energy, body heat from occupants and heat from electrical appliances will keep the place warm.

In addition, there is one fireplace in the centre of each house, which uses natural gas. Natural gas is plentiful in Iran: it is cheap and readily available, burns well and makes far less pollution compared to oil.

During summer, the houses are cooled passively – that is by shading on the south, east and west sides of the building, utilizing summer wind from the south and south-east (for cross-ventilation inside the buildings), chimney effect (through staircases) and evaporative cooling (from plants and pools in the yards and basements). Additionally, there is a water evaporative cooler (6000 cubic feet per minute) for each house, which cools all of the major spaces through ducts. Of course, it is used as a backup system when it is too hot outside and uncomfortable inside. Photovoltaic collectors are not used for these houses because the initial cost is far too expensive, compared to electricity from the grid. Low energy-use lighting fittings are used, which save energy and last much longer compared to regular fittings.

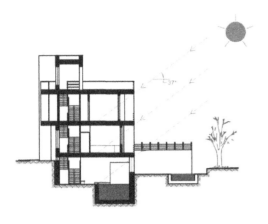

Winter Day Time

• Direct Solar Radiation

Section A-A

0 1 2 3 4 5

Winter Day Time

• Direct Solar Radiation

Section B-B

0 1 2 3 4 5

Source: Neda Taghi

5.26 Winter daytime; sections A–A and B–B of Akbari House

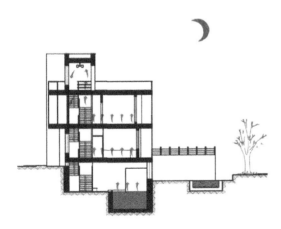

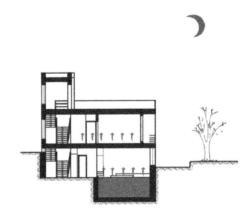

Winter Night Time

• Heat Recycle

Section A-A

0 1 2 3 4 5

Winter Night Time

• Heat Recycle

Section B-B

0 1 2 3 4 5

Source: Neda Taghi

5.27 Winter night; sections A–A and B–B of Akbari House

SOLAR ENERGY UTILIZATION

Having a large transparent southern façade and placing all of the major rooms on the south side of the houses provides solar heat in the winter and natural lighting throughout the year. As a result, solar energy is also used for natural lighting.

BUILDING HEALTH AND WELL-BEING

Using natural resources and energy sources creates a healthier and a more pleasant environment, with less noise, pollution and cost.

WATER RECYCLING AND CONSERVATION

Rainwater from the roofs and yards is directed towards flowerbeds in the courtyards. The sprinklers on top of the pools in the yards humidify and cool the air in the yards, as well as the summer wind before it enters the houses. Furthermore, in this way fresh water is added to the pools, while excess water from the pools feeds the flowerbeds.

CONTROLS

Controls are manual.

PLANNING TOOLS APPLIED

Dr Hasegawa Ken-Ichi (Akita Prefectural University) conducted the building simulation using the programs TRNSYS and COMIS. This work achieved internal air temperatures within the thermal comfort zone in winter and summer were achieved. For the evaluation of thermal comfort during winter, a comfort zone is defined as indoor temperature above 18°C. The case is calculated for closed windows during winter (November to February). In summer, windows are open during the night (12 am to 7 am). Calculations are for the three months of June, July and August.

PRINCIPLES, CONCEPTS AND TERMS

The main climatic principles and concepts used were as follows:

- large southern transparent exposure towards winter sun and summer breezes;
- protection against winter cold winds;
- thermal insulation of the building's outer skin;
- maximum use of solar energy for lighting and winter heating;
- maximum use of summer breezes;
- the chimney effect; and
- evaporative cooling and cross-ventilation for summer cooling.

Respect for local social traditions and building techniques, maximum use of recyclable building materials, and cost-effective methods for the construction and use of the building were also critical issues during the design of the two houses.

OTHER INFORMATION

Further information can be found at www.ifco.ir.

Chapter 6
Tokyo: A Warm
Temperate Climate

Nobuyuki Sunaga, Motoya Hayashi, Ken-ichi Hasegawa and Tamaki Fukazawa

INTRODUCTION

This chapter provides the information on solution sets for the warm temperate areas of Japan. First, a description of the climate of Japan is provided, followed by a description of the Tokyo area. The effectiveness of the solution sets are examined through computer simulation. Finally, three case studies are presented.

CLIMATE OF JAPAN

The country of Japan consists of many islands, is located between 25° and 45° north latitude and extends from the subtropical zone to the subarctic zone. The length of the country is about 3000km.

In most areas, a cycle of six seasons exists. Spring has a pleasant climate, but early summer is rainy, and throughout the summer it is hot and humid as in the tropics; the typhoons come at the end of summer. Autumn is pleasant; winter, of course, is cold, but the areas on the Pacific Ocean side (the south-east) of the islands have the benefit of sunshine. Along the north-west of the islands there is heavy snow in winter.

Most areas in Japan belong to the temperate zone, but the heating degree day, D_{18-18}, ranges from under 100 in Okinawa to over 4500 in East Hokkaido. In the energy standard, the insulation level and solar-shading level are specified for six zones, which are divided by the heating degree day. Figure 6.1 shows the characteristics of the climate at a typical point in zones I, IV and VI. In the following section, the characteristics of zone IV can be described as the typical climate of Japan because this zone contains 70 per cent of all the households in Japan.

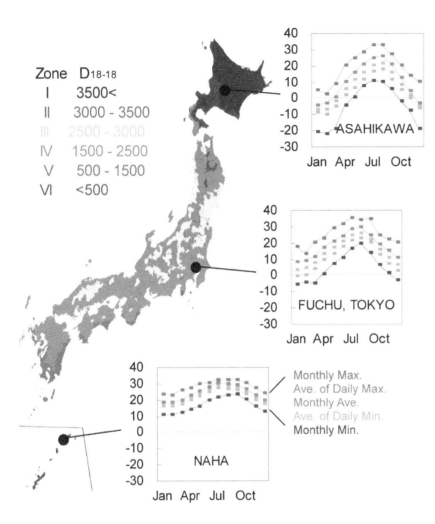

Source: IBEC (1999) and AIJ (2000)

6.1 The heating degree-day $D_{18\text{-}18}$ and the climate zones in Japan. The two figures of 18 are the limit of outdoor temperature for heating and room temperature.

Climate of the Tokyo area

In Tokyo, in climate zone IV, both heating and cooling are needed. The heating period is roughly from the end of November to the end of March. In winter, it is cold and dry; but there are many sunny days and so the climate is suitable for solar architecture. The dominant winds in winter are from the north-west; therefore, old farmhouses in this area have windbreaks on the north and west sides of the site.

The cooling period is from the beginning of July to the beginning of September. In July and August it is hot and wet, equal to the tropical climate. In June and September, it is very rainy and wet, so cooling is sometimes needed. In the cooling season, both the temperature and humidity are high in the daytime; the temperature falls a little at night, but the humidity becomes higher. Therefore, cooling and dehumidifying are needed on most days in this season. The dominant wind direction in the Tokyo area in summer is true

Source: chapter authors

6.2 Characteristics of the climate in zones I, IV and VI

Zone : I (D_{18-18} : 4237)	Zone : IV (D_{18-18} : 1955)	Zone : VI (D_{18-18} : 58)
Climate zone : Subarctic	Climate zone : Temperate	Climate zone : Subtropical
Location : Asahikawa, Hokkaido	Location : Fuchu, Tokyo	Location : Naha, Okinawa
Latitude : 43°46′ N	Latitude : 35°41′ N	Latitude : 26°12′ N
Longitude : 142°22′ E	Longitude : 139°29′ E	Longitude : 127°41′ E
Altitude : 112m	Altitude : 58m	Altitude : 28m
Ave. Temp. : 6.9 deg.C	Ave. Temp. : 14.6 deg.C	Ave. Temp. : 22.7 deg.C
Ave. Solar : 11.6 MJ/m² day	Ave. Solar : 12.2 MJ/m² day	Ave. Solar : 14.1 MJ/m² day
The lowest temperature in winter reaches -20°C, so sufficient thermal insolation is indispensable. The insolation in winter is not so much, but it is utilizable because of its sunny weather. The average of the daily highest temperature in Aug. is 26°C and the average of the lowest temperature is 18°C (at 4 am), so the night cross-ventilation is effective. The insolation in early summer is large but the wind power is weak.	The average of the hourly temperature chages from 0°C in Jan. to 30°C in Aug. and so both heating and the cooling are needed. In summer the night outdoor temperature is utilizable and the cross-ventilation by the south wind is effective. The insolation in winter is large and it is suitable for solar architecture. The annual average of outdoor temperature is 15°C, so ground cooling and heating are useful.	Cooling is needed for most of the year in this area. The shading and the cross-ventilation are fundamental strategies because this area abounds in solar power and wind. The hourly average temperature varies from 27 to 30°C. So the night outdoor temperature is not utilizable in mid-summer, but it is utilizable in May, June and Oct. The amounts of precipitation in May and Aug. are large because of the rainy season and typhoon.

Climograph

Climograph

Climograph

Precipitation and Insolation

Precipitation and Insolation

Precipitation and Insolation

average 1.4[m/s]   average 1.2[m/s]
Wind Vector

average 1.0[m/s]   average 1.5[m/s]
Wind Vector

average 3.5[m/s]   average 3.7[m/s]
Wind Vector

south and this is another reason why buildings have large windows on the south side.

The annual average earth temperature, equal to the outdoor temperature, is about 15°C, so earth cooling and heating can be used.

SOLUTION SETS

In the typical zone, the climate changes in the course of the year from requiring heating to requiring cooling and drying. Insulation and air sealing are basic solutions for this climate; and it is also very important to cope appropriately with solar radiation.

During the cooling season, shading and cross-ventilation are very important for cooling buildings and occupants. If these basic solutions are adequate, cooling systems may only be needed when the ambient temperature and humidity exceed comfort zone limits, so cooling energy will be saved.

In the heating season, the direct gains through the windows facing the south-east to the south-west will provide almost all of the heating energy on sunny days. Furthermore, solar collectors to supply hot water are effective in all seasons, as are photovoltaic systems, because of the high solar altitude and because, in this climate zone, even in winter, daylight hours are not very short.

The general recommendations are for the building envelope to have high insulation and good air tightness. The hot water system should use energy from solar collectors. A water tank is recommended in order to provide mains water conservation. A photovoltaic system for electrical generation is also suggested and can be placed on the roof.

Detached and row houses

The specific recommendations for detached houses and row houses are as follows:

- thermal hydronic strategies for the cooling season:
    - roof: high insulation, highly reflective surface; ventilation through the airflow space;
    - walls: ventilation through the airflow space;
    - floors: massive floor materials as heat storage;
    - windows: air entry routes allowing for wind direction in summer, upper openings for exhausting hot air and night ventilation; shades;
    - cooling equipment: heat-pump system for cooling and drying;
    - surroundings: plants and ponds for cooling the surroundings;

- thermal hydronic strategies for the heating season:
    - floor: massive floor materials as heat storage;
    - windows: large south-facing (± 45°) insulated glass windows for direct gain;
    - ventilation: a passive stack ventilation system with a fan and a sensor or a heat recovery ventilation system;
    - heating equipment: a heat-pump system for heating; and
    - heat recovery system: wastewater tank under floor for heat recovery;

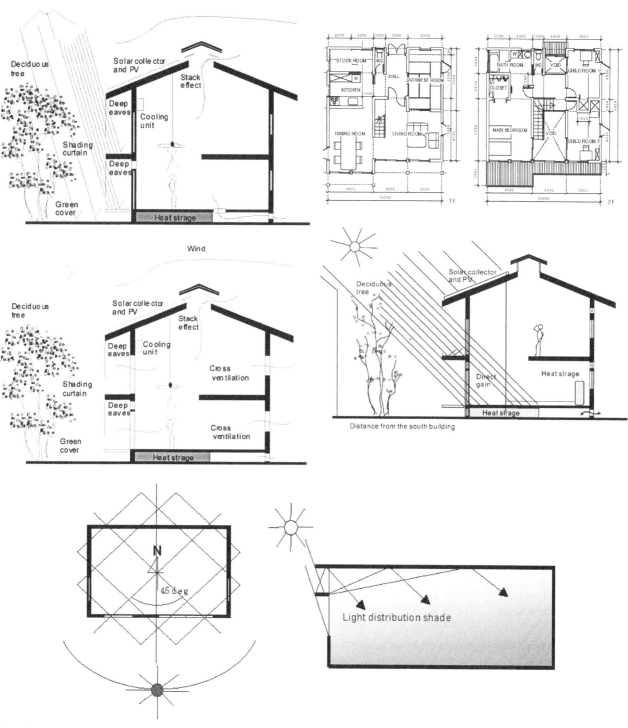

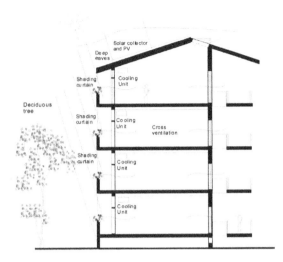

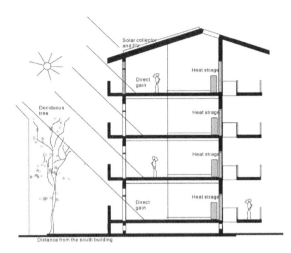

- lighting and acoustics strategies:
  - shading: light distribution shades and blinds; and
  - windows: screens to exclude insects and lattices to guard against intruders.

Apartment house

The specific recommendations for apartment houses are as follows:

- thermal hydronic strategies during the cooling season:
  - roof: high insulation, highly reflective surface; ventilation through the airflow space;
  - walls: shading, and balconies used as fins;
  - floor: massive floor materials for heat storage;
  - windows: air entry routes allowing for wind direction in summer, upper openings for exhausting hot air and night ventilation; shades;
  - cooling equipment: heat-pump system for cooling and drying;
  - surroundings: plants and ponds;

- thermal hydronic strategies in the heating season:
  - floor: massive floor materials for heat storage;
  - windows: large south-facing (± 45°) insulated glass windows for direct gain;
  - ventilation: heat recovery ventilation system;
  - heating equipment: heat-pump system for heating;

- lighting and acoustic strategies:
  - shading: light distribution shades and blinds;
  - window: screens to exclude insects and lattices to guard against intruders.

Source: chapter authors

6.4 Solution sets for apartments

SIMULATION STUDY OF THE PERFORMANCE OF THE TYPICAL SOLUTION SETS

For the evaluation of the Japanese typical solution sets (TSS), the simulation study has been conducted using the simulation methodology described in Chapter 2. The models used for calculation are a detached house and an apartment house. The TSS outlines are provided in the previous section on 'Solution sets'.

From the viewpoint of environmental consciousness, it is important to control the indoor thermal environment by means of natural heat flow through the building elements, without the use of heating and cooling equipment. Hence, it is necessary to design a building using good insulation and air tightness as a basis for its performance. In this simulation study, the strategies specified for the heating season comprise the effect of high insulation and large airtight south-facing insulated openings for direct gain, and massive floor/wall materials for heat storage. The strategies specified for the cooling season comprise the effect of high insulation and air tightness, massive floor/wall materials for heat storage, shading and natural cross-/stack ventilation.

Simulation methodology

To study the effect of building solution sets for cost-effective design on the improvement of thermal conditions, the calculation has been done using a model of the transient simulation program with a modular structure (TRNSYS), coupled with the multi-zone airflow model (COMIS). In the thermal building model TRNSYS, airflows between the rooms and from the outside are defined as input values. But in natural ventilation systems, these values depend upon external wind conditions and indoor and outdoor temperature conditions, so it is necessary to link the multi-zone airflow model COMIS with the thermal model TRNSYS (COMVEN-TRNSYS, 1997). In the following simulation study, this coupled model was used for analysing typical solution sets for a warm temperate climate.

For the calculation of multi-zone air infiltration, the air tightness of a whole house is expressed by an effective leakage area per floor area, αA, for the indoor–outdoor pressure difference, Δp, of 9.8Pa. In the following calculation, the leakages are distributed around wall openings. Furthermore, the leakages of the walls are distributed according to the proportion of windows. For the window openings, the multi-zone air infiltration through large openings can be calculated. The simulated house is assumed to be located in a place with no obstructions around it. Wind pressure coefficient data provided by the Air Infiltration and Ventilation Centre (AIVC, 1994) is used. It is assumed that there are four occupants (parents and two children) living in the house, and the schedule of their daily life related to heat generation is taken into account.

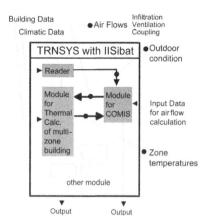

Source: COMVEN-TRNSYS (1997)

6.5   Diagram of coupling TRNSYS with COMIS

For the thermal evaluation, the thermal neutrality temperature (T_n) is used. This has been developed by a number of researchers (Nicole and Roaf, 1996; Humphreys, 1978, and others). It is defined as the temperature at which a person feels thermally neutral. For example, the relationship between the neutrality temperature and the monthly mean outdoor temperature is shown as follows:

$$T_n = 17.0 + 0.380 \cdot T_m \text{ (Nicole's equation)}$$

where T_n is the thermal neutrality temperature and T_m is the monthly mean outdoor temperature. Based on one of Nicole's equations, the comfort level is $T_n \pm 2°C$. The predicted neutrality temperatures for Fuchu city in Tokyo are as follows:

- July: 25.2°C /24.6°C–28.6°C
- August: 27.2°C/25.3°C–29.3°C (monthly average outdoor temperature/ thermal comfort zone from the Nicole's equation).

In evaluating indoor thermal comfort, it is desirable that daily averaged maximum and minimum temperatures are within the comfort zone.

Detached houses

Calculation model of TSS

The detached house model used for calculation has a TSS type and a standard type for comparing the effects of the TSS strategies. Each of the two types has 16 rooms, including an under-floor space. The living room faces south. The building features for the standard type and the TSS type are as follows:

- standard type:
  - structure, storeys: timber, two storeys;
  - floor area: 184 square metres;
  - insulation (GW32K): 55/35/45mm (ceiling/wall/floor);
  - openings: double-glazed windows; and
  - air tightness: effective leakage area: $\alpha A = 5.0\text{cm}^2/\text{m}^2$;

- TSS type:
  - structure, storeys: timber or reinforced concrete, two storeys;
  - floor area: 184 square metres;
  - insulation (GW32K): 160/90mm (ceiling/wall);
  - insulation on the foundation (polystyrene foam: 35mm);
  - openings: double glazing and low-emissivity glass;
  - thermal mass: floor in first-storey rooms (concrete: 20mm);
  - TSS RC type (concrete): 15/20mm (wall/floor); and
  - air tightness: effective leakage area: $\alpha A = 2.0\text{cm}^2/\text{m}^2$.

Table 6.1 Cases for calculating a detached house model, Fuchu City, Japan

	Standard		TSS				TSS RC			
Case number	1	2	3	4	5	6	7	8	9	10
Season	winter	summer	winter	summer	summer	summer	winter	summer	summer	summer
Structure	Timber						Reinforced concrete			
Shelter performance level	Level 1		Level 2							
Upper openings			✓	✓	✓	✓	✓	✓	✓	✓
Window opening 8:00–20:00		✓		✓	✓			✓	✓	
20:00–8:00		✓*		✓		✓		✓		✓
Shading (south openings)			✓	✓	✓	✓	✓	✓	✓	✓
Massive floors			✓	✓	✓	✓	✓	✓	✓	✓

Source: chapter authors
*All openings of ground floor are closed
TSS = Typical solution sets

Thermal insulation and air tightness of buildings of the TSS type (level 2) accords with the latest Japanese building code for energy efficiency; insulation and air tightness for a standard type (level 1) are set to the standard level. In the case of TSS RC type, the insulation is installed on the outside of the concrete walls, thus using the thermal mass of the inner space. All rooms are assumed to be ventilated at 0.5 air changes per hour (ACH) all day to keep the indoor air clean, and this ventilation is switched off when windows are opened. It is assumed that there are four occupants (parents and two children) living in the model. In this simulation, the heat generation related to their daily life is taken into account. The weather data for Fuchu city from METEONORM Ver.5.0 is used for the calculations. Based on strategies for TSS, ten cases are chosen for the calculations. Table 6.1 shows the conditions for each case.

Calculation results

The calculations were conducted for ten cases. To evaluate the effect of TSS, this calculation focused on the space-heating load during winter and on indoor temperature and thermal comfort during summer.

Figure 6.6 shows the total space-heating load during winter. Each room is heated for the whole day during winter and the room temperature is maintained

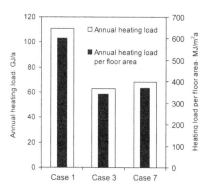

Source: chapter authors

6.6 Comparison of heating loads for three cases for the detached house solution set

at 20°C. The heating loads in case 3 and case 7 are less than that in case 1. Case 3 and case 7 have thermal insulation performance and air tightness according to the latest Japanese building code. On the other hand, case 1 is supposed to have the normal insulation level for a conventional Japanese

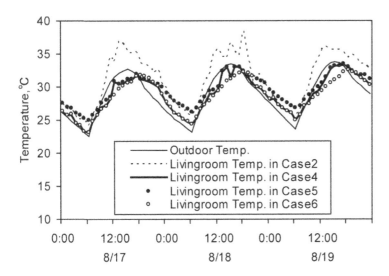

Source: chapter authors

6.7 Simulation of living room temperatures for the detached house solution set (cases 2, 4, 5 and 6)

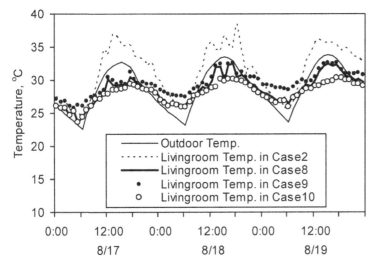

Source: chapter authors

6.8 Simulation of living room temperatures for the detached house solution set (cases 2, 8, 9 and 10)

house. It therefore appears that improving thermal insulation is effective in saving energy consumption for space heating. The TSS model is thus more effective than the standard model at saving energy.

Figure 6.7 shows the calculation results of the effect of Japanese TSS (cases 4, 5 and 6) on indoor temperature during summer. The results for a standard condition (case 2) are included. The differences in calculation conditions between cases 4, 5 and 6 are in the patterns of window opening as shown in Table 6.1. The temperature profile is for three days, including the hottest day of the year. The temperature of the standard condition (case 2) is higher than that of the TSS model throughout the day; it is significantly higher during the daytime because of the lack of shading on the southern openings. Ventilation in the night is effective in decreasing the indoor temperature while windows are opened; but the daily maximum temperature of case 6 does not decrease more than the other cases, which are ventilated during daytime and for the whole day. This is because the thermal mass is small for these buildings.

In Figure 6.8, the calculation results show the effect of TSS (cases 8, 9 and 10) on indoor temperatures during summer. The results also include the standard condition (case 2); the calculation conditions vary across cases 8, 9 and 10 according to different patterns of window opening, as shown in Table 6.1. As the construction of these cases is assumed to be reinforced concrete, the effect of thermal mass is expected to be greater than for cases 4, 5 and 6. The daily temperature swing of cases 8, 9 and 10 is smaller than that of cases 4, 5 and 6. The greater thermal mass enables the indoor temperature to be decreased during daytime for the cooling heat storage. Indoor temperature during daytime following night-time ventilation is lower than for any other cases.

In order to evaluate thermal comfort among these TSS conditions, an adaptive model is applied. Figure 6.9 shows the evaluation of thermal comfort in a living room and a main bedroom in August. For the standard condition (case 2), the difference between maximum and minimum temperatures is large and out of the comfort zone. On the other hand, in the case of the TSS model, the difference between maximum and minimum temperatures is smaller than that of the standard condition. The maximum temperature in the case of the TSS model with large thermal mass is lower than that of the TSS model with small thermal mass and approaches the comfort zone. According to the outdoor temperature, during July and August, 17 days have a maximum temperature within the comfort zone. For the standard condition, the indoor temperature of the living

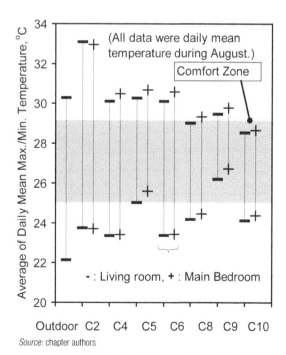

Source: chapter authors

6.9 Comfort study for the detached house solution set (cases 2, 4, 5, 6, 8, 9 and 10)

room and the main bedroom are within the comfort zone for only eight and nine days, respectively. On the other hand, in the case of the TSS model of timber construction, the indoor temperature of the living room is within the comfort zone for 19 days. And in the case of the TSS model constructed of reinforced concrete and with night ventilation (case 10), the indoor temperature of the living room and the main bedroom are within the comfort zone for 42 and 39 days, respectively. This indicates that adding adequate thermal mass to the inside of building walls, plus the use of night-time ventilation, is very effective in maintaining thermal comfort during summer.

Apartment house

Calculation model of TSS

As for the case of the detached house model, the apartment house model used for calculation has a TSS type and a standard type. This is a five-storey apartment house and there are five families on each floor. Each house has ten rooms, including a green room (conservatory) attached to a living room, which faces the south. The green room is an option. In the apartment house model, the basic building features for both the standard type and the TSS type are the same. These are as follows:

- structure, storey: reinforced concrete, the middle storey and apartment;
- floor area: 114 square metres;
- insulation (polystyrene foam): 45mm (south and north wall);
- opening: double glazing and low-emissivity glazing;

Table 6.2 Cases for calculating an apartment house model, Fuchu City, Japan

		Standard				TSS				
Case number		1	2	3	4	5	6	7	8	9
Season		winter	winter	summer	summer	winter	winter	summer	summer	summer
Window opening	8:00–20:00			✓*	✓*			✓	✓	
	20:00–8:00			✓*	✓*			✓		✓
Shading (south openings)					✓	✓	✓	✓	✓	✓
Green room		✓		opened		✓		opened		

Source: chapter authors

* The openings on the south are opened and there is the air inlet/outlet (10 x 40cm) on the northern windows

- air tightness: effective leakage area: $\alpha A = 2.0cm^2/m^2$; and
- thermal insulation and air tightness of buildings is according to the latest Japanese building code for energy efficiency.

The insulation is installed on the outside of the concrete walls. It is assumed that there are four occupants (parents and two children) and the heat generation related to their daily life is taken into consideration. The weather data for Fuchu city from METEONORM Ver.5.0 is used for the calculation.

Based on strategies for TSS, nine cases for calculation are chosen. Table 6.2 shows the conditions for each case.

Calculation results

Figure 6.10 shows the total space-heating load of an apartment house model during winter. Each room is heated for the whole day during winter and the room temperature is maintained at 20°C. In case 1, a sunspace is attached to the south room without shading for the promotion of solar gain from the openings. The heating load of case 1 is 3.5GJ per annum, lower than that of case 2 without the sunspace and shading, and is the lowest of any of the cases. This indicates that the sunspace is effective for decreasing the heating load. When the sunspace is attached and shading is installed in winter, the effect of sunspace disappears for decreasing heating load. Therefore, the best practice in winter seems to be with a sunspace and no shading. For cooling indoor space, shading devices should be attached to openings in summer. It would be better to control transmitted solar radiation through openings according to outdoor conditions by using movable shading device.

Figure 6.11 shows the calculated indoor temperature profiles for three days during summer in the case of an apartment house model. Figure 6.10 includes the results of a reference case (case 4). The differences in the calculation conditions between cases 7, 8 and 9 are the patterns of window opening, as shown in Table 6.2. The temperature in all cases is lower than the outdoor temperature during the daytime. But temperature difference between cases is not clear. In the detached house model, the night ventilation with thermal mass is the most effective for decreasing the indoor temperature; but its effect in an apartment house model are not so clear.

In order to evaluate thermal comfort among these conditions of the TSS, the adaptive model is applied. Figure 6.12 shows the evaluation of thermal comfort in a living room and a main bedroom in August. In the case of the reference case without and with shading (cases 3 and 4, respectively) there is a difference in maximum temperature, and the shading device is effective for decreasing the maximum temperature of the room. In the TSS model, the northern openings are opened to promote cross-ventilation. Therefore, the temperature difference between the living room and the main bedroom is very

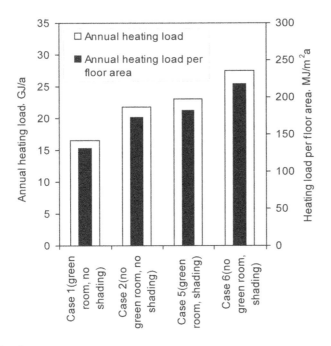

Source: chapter authors

6.10 Comparison of heating loads for the apartment house solution set (cases 1, 2, 5, and 6)

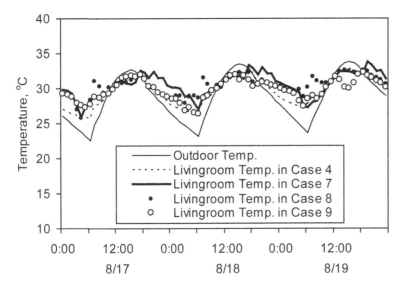

Source: chapter authors

6.11 Simulation of living room temperatures for the apartment house solution set (cases 4, 7, 8 and 9)

low. On the other hand, there is a difference for those reference cases in which northern openings are not used for cross-ventilation, and where the minimum temperatures of a main bedroom are higher than that of TSS models. TSS models (cases 7, 8 and 9) have different patterns of opening windows. The maximum temperature of case 7, which is the case of opening windows for the whole day, is the lowest among the TSS cases. According to the outdoor temperature, during July and August there are 19 days that have a maximum temperature within the comfort zone. In the case of the reference case (case 4), the indoor temperatures of a living room and a main bedroom show that they are within the comfort zone for 14 and 6 days, respectively. On the other hand, in the case of the TSS model with full-time ventilation (case 7), the indoor temperature of a living room and a main bedroom are shown to be within the comfort zone for 23 and 16 days, respectively. In the case of night ventilation, the days within the comfort zone in these two rooms are 17 and 11, respectively. This indicates that night ventilation may be a good way of maintaining the indoor thermal environment within the comfort zone during summer; but it is more effective to open windows in the night and during the daytime to promote decreased indoor temperature.

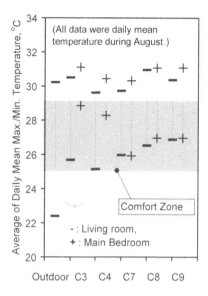

Source: chapter authors

6.12 Thermal comfort study for the apartment solution set (cases 3, 4, 7, 8 and 9)

Case study 6.1 Eco-village, Matsudo, Japan

Nobuyuki Sunaga

Source: Taisei Corporation

6.13 Eco-village, Matsudo, Japan

PROFILE

Table 6.3 Profile of the eco-village in Matsudo, Japan

Country	Japan
City	Matsudo
Building type	Apartment (family, single)
Year of construction	2000
Project name	Eco-village Matsudo
Architect	Taisei Co

DESCRIPTION

This project achieves energy conservation and peak-load shift of electric power consumption in apartment houses by integrating architectural passive design methods and low-energy mechanical equipment. The site plan is designed so that two apartment buildings, a family type and a single person type, are arranged to make a courtyard and to preserve the large green space

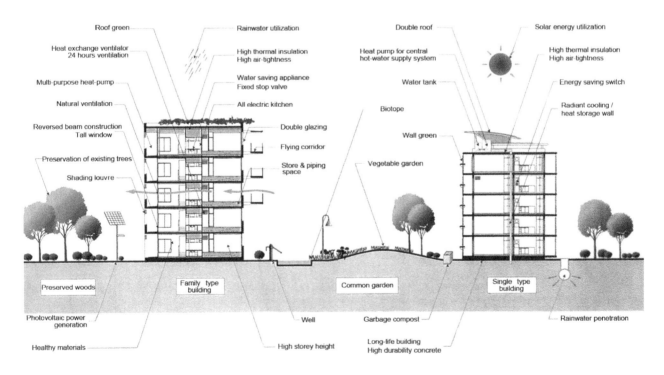

Source: Taisei Corporation

6.14 Section showing environmental strategies in the eco-village, Matsudo, Japan to the south of the site. The major passive design elements are the sunshades, the roof garden, the natural ventilation and the high extent of thermal insulation. Low energy mechanical components include a heat exchange ventilator, a multi-purpose heat-pump system with ice storage, and a wall radiant cooling/storage system utilizing the waste cooling air by way of a hot water heat-pump system. The residents set a high value on good cross-ventilation, and this project won a low energy building prize from the Institute for Building Environment and Energy Conservation in 2001.

CONTEXT

The winner of the first prize in the institute's invited design competition, Eco-village Matsudo, is planned as a model for a new type of apartment suitable for the 21st century. It aims to achieve a comfortable residential environment that is strongly conscious of energy conservation, more effective use of natural resources, and a good relationship with the local environment and the populace. It is intended to preserve the amenity throughout the life cycle of the buildings.

These apartment houses, with one building for families and one for single adults, are built as dormitories for a business enterprise. Both buildings are five storeys; in planning the site, computer simulations of the effects of wind were carried out and the preservation of existing trees on the south side of the site were taken into account.

ECONOMIC CONTEXT

Since these buildings are company dormitories, the company provides all of the funding. Low energy consumption and low operation cost are planned for each of the two buildings by integrating architectural passive design methods and high performance mechanical services. Passive techniques, such as cross-ventilation and natural lighting, are used for both buildings; but different mechanical systems are adopted for each building to allow for the inhabitants' varying residence patterns.

A new multi-purpose heat-pump system with ice storage, named Home Ice, has been developed for the family-type building to shift the electric consumption of air conditioning to the night. For the singles' building, a new wall irradiative cooling/storage system has also been developed, using waste cooling air from the hot water heat-pump system.

SITE DESCRIPTION

The site is located in the north-eastern suburbs of the Tokyo city area. The site is flat and has many existing trees at its southern corner. There are some office buildings in neighbouring sites to the north-west and north-east, as well as an apartment house in the south-east and some detached houses in the south-west. There is no boundary fence on the south-east and south-west of the site, allowing for good communication between neighbours.

BUILDING STRUCTURE

Both buildings are five storeys. The family building has 50 flats and its total floor area is 4500 square metres. Each flat is about 80 square metres in

(a)

(b)

6.15   Eco-village, Matsudo, Japan: (a) elevation of the family buildings; (b) single person accommodation

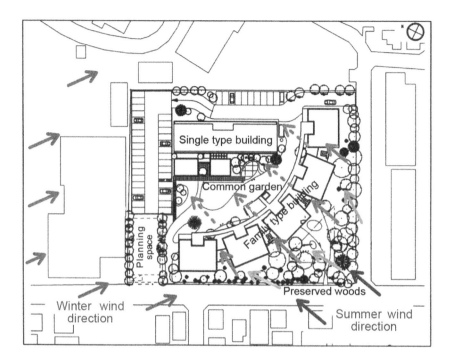

Source: Taisei Corporation

6.16 Site plan of the eco-village, Matsudo, Japan

extent, consisting of a living/dining room, a kitchen, a Japanese-style room, two bedrooms, a bathroom and a toilet. The living/dining room has south-facing balconies. The building is separated into five blocks so that all flats have three façades in contact with outside air. The roofs are planted with lawn and there are louvres on the upper side of the balconies. Each flat has a multi-purpose heat-pump system with ice storage, named Home Ice.

The singles' building has 65 flats and its total floor area is 1895 square metres. Each unit is about 25 square metres in extent, consisting of a main room, a kitchen, a bathroom and a toilet. This building is rectangular in shape, with an east–west axis. There is an extra roof at the top of the building for roof shading, and the main rooms have south-facing balconies with egg crate-type shading. Each flat has a conventional air conditioner and a radiant cooling wall that is cooled by the waste cooling air from the hot water heat-pump system.

Source: Taisei Corporation

6.17 Earth roof system of the eco-village, Matsudo, Japan

BUILDING CONSTRUCTION

Both buildings have a reinforced concrete structure. The family building has a reversed beam structure and a storey height of 3.2m, making for a light living room and good ventilation conditions, as well as a double floor. The double floor makes it possible to install storage spaces and spaces for piping. High durability concrete is used for the building frames, and materials with a low emission of volatile organic compounds (VOCs) are used. The material of the thermal insulation is polyurethane foam, with a thickness of 35mm for roofs and 30mm for walls. All windows are double glazed.

VENTILATION SYSTEM

The building uses cross-ventilation with south-facing and north-facing windows. All flats have a total heat exchanger. The family-type flats have a smaller size of kitchen fan because they use a microwave oven.

APPLIANCES

All of the family-type flats have a multi-purpose heat-pump system with an ice storage system, called Home Ice. The output of the heat-pump system is 1.1 kW and the capacity of the ice storage system is 150 litres. The system's coefficient of performance (COP) is 2.9 for cooling only, 2.3 for heating only, 2.8 for hot water supply only, and 5.5 for cooling and hot water supply. This system reduces the energy consumption by 35 to 44 per cent and the annual cost by 92,000 yen, compared with the normal system. The peak shift of electrical power consumption is 26 per cent.

The singles' building has a wall radiant cooling/storage system cooled by the waste cooling air from the hot water heat-pump system. The output of this heat-pump system is 14.7kW and the capacity of the hot water storage tank is 10,000 litres. This system is expected to reduce energy consumption by 60 per cent and running costs by 75 per cent through the use of night power tariffs and the high COP heat-pump system. Furthermore, this system cools the waste air more than 5 degrees relative to the inlet air temperature. That means that the temperature of the waste air is below 20°C when the outdoor temperature is 25°C. The cool air is sent to ducts that pass behind the wall of each flat, cooling the wall in the night. This wall keeps the room cool all day. A conventional air conditioner of 2kW is installed in each singles' flat.

ENERGY SUPPLY SYSTEM

This building uses conventional energy supply systems for electricity and water. There is no gas system.

SOLAR ENERGY UTILIZATION

These buildings use passive solar energy from the south-facing windows.

BUILDING HEALTH AND WELL-BEING

The buildings conform to Japanese air quality standards. Low VOC emission materials are used and the building has a natural and mechanical ventilation system.

CONTROLS

The mechanical systems are automatically controlled.

PLANNING TOOLS APPLIED

Computer fluid dynamics simulation was used for examining the wind effect on both the outside environment and the inside air distribution.

INFORMATION

For further information please contact Tatsuyuki Yamada at yamada@ arch.taisei.co.jp and Hikaru Kobayashi at hkobayas@arch.taisei .co.jp. The Taisei Corporation website also provides information: www.taisei.co.jp.

(a)

(b)

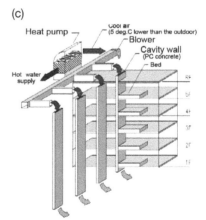

(c)

Source: Taisei Corporation

6.18 Eco-village, Matsudo, Japan: (a) shading system; (b) heat-pump system; (c) schematic diagram of the cooling system

Case study 6.2 A house in Motoyama, Kochi, Japan

Nobuyuki sunaga

Source: Prof. Yuichiro Kodama, Kobe Design University

6.19 A house in Motoyama, Japan: front elevation

PROFILE

Table 6.4 Profile of a house in Motoyama, Kochi, Japan

Country	Japan
City	Kochi
Building type	Detached
Year of construction	2003
Project name	House in Motoyama, Kochi
Architect	Yuichiro Kodama/ESTEC Design Co Ltd

Source: Nobuyuki Sunaga

DESCRIPTION

The theme of this project is being on close terms with nature. The architects started by examining and analysing the natural features and microclimate of the site; in due course, the building was designed as a typical passive house for this climate zone. The building has an east–west axis, large south windows with eaves and small north windows, complying with the basic principles of building design. It is well insulated and has heat storage floors and walls for a direct gain system in winter. Sun shading occurs through a double roof with a reflective surface, and the cross-ventilation is designed for summer conditions. The owners enjoy life in this house without mechanical cooling.

Source: Prof. Yuichiro Kodama, Kobe Design University

6.20 Interior views of a house in Motoyama, Japan

CONTEXT

This house is a new, privately built detached house, constructed in the vicinity of the Yoshino River. The building has a piloti structure with a bridge leading to the entrance as a precaution against river flooding. The main theme of this project is to integrate the built environment with nature and to create a home where the inhabitants can be aware of the changes of season, weather and time.

ECONOMIC CONTEXT

The construction costs of this building are no more than those of conventional buildings in the region. However, cooling and heating costs are low, so the life-cycle cost is at a very low level. Moreover, the inhabitants have a comfortable indoor climate. Since the building is in private use, the owner provides all of the funding.

SITE DESCRIPTION

The site is located in the country in the centre of Shikoku, on the junction of two rivers. The land is slightly inclined toward the south at the river junction and is covered by chestnut trees. In summer, trees hide the building and in winter, when the leaves have fallen, the house appears through the branches.

BUILDING STRUCTURE

The building is a two-storey piloti construction. There is only a storeroom and a machine room on the ground floor, with a living, dining and kitchen area, a bedroom and a children's room on the first floor. A bridge connects the entrance on the first floor to the upper level of the site as an evacuation route in case of flooding.

The building has a rectangular shape with an east–west axis, large south windows with eaves and small north windows. There are no windows in the east and west walls. It is well insulated and has heat storage floors and walls for a direct gain system in winter. The double roof, with its reflective surface and adequately sized eaves, is designed for sun shading in summer. Jalousie windows are installed on the north and south side for cross-ventilation. This building is heated mainly by solar energy and cooled by night-time ventilation. It has an auxiliary floor heating system and no mechanical cooling system.

BUILDING CONSTRUCTION

Taking seasonal floods into account, a reinforced concrete structure was selected for the lower part of the building, from the basement up to the floor of the first floor. The upper floor consists of a steel frame structure and a timber curtain wall.

(a)

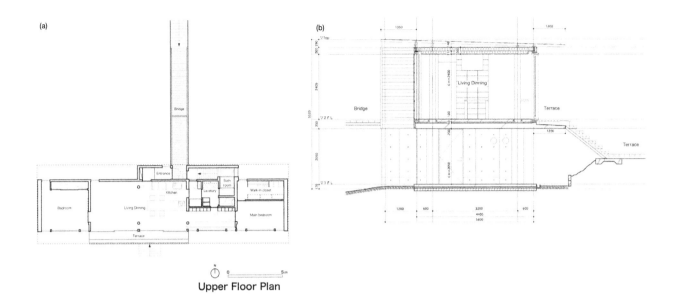

Upper Floor Plan

(b)

(c)

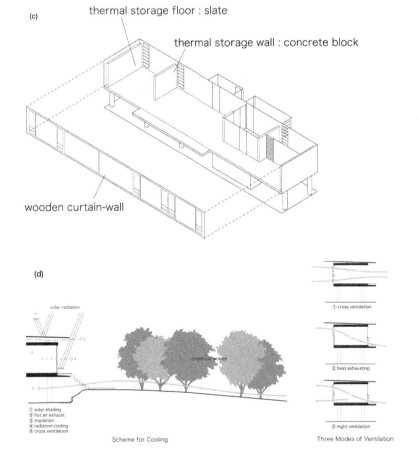

thermal storage floor : slate

thermal storage wall : concrete block

wooden curtain-wall

Source: Prof. Yuichiro Kodama, Kobe Design University

6.21 A house in Motoyama, Japan: (a) floor plan; (b) section; (c) thermal storage system; (d) site section showing passive cooling

(d)

solar radiation

① solar shading
② hot air exhaust
③ insulation
④ radiation cooling
⑤ cross ventilation

Scheme for Cooling

① cross ventilation

② heat exhausting

③ night ventilation

Three Modes of Ventilation

The roof is a double structure and the outer roof is made of galvanized steel with a reflective surface. This allows wind to pass between the two layers of the roof in order to cool it during summer. The roof insulation is 250mm thick cellulose fibre, and the wall has 100mm of glass wool. The floor of the upper storey consists of a 200mm concrete slab, 60mm of polystyrene foam and 90mm of heat-storage concrete. Both transverse walls and three partition walls are made of concrete blocks for heat storage. The windows are double glazed, except for the jalousie windows.

VENTILATION SYSTEM

The building uses cross-ventilation and there are three methods of ventilation. First is the so-called cross-ventilation achieved by opening the windows wide during the day. Second, when the outdoor temperature is higher than the room temperature, windows are opened slightly to discharge internal heat production. Third, at night, the jalousie windows are fully opened for night ventilation for cool storage. The jalousie windows play an important role, not only for cross-ventilation, but also in the design of the façade, especially on the north side.

APPLIANCES

The building has a hot water floor-heating system on the first floor for auxiliary heating. There is no mechanical cooling system.

ENERGY SUPPLY SYSTEM

This house uses conventional supply systems for electricity, oil and hot water. In winter, solar and internal heat gain warms the building, and the heat is stored in the concrete floor and walls. In summer, the building is cooled by cross-ventilation, and night ventilation cools the interior and the heat-storage floors and walls.

SOLAR ENERGY UTILIZATION

This building is a typical passive solar house. It utilizes the direct heat gain from the south windows with eaves. The heat is stored in the floor, transverse walls and three partition walls. The eaves are of an appropriate size to protect the windows from intense direct solar radiation in summer.

BUILDING HEALTH AND WELL-BEING

This building conforms to Japanese air quality standards. Low VOC emission materials are used and the building has a natural and mechanical ventilation system. Lifestyle spaces, such as terraces, are created to provide functional and amenity spaces outside the building.

CONTROLS

Manual controls are used.

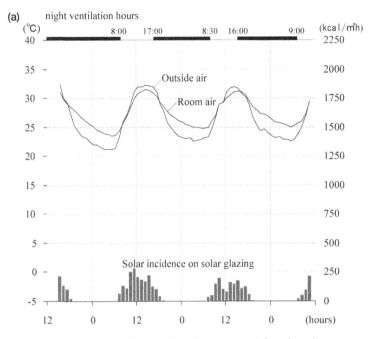

(a)

Temperature fluctuation in August (simulated)
(without aux. cooling)

Source: Prof. Yuichiro Kodama, Kobe Design University

6.22 A house in Motoyama, Japan: (a) simulation results for summer; (b) simulation results for winter

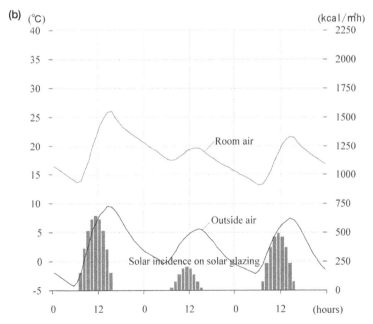

(b)

Temperature fluctuation in January (simulated)
(without aux. heating)

PLANNING TOOLS APPLIED

Thermal simulation was carried out using a computer simulation programme named Solar Designer. The results show that the winter room temperature is kept at 15°C to 25°C by solar energy alone, and the summer room temperature is kept under 28°C by cool storage from night ventilation and by closing windows during the daytime. Dr Yuichiro Kodama, the architect of this house, developed the programme.

INFORMATION

Further information on the programme (the design tool) Solar Designer from Quattro Corporate Design can be obtained from the website www.qcd.co.jp/sd/index.html. Dr Yuichiro Kodama may also be contacted at y-kodama@kobe-du.ac.jp and on the website (Japanese) www.kobe-du.ac.jp/env/kodama/.

Case study 6.3 Environmental Symbiosis House Project, Nara, Japan

Yoshinori Saeki

PROFILE

Table 6.5 Profile of the Environmental Symbiosis House Project

Country	Japan
City	Nara
Building type	Detached
Year of construction	1994
Project name	Environmental Symbiosis House Project
Architect	Yoshinori Saeki, Daiwa House Industry Co Ltd

Source: Yoshinori Saeki

DESCRIPTION

With this experimental house, we seek to create an environmental symbiosis house from every viewpoint and on the basis of three themes: preservation of the global environment; harmony between the house and its surrounding environment; and creation of a healthy and pleasant living environment. In our research activities, the aim was to use passive construction methods to make the most of location and climate conditions. The aim was also to use the most advanced technologies for saving resources and energy. The team worked diligently on experiments and verification activities to achieve a good balance in combining traditional methods and advanced high technologies.

(a)

Source: Daiwa House Industry Co. Ltd

6.23 Environmental Symbiosis House: (a) front elevation; (b) north elevation

(b)

CONTEXT

In order to take into consideration the Earth's precious resources, such as sunlight, breezes, water and so on, we have conducted various experiments and verification activities relating to environmental symbiosis.

BUILDING STRUCTURE

This house comprises two storeys, with a basement, and has eight rooms: a Japanese-style room, living room, dining room and kitchen on the ground floor; three bedrooms on the first floor; and an audio room in the basement.

BUILDING CONSTRUCTION

The Environmental Symbiosis Experimental House is a prefabricated house: a high-quality stable structure formed from steel frames and proof-stress panels. All panel frames, exterior wall materials, heat-insulating materials and window sash frames are prefabricated in the factory.

Source: Daiwa House Industry Co. Ltd

6.24 Environmental Symbiosis House: sun room

Painting of exterior walls is also carried out in the factory in order to avoid possible air pollution of the surrounding environment. Air tightness is maximized by joining panels with highly efficient bolts and by scrupulously patching and taping sheets. Light gauge steel is used for the frames, ceramics for the exterior walls, and highly efficient glass wool for heat-insulating materials. In cold districts, in order to improve efficacy, a greater quantity of insulating materials is utilized. Long eaves protect against the fierce sunlight of summer and reduce the energy needed for air conditioning. The rays of sunlight in winter are allowed to shine into the rooms to heighten heating effectiveness.

VENTILATION SYSTEM

Well hole

Just like the human body, a house needs to breathe. Here, at the centre of the house, a small roof portion (well hole) is provided in order to allow heat from

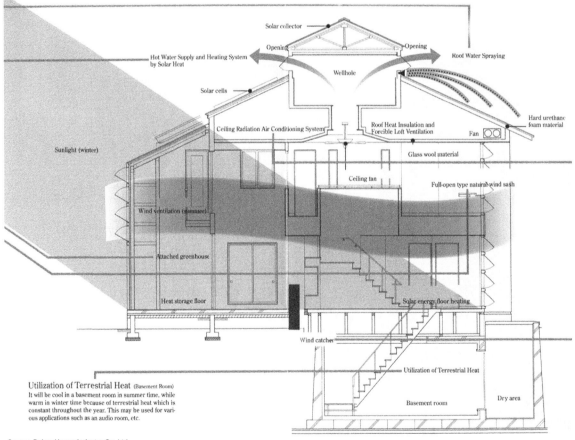

Utilization of Terrestrial Heat (Basement Room)
It will be cool in a basement room in summer time, while warm in winter time because of terrestrial heat which is constant throughout the year. This may be used for various applications such as an audio room, etc.

Source: Daiwa House Industry Co. Ltd

6.25 Environmental Symbiosis House: section

each room to naturally exit within the house by taking advantage of air temperature differences (this may be even more effective if a ceiling fan is used).

Natural air conditioning by wind ventilation (full-open type natural wind sash)

Here in the northern part of Nara Prefecture, the north wind blows during summer. In order to take advantage of this natural phenomenon, full-open type natural wind sashes are placed on the north and south sides of the building. Using the north wind for natural air conditioning is referred to as a 'passive cooling system'.

In order to support natural air conditioning through the use of wind ventilation, a wing-shaped projecting wall is placed at both the east and west sides of the house. These walls catch the north wind and deflect it into the house.

APPLIANCES

Sunlight control (pergola and vine screen)

A pergola full of deciduous plants is arranged at the south side of the house. It prevents sunlight from going into the room in summer, while allowing sunlight into the room in winter. Furthermore, a vine screen is positioned to shut out the western sunlight and prevent the wall surface temperature from rising.

UTILIZATION OF PLANTS

Trees and plants have various functions for environmental symbiosis, such as transpiration to prevent the temperature of materials around them from rising, atmosphere purification and so on. Evergreen trees planted on the north side shut out the north winds of winter, while deciduous trees planted on the south side offer shade in summer, but allow sunlight to come in during the winter.

SOLAR ENERGY UTILIZATION

Hot water supply and heating system using solar energy

This trial system utilizes solar energy as part of the energy required for daily living, such as hot water and heating. A solar collector is arranged on the southern rooftop, and the solar energy collected is used to generate hot water for the kitchen, bathroom, washroom and for part of the floor heating.

Attached greenhouse

This is used as a sun room in wintertime and also has the effect of auxiliary heating by allowing solar heat to be stored in the floor. When opened in summertime, it becomes a well-ventilated space.

Utilization of terrestrial heat (basement room)

The basement is cool in summer and warm in winter because of terrestrial heat, which is constant throughout the year. This room may be used for various applications, such as an audio room.

Source: Daiwa House Industry Co. Ltd

6.26 Environmental Symbiosis House: wind catcher

WASTE AND WATER SYSTEMS

Utilization of rainwater

In order to make the most of water resources and rainwater, rainwater ponds have been arranged on the west and east sides of the building. Rainwater is purified and sterilized by an outdoor rainwater-processing unit, then pumped out for use in spraying the roof, watering the garden and flushing the toilet.

Roof water spraying

This is a contemporary version of the water sprinkling that has been carried out since earlier times. On hot summer days, this system sprinkles water onto the rooftop to prevent room temperatures on the second floor from rising. Purified and sterilized rainwater is used in this system. After sprinkling, the remaining water is collected in the pond and recirculated.

SUMMARY

Traditional Japanese houses have a post-and-beam structure, which is similar to the structure of vernacular houses in tropical zones such as south-east Asia. There are many air leakages in these houses; the structure is suitable for humid and hot weather. After the oil crisis in the 1970s, the structure was improved to be airtight and insulated using films and other materials. The prefabricated houses with original structures and the imported houses (such as 2 × 4) expanded all over Japan.

In these situations, the thermal performance of normal Japanese houses has become higher and the demand level for the indoor climate has also become higher. Especially, the energy consumption for cooling has become higher because of the diffusion of air conditioning and the loss of traditional habits such as opening windows and sprinkling water on the ground before windows. As a result, both of the energy consumption for heating and that for cooling are increasing continuously.

For a better living environment and for energy saving, the solution set considering both winter and summer is very important in Japan. The results of the simulations showed the necessity of the design concept for cooling and heating. This concept can be found in these case studies. The variety of solution sets in the case studies showed that the hybrid of the passive solution methods and the active solution methods is not easy. But the typical solution sets shown above will be a guide to a new design concept and the simulation technologies will adjust the design details.

REFERENCES

AIJ (Architectural Institute of Japan) (2000) *Expanded AMeDAS Weather Data*, AIJ, Japan

AIVC (Air Infiltration and Ventilation Centre) (1994) *Numerical Data for Air Infiltration and Ventilation Calculations*, Technical note AIVC 44

COMVEN-TRNSYS (1997) *Multi-Zone Air Flow Model COMVEN-TRNSYS*, IEA-ECB Annex 23

Humphreys, M. A. (1978) 'Outdoor temperatures and comfort indoors', *Building Research and Practice*, vol 6, no 2, pp92–105

IBEC (Institute for Building Environment and Energy Conservation) (1999) *Standard and Guidelines for Energy Conservation of Residences*, IBEC, Japan, p57

Nicole F. and Roaf S. (1996) 'Pioneering new indoor temperature standard: The Pakistan project', *Energy and Buildings*, vol 23, pp169–174

Chapter 7
Brisbane:
A Subtropical Climate

Richard Hyde, Luke Watson, Katherine Khoo, Nardine Lester and Joel Kelder

INTRODUCTION

The underlying principle behind climate responsive design is understanding the climatic parameters in which the building is situated. Climate, by definition, is related to the atmospheric conditions of temperature, humidity, wind, vegetation and light specific to a geographical location. Within this location, a series of climate conditions can be found. These can be categorized into three levels. First, there are the global conditions of the region created by dominant geographical features of land, sea, sun and air. Next, these are modified by local conditions that are dependent upon dominant features of water, topography and vegetation. Finally, there are the site conditions and building context, which are an interaction of local conditions and the building. These three levels of climatic conditions combine to create a complex inter-relationship between both macro- and microclimatic conditions. This complex interrelationship is what designers have to deal with to produce a building design that is both functional and comfortable for the climate in which it is situated. This becomes a complex task because of the nature of the ever-changing and unpredictable environmental patterns, associated with different environmental zones throughout the world.

SUBTROPICAL CLIMATES
Brisbane climate

Located at latitude 27.4° south, longitude 153° east and at an elevation of 38m above sea level, Brisbane enjoys a subtropical climate, with hot humid summers and balmy nights. Summer can have a number of very hot days. The winters are mild and dry, with most days being sunny. Summer averages a mean temperature of 25° and winter a mean temperature of 15°, with average diurnal temperature variation around 9°–10°. Humidity and rainfall are high over the summer months but low in winter, with July exhibiting 24 dry days on

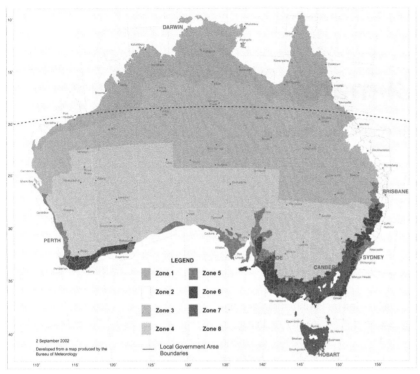

Source: Cement and Concrete Association, Bureau of Meteorology

7.1 Brisbane location and climate: zone 2 subtropical climate; Latitude 27° 28' S; longitude 153° 2' E; reference longitude 150° 0'

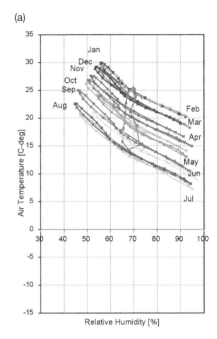

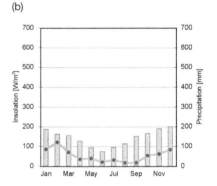

Source: Tamaki Fukazawa

7.2 Brisbane: (a) climograph the climate creates both under-heating (250-500 heating degree days) and overheating periods (50-100 cooling degree days); the extreme weather patterns can create high temperatures in summer; (b) rainfall (shown as a line graph) and solar radiation (called insolation and shown as a bar graph)

average. Annual precipitation is recorded as being, on average, between 650mm to 1200 millimetres. Relative humidity ranges between 60 to 70 per cent. The coastal areas enjoy trade winds all year round, while inland areas receive mainly southerly and westerly winds in winter, but lack breezes in summer, causing thermal stress during the daytime. Coastal areas are frequently visited by severe tropical storms that cause flash flooding and are at risk of cyclonic activity mainly in the summer period.

Figure 7.2 (b) shows the solar radiation and rainfall pattern. The availability of solar radiation creates a significant potential for solar thermal and solar electric systems, with, on average, seven hours of sunshine per day. Rainfall, on the other hand, is problematic for water collection and is not consistent throughout the year. Rainfall is delivered either as a monsoonal event with large amounts of water over a short period or through coastal showers. Winter is a dry season with many months without rain.

Figure 7.2 (a) shows the climate data in a climograph, emphasizing the main issues. The comfort zone is in the centre, with temperatures and relative humidity polled for specific times of the year. Yet, from the statistics on degree days for heating and cooling, it would appear that this is not an over-heating but an under-heating climate, although intuitively the climate is perceived as one where the main issue is cooling due to the high humidity and elevated summer temperatures that are clearly evident.

Figure 7.3 shows the number of days when temperatures are above the comfort zone. It is these extremes of temperature, particularly in summer, that present the most challenges for designing climate responsive buildings.

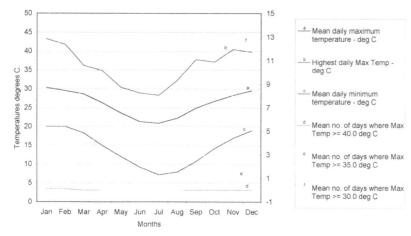

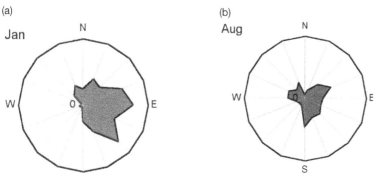

Source: Tamaki Fukazawa

7.3 Days of discomfort provided by the climate present a challenge to passive design in summer and require particular solution sets if comfort is to be maintained

Source: Tamaki Fukazawa

7.4 Wind roses for Brisbane. Summer sea breezes from the south-east and north-west provide one source of cooling in summer. Cold south-easterlies and westerlies create heating problems in winter. Designing with respect to the meso-climate and microclimate conditions is important if the effects of airflow are to be accommodated effectively.

SOLUTION SETS

Typological solution set for subtropical climates (southern hemisphere/assumed north–north-east wind vector)

The subtropical winter months are dominated by south-east winds and mostly fine weather, with blue skies, warm days and cool nights. Summer months are hot and humid, with thunderstorms starting in late October or November. Bursts of monsoon rains from late December through to early April deliver highest rainfalls, and this is also the season when tropical cyclones can threaten.

The success of a building's natural ventilation performance, particularly in the tropics, is related to the thermal performance of the building's construction, materials and wind interactions around the building's façades. To ensure that this occurs, solutions must look at whether the construction materials and form of the building will enhance natural ventilation and control solar heat gain. In order to enhance these naturally occurring breezes, it is imperative to site the building to capture the prevailing breezes and ensure that the location does not restrict the airflow to other buildings within the site or create wind shadows.

The solution sets can be rationalized from three (detached, row house and apartment) to two types when criteria of both low cost and passive building design are considered. Each solution must provide effective natural ventilation that removes internal heat loads and supplies air movement and fresh air to the occupants.

Two housing typologies have been decided upon as suitable applications within a tropical location. One of them is primarily a low-rise building set at 1m off the ground. This allows wind ventilation underneath the building to pass through to the adjacent spaces. The building has a narrow plan, being one room deep. This allows cross-ventilation through the building, providing passive cooling to internal spaces. The building is oriented to the north-east to capture the prevailing breezes. A high level of openings at the façade ensures that there is an adequate amount of ventilation. The use of an efficient roof construction system ensures minimum radiant heat into the spaces below.

The other is a higher density building that is placed in clusters. It is imperative that within these clusters breezes are able to pass through and around the building masses. It is therefore very important that the orientation of the buildings corresponds to the wind conditions onsite. The form of the building is primarily of narrow plan with adjacent outdoor spaces that are used as a communal living area. The structure provides sun shading to the windows, as well as to the majority of the western walls.

Suggested typology set for a subtropical climate

A suggested typology set for a subtropical climate is as follows:

- a rectangular shape for detached houses, stretched along an east-west axis, with the main orientation towards the north/south;
- for higher density multistorey developments (up to three storeys), a narrow plan with associated external communal spaces is recommended (it is important that these buildings are located to allow ventilation through to other buildings, preventing wind shadows);
- a narrow plan, oriented along the long axis towards the north-north-east to capture prevailing breezes;

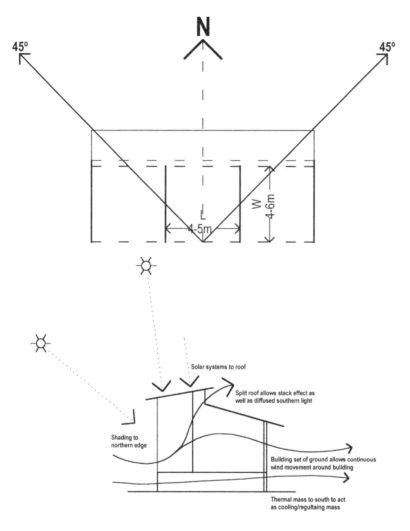

Source: Luke Watson and Joe Kelder

7.5 Typical solution sets for a compact single storey unit

- sufficient overhangs to the northern façade to exclude summer sun radiation but to allow winter sun penetration;
- a pitch of about 30° to the north to provide sufficient provisions for solar hot water systems;
- insulated walls and roof to prevent solar radiation through light-coloured materials and voids; use of radiant barriers important to resist radiant heat flux, which is the largest percentage of the total heat transfer (radiation 40 per cent, convection and conduction 30 per cent);
- a thermal mass to the southern side of the building to allow it to act as a cooling mass (it is important that this is not exposed to solar radiation since it will act as a warm mass, radiating heat back into the space);
- a construction of lightweight materials, as well as heavyweight thermal cooling mass to the southern side (reverse brick veneer construction);

- thermal mass (concrete) to floors (these are to be protected from direct solar gain); and
- sun shading and insulation to elements on the western and eastern side of the building.

Typological solution set:
A detached house in response to a warm, humid climate

Table 7.1 summarizes the main concepts regarding the most suitable typology for a subtropical climate.

Table 7.1 Characteristics for a detached house

Form: small compact unit

Floors: one maximum

Dimensional ratio (length/width): 0.83–1

Solar orientation (0° = north): main façade variable from –45° to +45°

Wind orientation: cool north–north-east from 0° to +45°

Roofing: insulated pitched, about 30° pitch

Solar protection: façade-shadowing system to the north, east and west

Active systems: photovoltaic cells, solar hot water system

Passive systems: passive ventilation, solar radiation; night ventilation/day closure, cooling mass

Glazed/opaque surfaces ratio: 15–20%

Living rooms: oriented to the south

Stack limitation: internal/external temperature = 1°–3° difference

Daylighting/skylight orientation: 180° (south) to avoid direct solar gain

Insulation: required for all roof structures, as well as unshaded external walls

Source: Luke Watson and Joe Kelder

The main solution set for this climate is to use a *highly interactive building* to accommodate a variety of different heat flux and ventilation requirements during the year. The buildings are essentially skin loaded – that is, insolation is high and the main heat gain is from the exterior, rather than from the heat gain from internal functions (building skin heat loads of 1000W/m^2 are not uncommon). The role of the building envelope, microclimate and the area surrounding the building (meso-climate) should be viewed as *linked components* to facilitate the effective operation of the solution sets proposed.

Multi-storey units (on ground)

Source: Luke Watson and Joe Kelder

7.6 Typical solution sets for a multi-storey unit

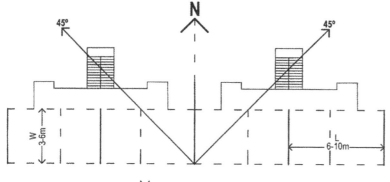

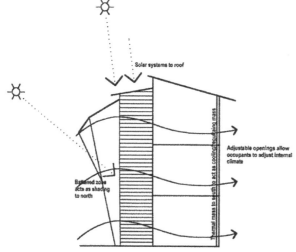

Typological solution set: A row house in response to a subtropical climate

Table 7.2 Characteristics for a row house

Form: row of three-storey houses

Floors: three maximum

Dimensional ratio (length/width): 0.6–2

Solar orientation (0° = north): main façade variable from −45° to +45°

Wind orientation: cool north–north-east from 0° to +45°

Roofing: insulated pitched, about 30° pitch

Solar protection: façade-shadowing system to the north, east and west

Active systems: photovoltaic cells, solar hot water system

Table 7.2 Characteristics for a row house (Cont'd)

Passive systems: passive ventilation, solar radiation; night ventilation/day closure, cooling mass

Glazed/opaque surfaces ratio: 15–20%

Living rooms: external associated spaces

Stack limitation: internal/external temperature = 1°–3° difference

Daylighting/skylight orientation: 180° (south) to avoid direct solar gain

Insulation: required for all roof structures, as well as unshaded external walls

---

Source: Luke Watson and Joe Kelder

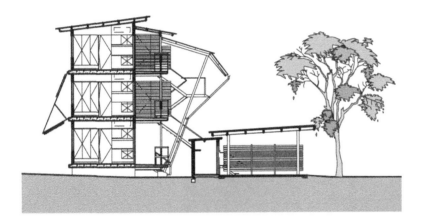

Source: Bligh Voller Nield and Troppo architects

7.7 Section through Lavarack Barracks: this project was used as a precedent for the solution set developed for apartments

Typological solution set:
An apartment in response to a subtropical climate

Table 7.3 Characteristics for an apartment

Form: row of four-storey units

Floors: four maximum

Dimensional ratio (length/width): 0.6–2

Solar orientation (0° = north): main façade variable from –45° to +45°

Wind orientation: cool north–north-east from 0° to +45°

Roofing: insulated pitched, about 30° pitch

Solar protection: façade-shadowing system to the north, east and west

Active systems: photovoltaic cells, solar hot water system

Passive systems: passive ventilation, solar radiation; night ventilation/day closure, cooling mass

Glazed/opaque surfaces ratio: 15–20%

Table 7.3 Characteristics for an apartment (Cont'd)

Living rooms: external associated spaces

Stack limitation: internal/external temperature = 1°–3° difference

Daylighting/skylight orientation: 180° (south) to avoid direct solar gain

Insulation: required for all roof structures, as well as unshaded external walls

Source: Luke Watson and Joe Kelder

Three case studies are presented in this chapter that provide examples of houses that have utilized the principles of subtropical design. The case studies respond to the three main types of solar sustainable housing: detached housing, row housing and apartments. In this region, the use of row housing is less common. The detached house is represented by the Design Studio, Brookwater, Brisbane, the row house by Lavarack Barracks, Townsville, Queensland, and the apartments by Cotton Tree Housing, Queensland.

Case study 7.1 The Design Studio, Brookwater, Queensland, Australia

Katherine Khoo and Nardine Lester

PROFILE

Table 7.4 Profile of the Design Studio, Brookwater, Queensland, Australia

Country	Australia
City	Brookwater Estate, Ipswich
Building type	Detached
Year of construction	2002
Project name	Design Studio
Architect	McKenzie Building (Brett McKenzie)

Source: McKenzie Building

PORTRAIT

In 2002, developers of the burgeoning Brookwater residential estate financed a purpose-built display house that endorsed options for sustainable living. The project was named the 'Design Studio' in an attempt to avoid the associated stigmas of display housing, this initiative demonstrates to consumers that sustainable living within mainstream urban environment confines is achievable without adopting extreme measures, without incurring

(a)

Source: McKenzie Building

7.8 Brookwater House: (a) pavilion design steps down the hillside to reduce cut and fill on the site; (b) front elevation showing passive and active systems

(b)

substantial financial expenditure or forgoing comfort and style. These objectives have been realized by integrating Housing Institute of Australia (HIA) GreenSmart strategies with architectural form in the design.

The designers, Brett McKenzie and his company Sustainable by McKenzie Building, have devised the house in accordance with the developer's instructions, which specified that its functional arrangement should be suited to either empty nesters or young urban professionals (primarily, people recognized as leading simplistic but quality lifestyles). However, the building designer maintains that the design principles are adaptable to the needs of most clients.

Ecological sustainable development (ESD) strategies comprise the studio's sustainable constituent, with holistic strategies included to contemplate building sustainability. These consist of passive and active strategies such as environment preservation; site management; strategically designed elements; abundant natural lighting and ventilation provision; efficient energy usage; passive heating and cooling; and water conservation.

In July 2003 the Design Studio was awarded the 2003 Australian Building of the Year award at the Housing Industry of Australia's National GreenSmart Awards, as well a Finalist Medal for National Energy Efficiency.

CONTEXT

The house has been designed as a prototype detached dwelling. It is positioned almost centrally and without apparent influence or constraints within the immediate site. It was built on an allocated site chosen by the estate's management as it lies within walking distance of the Brookwater Information and Directory Centre and alongside other housing displays. For the project, the developer commissioned Brett McKenzie, a prominent housing designer with a respected knowledge and experience of building sustainable houses. The resultant design addresses the described client scenario and creates a feeling of 'hanging among the gumtree vegetation', thus achieving a form that merges with the landscape and appearing as if it had grown from the site like a native plant. The house's sustainability objectives are accomplished with effective integration and coordination of strategies and devices recognized by the HIA GreenSmart programme.

ECONOMIC CONTEXT

The building was costed at approximately Aus$1250 per square metre and comprises a floor area of just over 300 square metres. Ten photovoltaic panels at an initial cost of Aus$17,000 were the most expensive of all the technical strategies employed at the Design Studio. However, the initial cost outlay is gradually reimbursed through the efficient use of the building and the selling of excess energy back to the main energy grid system.

SITE DESCRIPTION

The site is an average size for a development of this kind and is located centrally within the residential estate setting. It has a 14m drop in its topography that falls towards the 17th hole of the Brookwater golf course. Its neighbours are the estate's enquiry centre and another display house. The site is not imposed upon by the influences of the immediate surroundings, receiving abundant uninhibited breezes and daylighting.

The natural vegetation ensures that the area is partially shaded and that cool air pervades it. Its removal would have damaged the landscape by disturbing the ground and displacing the natural drainage system, instigating continuous soil erosion.

BUILDING STRUCTURE

The building is split into two pavilions connected by a breezeway. The building has an open plan and section encouraging maximum advantage of passive strategies. The northern pavilion is two storeys; the main bedroom is on the second level with its own bathroom. The stair leading back down to the

(a)

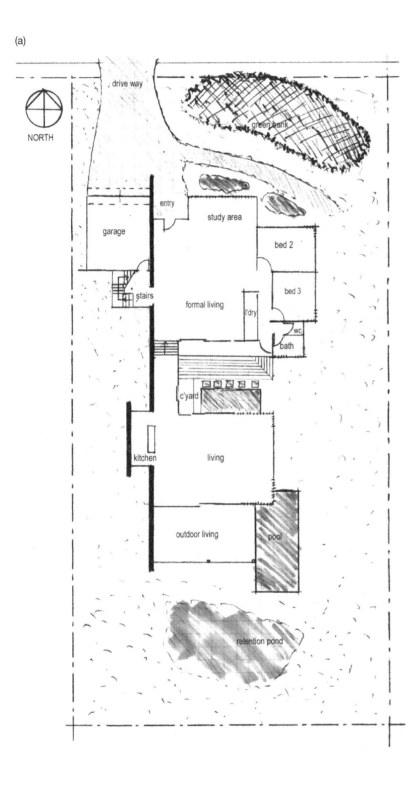

Source: McKenzie Building

7.9 Brookwater House: (a) concept plan – masonry walls are used as buffers to the western sun; the masonry is thick in order to provide capacitive insulation

(b)

Source: McKenzie Building

7.9 Brookwater House: (b) final lower floor plan; (c) final upper floor plan

(c)

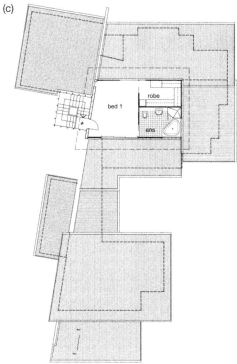

first level is a heat-attracting mechanism that draws heat through the house and out of a mechanical vent (thermal chimney effect). Two bedrooms, a living area (informal), a general bathroom and laundry are located on the lower level. The southern pavilion is accessed via a breezeway and courtyard. The kitchen is located on the western side to allow maximum ventilation through the adjoining living areas. The veranda opens out to the view of the Brookwater golf course, and the pool provides a cooling pool of water that wind passes over and naturally ventilates the interior. The house is passively cooled and

(a)

Source: McKenzie Building

7.10 Brookwater House: (a) planning on the site ensures areas for habitat conservation; (b) eastern elevation – cost-effective lightweight construction is used

(b)

(a)

(b)

Source: McKenzie Building

7.11 Brookwater House: (a) the courtyard provides amenity and a source of cool air for cross-ventilation; (b) cross-ventilation is used through the building; (c) the stove is placed on an external wall to reduce heat gain into the living spaces

(c)

heated using a variety of different strategies. Its flexible planning means that the house can be opened or closed, depending upon the situation, allowing better control over individual spaces.

The surrounding natural bushland creates a microclimate that cools air before it even reaches the building's envelope. As the Brookwater Estate development grows, this may either decrease or remain the same. If it decreases, the combination of efficient site planning and planting is sufficient enough to filter the hotter airflow.

BUILDING CONSTRUCTION

The house is composed of two forms, both comprising a lightweight building envelope. The constructed framing includes a combination of recycled timbers and plantation timber. Cladding the timber frame is a mixture of fibre-cement panels and reflective corrugated metal panelling; operable louvres on corners with exposure to prevailing winds also enhance the light/hard weighted material choice. Internally, walls feature plasterboard lining and plywood ceilings. The garage and stairwell are similarly constructed, but clad with polycarbonate sandwich sheeting. Internally, steel structures support the stairs.

Constructional difference occurs in the resolution of foundations. The upper entry pavilion uses a suspended concrete slab supported by a retaining wall utilizing stone and rubble from site spoil. The polished slab top serves as flooring for these areas. In the lower pavilion, footings are used (concealed from view) to support recycled timber flooring.

Lightweight sheet roofing with moderate-sized overhangs made of corrugated iron is used and has been designed to extend further on western aspects. Extending along the building's western side is a masonry wall performing as a passive device and providing the majority of thermal massing.

VENTILATION SYSTEMS

The site's longitudinal exposure towards the northern sun allows easy orientation of studio forms and spaces for receiving an abundant amount of daylight, using natural lighting efficiently throughout the entire day. Consequently, window design and positioning, such as louvre banks, are located on each of the north-facing corners, and ceiling-positioned horizontal strip windows encourage more natural lighting.

Passive heating and cooling measures were significant in the studio's spatial arrangement. The architectural form is arranged to use natural energy yielded by breezes effectively in order to ventilate. The site topography sweeps cooling breezes uphill. The location of the east-situated pool immediately outside a bank of louvre windows conducts evaporative cooling on breezes before their induction through the louvres, activating cross-ventilation of the lower pavilion. The northern pavilion uses the courtyard pond in a similar way.

A masonry spine wall features along the western side. This cooling element provides the majority of the building's thermal massing. It shields the house from western sun exposure, protecting inner spaces from excessive heat and thus assists the cross-ventilation strategy. It demonstrates how

(a)

(b)

(c)

Source: McKenzie Building

7.12 Brookwater House: (a) the stairway is used as a thermal flue; (b) external water feature to assist with cooling (c) stair details

elements can be sustainable and architecturally stylish. The polished concrete flooring in the upper pavilion also contributes the same benefits.

An internal stairwell protrudes past the masonry blade wall in the upper pavilion and also performs as a passive heating and cooling device. This feature assists temperature control by channelling hot air into it. The space produces a chimney effect, operating by a roof-fitted mechanical vent. In summer, an open vent releases the accumulating heat, relieving the studio of excessive heat loads and encouraging cooling breezes into the studio for ventilation. In winter, the closure of the vent reverses the effect by directing heat into the studio for circulation. This element's predominant construction of polycarbonate sandwich sheeting increases the beneficial effects.

APPLIANCES

Household operation efficiency is achieved by, and benefits from, using energy saving appliances and energy efficient light fittings and controls. Lighting use is minimized during the day due to the design of the building envelope, intended to allow maximum daylight penetration of spaces. Use of the 5-star-rated appliances reduces electricity consumption in comparison to other households. As a result, excess energy generated by the photovoltaic panels is dispensed to south-east Queensland's electricity grid and converted into credit on the household occupant's electricity bill. This decreased reliance on grid-supplied energy contributes overall to reducing energy production, which is the largest single source of Australia's greenhouse gas emissions according to the Sustainable by McKenzie Building website (www. sustainablebuildings.com.au).

ENERGY SUPPLY SYSTEM

Efficient energy is created by ten photovoltaic (or solar) panels installed on an extensive north-facing roof over the front entry. Initially costing Aus$17,000 (Derrick Jones, personal communication, 2000), the panels generate sufficient energy to power all household appliances (air conditioning does not apply as it is not utilized in this project). The panels feed surplus generated energy into the electricity power grid, with a grid electricity system operating to service night and cloudy day power requirements. This feature instigates a debit and credit incentive system, encouraging reduction of energy usage. While government reimbursement is available upon purchasing the panels, their cost is still their major disadvantage or setback. Credit from selling power back to the grid should outweigh the cost factor. On the day we were visiting, the solar panels had saved 2.17kg of carbon dioxide (CO_2), and the power usage was about 378W to 380W. The system used is the Plug and Power System by Pacific Solar.

SOLAR ENERGY UTILIZATION

With illuminance values from clear sky conditions prevailing with levels in the range of 100,000–250,000 Lux, equating to solar loads as high as

$1000W/m^2$, solar-generated electricity is used as above. Passive solar heating is employed in winter and avoided in summer. In winter, direct solar heat gain is permitted through penetration of the louvres and the ceiling height strip windows. Features such as the stairwell and masonry store solar heat gain and release heat into the house. The house is protected from solar heat gain in summer: on the north façade, there are moderate- to large-sized roof overhangs and on the east and west façades the site's natural gum-tree vegetation provides shade and the window area is minimized. Radiant barrier insulation and reflective material panels are used on the west walls to negate solar loads. Within the western zone of the building, non-habitable and limited use spaces, such as the garage and stairwell, are located.

BUILDING HEALTH AND WELL-BEING

The use of renewable resources – that is, plantation and recycled timbers – have been used for frame construction and flooring. Cutting and filling have been kept to a minimum to avoid interfering with natural drainage, which causes instability and destroys natural habitats.

The project works with existing trees and retains as many as possible to assist with the passive design strategies. By retaining the natural slope and land flow, there is minimal disturbance to the local ecosystem.

Volatile organic compounds (VOCs) are minimized by using healthy paints, carpets, textiles and furnishings. Doors and windows assist in the ventilation process, which keeps air fresh and VOCs to a minimum level of build-up. The garage location takes advantage of the ventilation flow and is at the end of the flow path so that harmful gases are not carried into the living areas.

Recycling grey water waste for toilets and irrigation ultimately saves precious fresh water. The natural site layout and storm water catchment at the lower end of the site act as a natural irrigation system and return excess water back to the water table. Photovoltaic panels are use to take advantage of solar radiation. Lifestyle spaces, such as the deck and courtyard, are created to provide functional and amenity spaces outside the building.

WATER RECYCLING AND CONSERVATION

A 25,000 litre storage capacity is provided, with five tanks (underneath the lower pavilion) for the collection of rainwater. Rainwater is collected on two-thirds of the roof surface area – the optimum to alleviate (heavy) reliance on council infrastructures and reserves. This saves approximately 400,000 litres of treated drinking water. The waste management system recycles sterilized grey wastewater by relegating it for recycled use in toilet systems and for yard irrigation. This process is calculated by Sustainable by McKenzie to save up to 50 per cent of home water requirements and approximately 350,000 litres of sewage water. Water-saving fittings and devices are installed on all tap ware to reduce water wastage. The EcoShower shower rose that uses 78 per cent less water than other showers alone saves 180 litres per shower.

PRINCIPLES, CONCEPTS AND TERMS

The building's premise was for the sole purpose of displaying and demonstrating strategies applicable to mainstream residential situations. Thus, its pivotal concepts are (in accordance with HIA GreenSmart principles) energy and water efficiency, water and waste management and recycling, as well as site and environmental management and biodiversity. With regard to occupant comfort, priorities are ventilation, appropriate daylighting and the conduction of natural cooling and heating. It is possible to achieve this through balancing purpose-massed and lightweight building components, installing and using efficient equipment and appliances, and making use of site characteristics to capture breezes, absorb solar heat, drain storm water and cool air temperatures.

INFORMATION

Brisbane City Council (2004) *Green Home Guide: Living in Brisbane 2010*, Brisbane City Council, Australia

Hay, P. R. (2002) *Main Currents in Western Environmental Thought*, UNSW Press, Sydney

HIA GreenSmart Magazine (2004) 'Passive perfect', *GreenSmart Magazine*, pp39–41

Hyde, R. (2000) *Climate Responsive Design: A Study of Buildings in Moderate and Hot Humid Climates*, E & F N Spon, London

Owen Lewis, J. (1999) *A Green Vitruvius: Principles and Practice of Sustainable Architectural Design*, James and James, London

Royal Australian Institute of Architects (2003) *BDP Environment Design Guide*, RAIA, Canberra, Australia

Websites

Australian Greenhouse Office, www.greenhouse.gov.au and www.yourhome.gov.au/index.htm (accessed 5 May 2004)

Commonwealth Bureau of Meteorology, www.bom.gov.au (accessed 21 May 2004)

HIA GreenSmart Magazine, www.greensmart.com.au

Queensland Government Housing, www.housing.qld.gov.au/initiatives/smarthousing/

Sustainable by McKenzie, www.sustainablebuildings.com.au (accessed 9 May 2004)

ACKNOWLEDGEMENTS

Acknowledgements are due to Derrick Jones, manager of Brookwater Design Studio.

Case study 7.2 Lavarack Barracks, Townsville, Queensland, Australia

Luke Watson and Joel Kelder

Source: Bligh Voller Nield and Troppo Architects

7.13 Lavarack Barracks, Townsville, Queensland, Australia

PROFILE

Table 7.5 Profile of Lavarack Barracks, Townsville, Queensland, Australia

Country	Australia
City	Townsville
Building type	Units
Year of construction	2001
Project name	Lavarack Barracks
Architect	Bligh Voller Nield and Troppo Architects

PORTRAIT

This project is a redevelopment of Lavarack Barracks in Townsville, Australia, completed by Bligh Voller Nield in conjunction with Troppo Architects. The redevelopment of Lavarack Barracks consisted of extensive master planning studies, followed by the progressive construction of over 1000 single-person accommodation units. The project was to be a low cost development that utilized prefabricated items to facilitate the speed of construction.

CONTEXT

The existing prefabricated steel buildings, dating from the Vietnam War, did not meet the current operational requirements of the defence force. Difficulties in the existing barracks included lack of acoustic and visual privacy, no individual bathrooms and inadequate environmental control. The organizational planning of the site also had failures in the lack of a functional hierarchy, the overlapping of working and housing facilities and sleeping quarters directly adjacent to heavy artillery parade grounds. Overall, the main intention of the design team was to produce new barracks with improved climatic performance and responsiveness, thus minimizing reliance on air conditioning to maintain comfortable internal temperatures. Particular

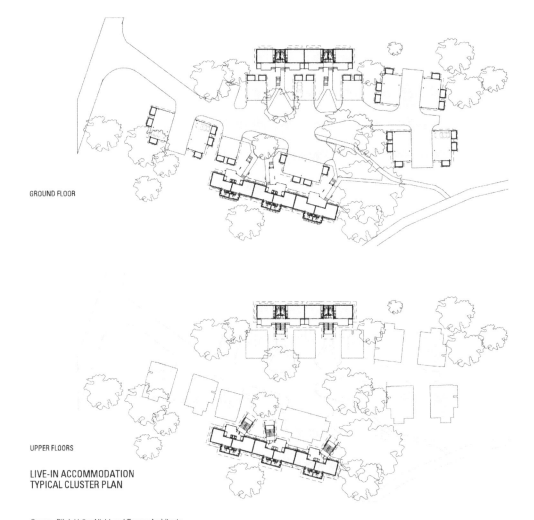

GROUND FLOOR

UPPER FLOORS

LIVE-IN ACCOMMODATION
TYPICAL CLUSTER PLAN

Source: Bligh Voller Nield and Troppo Architects

7.14 Lavarack Barracks: site plan

UNIT TYPE 1A - 18no.UNIT BLOCK
FRONT ELEVATION WITH CARPORTS

UNIT TYPE 1C - 18no.UNIT BLOCK
FRONT ELEVATION

TYPICAL CLUSTER SECTION

LIVE-IN ACCOMMODATION
TYPICAL 'OTHER RANKS' ELEVATIONS AND CLUSTER SECTION

Source: Bligh Voller Nield and Troppo Architects

7.15 Lavarack Barracks: elevations

attention has been directed towards building orientation in order to maximize exposure to cooling north to north-easterly breezes. The new barracks also aimed to improve the relationships between public and private space, and work and recreational areas.

ECONOMIC CONTEXT

Due to the strategic location of the port Townsville, Lavarack is the largest military installation in northern Australia. It was to be of a low construction and maintenance cost. With this in mind, a robust materials, as well as a kit of parts, construction approach was adopted.

SITE DESCRIPTION

Townsville defies the stereotypical tropical image. It lacks the lush green foliage, waterfalls and landscape generally associated with the tropical north.

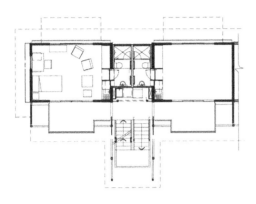

UNIT TYPE 1A

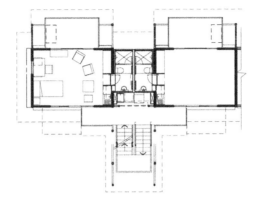

UNIT TYPE 1B

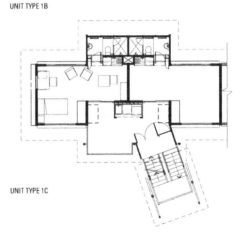

UNIT TYPE 1C

TYPICAL FLOOR PLANS - 'OTHER RANK' UNITS

Source: Bligh Voller Nield and Troppo Architects

7.16 Lavarack Barracks: plans

Townsville is situated on an ancient seabed in the rain shadow of the Great Dividing Range. The landscape bears little resemblance to what is usually associated with the tropical north, being more akin to the qualities of Central Australia. The site of Lavarack Barracks is a large flat plain bordered by Mount Stuart with its range of foothills to the south.

BUILDING STRUCTURE

The buildings have narrow cross-sections, using cross-ventilation for cooling purposes. The main living area is one room deep with both bedrooms and bathrooms accessed off this central space. Located to the north is an external space that acts as a buffer to solar penetration. There are also tilt-slab walls, precast floors and clipped-in-place prefabricated units.

BUILDING CONSTRUCTION

The buildings are steel framed structures as follows:

- external walls: concrete tilt-up panels; insulated timber with a 50mm air gap;
- internal walls: two by lightweight plaster with 50mm air gap;
- floor slab: precast floor slabs;
- ceiling: plasterboard ceiling with insulation;
- glass: clear float, 6mm;
- louvres: aluminium;
- roof: insulated roof with foil-radiant barrier;
- doors: wooden doors.

Construction was highly prefabricated in order to meet budgetary constraints and to provide an efficient means of adapting the basic unit design to each site. Roofs, stairwells, sun shading devices and bathrooms were prefabricated and simply clipped in place to tilt-slab walls and precast floors.

VENTILATION SYSTEM

The buildings use the cooling north–north-east breezes for cross-ventilation. To maximize the usage of the site's predominant breezes, studies were conducted at an early stage in the design process. Advanced environmental concepts provided design advice relating to general building form and site planning issues. The studies evaluated and quantified thermal and ventilation properties for a variety of building types and site planning arrangements; detailed thermal analysis and computation fluid dynamic (CFD) modelling was conducted and recommendations were made. Natural ventilation is maximized by achieving good cross-ventilation by a narrow cross-section and by opening the building to prevailing breezes. The building further enhances ventilation effectiveness by funnelling wind through the building and ensuring that air passes below the building, to be available for adjacent buildings.

APPLIANCES

While it is not evident from the study material reviewed what appliances and plumbing components were used in the units, it could be assumed that little energy consideration in terms of the 'star-rating' system was taken into account.

ENERGY SUPPLY SYSTEM

Energy supply systems for this development have lacked a degree of thought. The way in which active systems are delivered to the project also lacks thought and execution. However, much effort has been applied in terms of passive design outcomes. The level of good design and pre-planning is of a high standard. The site itself has been innovatively used to reduce cooling loads within the buildings. The best position for each building was found by assessing the interior demands of the building in relation to the data obtained by the detailed site analysis. Critical relationships, such as the overall

Source: Bligh Voller Nield and Troppo Architects

7.17 Lavarack Barracks: construction uses tilt-up concrete systems involving site prefabrication

arrangement between buildings, vegetation and site topography, have been carefully considered. Internal and external spaces have been designed with bioclimatic aims in mind; hence, buildings and the space surrounding them work together to control and enhance internal and external environments, as well as protecting the site and local ecosystems and biodiversity. Overall, the design exploits and manipulates site characteristics to reduce the energy consumption of the buildings and endeavours to create the best possible microclimate for the buildings and their occupants.

SOLAR ENERGY UTILIZATION

All north-facing windows have significant fixed external overhangs to minimize solar gains in summer. The need for space heating is minimal. The scope for utilizing a soft energy path through use of green technologies has been largely neglected. It was evident that no solar hot water system was utilized within the site and this could be looked at as a way of greatly improving the energy efficiency of the buildings. In a climate where there is an abundance of natural light, the option of photovoltaic cells could also be considered for their environmental and economic benefits. Furthermore, green technologies such as wind turbine technology have not been utilized despite the site's adequate wind exposure.

BUILDING HEALTH AND WELL-BEING

The brief called for no maintenance, robust finishing and detailing. Painted surfaces have been kept to a minimum. In addition to new landscaping works, existing trees and shrubs have been retained. Trees and shrubs absorb CO_2 and can remove up to 75 per cent of dust, lead and other particles from the air. The native vegetation has a high canopy with tall trunks, providing shade from the sun but at the same time allowing cooling air circulation at ground level. The existing vegetation was carefully examined to identify those plants and trees that would be important in achieving energy conserving landscaping. Mature trees require much less effort to maintain and are generally of a larger size than newly planted varieties. In addition, the slow-growing native vegetation is highly resilient to storm damage, as well as to insects and disease. The roads onsite are frequently used by military vehicles, many of which are large trucks or combat machinery; vegetation plays a major role in maintaining good air quality, as well as creating natural sound buffers.

WATER RECYCLING AND CONSERVATION

All rainwater is directed onsite into dry rock-lined gullies/swales that feed into a larger water catchment area within the army barracks site. This water is then reused as irrigation water for the site vegetation and for cleaning exterior surfaces, vehicles and combat machinery. Water collection requires no use of ultraviolet (UV) filters: the water is aerated as it passes through the rock-lined gullies and impurities are filtered out through reed filtration and settling ponds within the dams. The drainage system mimics nature and costs substantially less to build than a conventional drainage system. The natural drainage patterns of the site have been retained and followed as much as possible. Impervious surfaces such as driveways, parking lots and paving have been minimized, thus slowing water runoff and greatly reducing damage to the surrounding land and waterways. The drainage system ensures that as much rainwater as possible finds its way back into the soil in a clean condition. The possibility of grey water treatment and recycling has not been utilized; a variety of systems could have been used successfully on the site, such as septic tanks, rotating biodiscs, reed beds or dry toilet systems. An effective grey water treatment system would have provided additional water for irrigation and general outdoor cleaning, particularly during the dry season.

CONTROLS

Manual controls are used.

PLANNING TOOLS APPLIED

Thermal comfort modelling was employed to accurately predetermine the thermal comfort levels within the building structure. First, a three-dimensional thermal model of the building was constructed for each option. The model described:

- material constructions;
- windows, doors and air-transfer openings required to describe the ventilation paths;
- internal diversified load profiles for people, lights and equipment;
- schedules for the opening and closing of windows and exhaust louvres; and
- shading and overshadowing for each hour of the day for each day of the year.

The second stage required the simulation of the three-dimensional (3D) building model against hourly recorded local weather data. This allowed for the assessment of each option's thermal performance with the various building materials and shading configurations. The simulation provided detailed results of predicted room temperatures, air movement and radiant loads for each day in a typical year. Finally, thermal comfort levels were calculated using inputs that described the clothing that people wear and their activity rate. Any level of the expected occupants' discomfort was minimized.

PRINCIPLES, CONCEPTS AND TERMS

The basic principles, concepts and terms for the project are outlined as follows:

- *Heat sink*: an area of the building or site that absorbs and stores solar heat through its mass, and which continues to re-radiate to the immediate environment even after the solar load is removed.
- *Biodiversity*: the natural pre-existing flora and fauna of an area.
- *Cross- and stack ventilation*: in moderate warm climates, the purpose of ventilation is to provide fresh air to the occupants, to cool the building

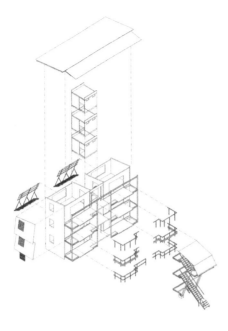

Source: Richard Hyde

7.18 Lavarack Barracks: exploded axonometric showing the construction components

fabric and to cool the occupants through air movement and heat loss. Cross-ventilation is horizontal and is usually wind driven; stack ventilation is usually vertically driven by thermal difference.

- *Soft energy path*: this entails using energy much more efficiently than is common ('doing more with less'), obtaining energy from soft technologies and intelligently using fossil fuels for the transition.
- *Soft technologies*: these technologies are renewable, running on sun, wind, water, and farm and forestry wastes.
- *Microclimate*: this is a local zone where the climate differs from the surrounding area.

INFORMATION

Advanced Environmental Concepts (2002) *Environmental Performance Analysis*, May, Bligh Voller Nield, internal publication, Brisbane
Bligh Voller Nield (2003) *Environmental Sustainability in Practice*, May, Bligh Voller Nield, internal publication, Brisbane
Bureau of Meteorology, www.bom.gov.au
Owen Lewis, J. (1999) *A Green Vitruvius: Principles and Practice of Sustainable Architectural Design*, James and James, London
Lovins, A. (1977) *Soft Energy Paths: Towards a Durable Peace*, Penguin, London
Noble L. (2002) 'Lavarack Barracks', *Architecture Australia*, March/April
Shane Thomson, Bligh Voller Nield

Case study 7.3 Cotton Tree Housing, Queensland, Australia

Richard Hyde

PROFILE

Table 7.6 Profile of Cotton Tree Housing, Queensland, Australia

Country	Australia
City	Cotton Tree, Sunshine Coast
Building type	Units
Year of construction	1994/1995
Project name	Cotton Tree Housing
Architect	Clare Design (Lindsay Clare and Kerry Clare)

Source: Richard Hyde

Source: Richard Hyde

7.19 Cotton Tree Housing: northern façade opens to a courtyard showing balconies that are enlarged to form external rooms

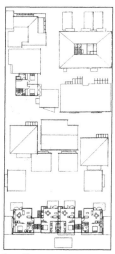

First Floor

Second Floor

N (equator)

Ground Floor

Source: Clare Design (Lindsay Clare and Kerry Clare)

7.20 Cotton Tree Housing: site plan

PORTRAIT

The architects for the Cotton Tree Housing project, Clare Design, aimed to demonstrate the use of design principles that can be used in ecological sustainable development (ESD) for multi-residential construction in warm climates.

A strong feature of the design is the attitude to ESD in selecting building materials and building by-products and in the assembly. The scheme follows the 'timber-and-tin' typology found in Queensland architecture; the use of corrugated iron fibre cement for cladding and roofing is common, juxtaposed with hardwood timber. The logic of this type of construction has stood the test of time, demonstrating sensitivity to the eco-efficiency of construction.

Source: Richard Hyde

7.21 Cotton Tree Housing: south (polar-facing) façade. Three storey units are placed to the southerly zone on the site to facilitate ventilation to the northern zones

Source: Richard Hyde

7.22 Cotton Tree Housing: permeable stairways in the three storey units facilitate airflow across the site

CONTEXT

The site is located only a few hundred metres from the Pacific Ocean, and although not directly on the beach, the site receives benefits from the ocean effect on the climate. The ocean effect moderates the macroclimatic conditions, which are characterized by cool, dry winters and summers that are hot, wet and humid.

An unusual aspect of the project involved the pooling of adjoining land owned by the then Department of Housing, Local Government and Planning and a private owner. The pooling of the available land enabled a single overall design to be prepared by a single consultant (employed separately by the two parties). This led to a radical redrawing of the boundary between the two ownerships and then to two separate building contracts undertaken by one builder. The project has two parts: a public housing component and a private component. The public housing component consists of one-bedroom apartments and two-bedroom attached houses, facing Kingsford Smith Parade, and the private component consists of attached houses, facing Hinkler Parade.

With this building type, high-density solutions are often used that invariably compromise the climate responsive character of the design. The result is the use of excessive space heating and cooling as a Band-Aid to poor design. In this case, the site planning and the massing of the building form enable the units to draw power from renewable sources. The use of passive strategies for heating and cooling are employed effectively, and the benefits with regard to the lifestyle of the users are apparent. The design of the external space has been given central importance in the scheme, which provides social and community benefits.

ECONOMIC CONTEXT

The building was designed to meet cost criteria for commercial and public development. The project focused on material issues in terms of increasing durability and reducing the maintenance of components, while still achieving design objectives.

SITE DESCRIPTION

One of the main opportunities of master planning the two subdivisions into one development was to facilitate the use of microclimate design principles for the form and position of the development's units. The resulting design

Source: Richard Hyde

7.23 Cotton Tree Housing: two-storey 'narrow frontage' units are located on the northern edge of the site; these have a deep plan and hence roof ventilations are used to facilitate ventilation

Source: Richard Hyde

7.24 Cotton Tree Housing: single-storey units and permeable fencing maximize airflow through the site

South/North Section (public housing at left)

Source: Clare Design (Lindsay Clare and Kerry Clare)

7.25 Cotton Tree Housing: site section

layout shows the permeability of the building massing to promote airflow through the site. The geometry of the site is not favourable to this approach; it has it longest boundaries facing east and west, with its smallest boundary facing the northerly favourable solar and wind vectors. This influenced the massing and location of units, with two-storey and single-storey units placed on the northerly area of the site, and the three-storey units to the southerly area. Access roads, paths and courtyards were used to funnel and direct the airflow to units in order to maximize summer cooling. Only a few units do not avail themselves of direct access to the sea breeze in summer.

For winter, most units have a northerly orientation. Each unit has been provided with a large balcony and deck spaces of generous proportions, allowing for a variety of activities that are characteristic of the informal Queensland lifestyle. Generally, these external areas are shaded, thus enhancing their functionality in summer; but the shading may provide problems in winter with solar access to the living rooms.

Careful selection of landscaping materials and fencing allows airflow through the site. Existing trees are maintained for shade and covered car parking is provided.

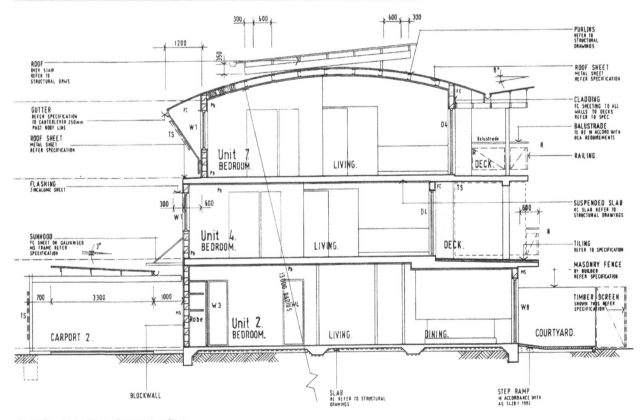

Source: Clare Design (Lindsay Clare and Kerry Clare)

7.26 Cotton Tree Housing: building construction

BUILDING STRUCTURE AND CONSTRUCTION

The external appearance of the buildings follows a lightweight timber-and-tin aesthetic, characteristic of the regional architecture. Interestingly, the buildings are a mixture of heavy and lightweight materials and systems: the low-rise buildings are timber with suspended timber floors and masonry cross-walls, while the three-storey units are load-bearing masonry walls with concrete slabs and lightweight external veneer systems. The buildings use corrugated iron cladding to the first and second floor, with fibre cement for the top floors. Bulk insulation and radiant barriers are used in the roofs, with only radiant barriers and air cavities employed as thermal protection in the walls.

The materials and detailing demonstrate a life-cycle perspective, as well as a clear tectonic in the architecture. The honest expression of materials and the differentiation in the façade respect the need for a robust durable material at the ground plane. The composite wall construction optimizes the low embodied-energy potential, while reducing heat gain and loss. Minimal corrective maintenance is required for the envelope.

VENTILATION SYSTEM

Rather than two separate blocks, the designers were faced with a larger rectangular site aligned north and south. The need to create a massing effect that allowed cooling breezes from the north to flow through the site became a major challenge.

Furthermore, the challenges of this building type are also bound inextricably to the design of the external spaces. Indeed, the building form creates a series of courtyards, breezeways and light wells, which are integral with the functionality of the internal spaces and require similar consideration. The microclimate is an important challenge for this type of development, with the need for shade and cool pools. The ocean effect moderates the macro-climate conditions, which are characterized by cool, dry winters and summers that are hot, wet and humid.

In winter, the long under-heating period is moderated by reduced minimum temperatures on the coast. The clear dry days result in comfortable and bright weather. The cool air temperatures during the day can be overcome by the solar effect. The local architecture responds to these winter conditions by controlling glare and admitting early morning and evening sun to provide warmth. Sheltering from cool westerly winds is also a necessary response; once sheltered, there is enough solar energy in the sun to combat the lower air temperatures.

In summer, the weather is characterized by humidity that is often in the range of 60 to 100 per cent, making a still air temperature of 27 to 30°C uncomfortable. The need for ventilation for cooling of both the building fabric and the occupants is a major priority. A further complication is that with residential construction, differing comfort levels apply for daytime and night-time activities. This is particularly important in summer when elevated night-time temperatures and low wind speeds coincide with the need for lower comfort levels. Hence, the design of bedrooms becomes a problem for residential construction – particularly for unit construction.

Source: Richard Hyde

7.27 Cotton Tree Housing: external rooms are constructed of a lightweight elemental system (bolt on and off) for ease of maintenance

The passive building solution involves utilizing microclimate, envelope control and site planning to maximize the prevailing breeze; thus, the location of the building and its access to natural resources of breeze and sun are of central importance to the effectiveness of these strategies. Yet, for hot windless nights, backup active systems, such as ceiling fans, are crucial, particularly in the bedroom area. The electrical consumption of a ceiling fan is equivalent to a small light fitting and provides effective cooling for minimum greenhouse gas emissions. It is important to note that this is conditional upon the building fabric remaining at ambient temperature. Poor passive design of buildings often results in the transfer of external heat to the bedroom area. The radiant heating effect from the walls and the ceilings on the occupants is a problem. This type of heat load cannot be controlled effectively by the use of ceiling fans. It is common in this situation for owners to consider using a split system to air condition bedrooms. There is a debate about the effectiveness of this approach. Owners should consider the added system cost, ongoing maintenance and air quality issues, as opposed to improving the quality of the building envelope. Generally, split systems do not have a fresh air intake and therefore recycle internal air, which can result in further problems, including odours and user health.

APPLIANCES

While it is not evident from the study material reviewed as to what appliances and plumbing components have been used in the units, it could be assumed that little energy consideration in terms of the 'star-rating' system was taken into account.

Additional savings could be made if the buildings used energy saving features, such as energy efficient electric lighting. This is difficult to achieve in housing developments such as this where the selection of fittings is largely controlled by the occupants. Additional energy-production strategies, such as solar hot water heating and photovoltaics, have also become more affordable since the introduction of government subsidies. These can be easily retrofitted to the design and would further reduce the environmental footprint.

ENERGY SUPPLY SYSTEM

The approach to energy efficiency is primarily through the use of climate responsive strategies and passive design features in the buildings. This means that the environmental loads from the sun and other agents are minimized and the need for active environmental control is avoided. An indication of the effectiveness of the approach is seen by the absence of air-conditioning systems, which are often used as a Band-Aid to poor passive design.

Energy supply systems for this development have lacked a degree of thought. Furthermore, the ways in which active systems are delivered to the project lack consideration and efficient execution. However, much effort has been applied in terms of passive design outcomes. The level of good design and pre-planning is of a high standard. The site itself has been innovatively used to reduce cooling loads within the buildings. The best position for each building was found by assessing the interior demands of the building in relation to the data obtained by the detailed site analysis. Critical relationships, such

Source: Richard Hyde

7.28 Cotton Tree Housing: details of the façade showing the fibre cement and steel sheeting which acts as cladding to the reinforced concrete masonry used to form the block construction)

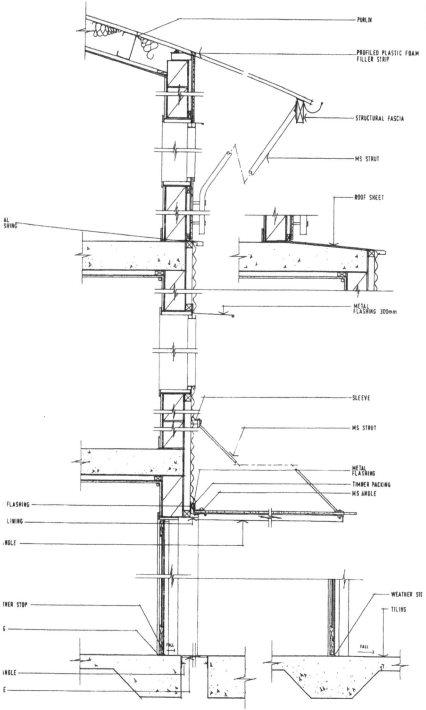

PURLIN

PROFILED PLASTIC FOAM
FILLER STRIP

STRUCTURAL FASCIA

MS STRUT

ROOF SHEET

AL
SHING

METAL
FLASHING 300mm

SLEEVE

MS STRUT

METAL
FLASHING
TIMBER PACKING
MS ANGLE

FLASHING

LINING

ANGLE

WEATHER STO

TILING

THER STOP

G

FALL

FALL

ANGLE

E

Source: Clare Design (Lindsay Clare and Kerry Clare)

7.29 Cotton Tree Housing: reverse brick veneer construction in this development provides thermal mass, which is insulated from the external heat flux, providing a way of moderating thermal performance

as the overall arrangement between buildings, vegetation and site topography, have been carefully considered. Internal and external spaces have been designed with passive principles in mind; hence, buildings and the space surrounding them work together to control and enhance the internal and external environments, as well as protecting the site and local ecosystems and biodiversity. Overall, the design exploits and manipulates site characteristics to reduce the energy consumption of the buildings and endeavours to create the best possible microclimate for the buildings and its occupants.

SOLAR ENERGY UTILIZATION

The high levels of solar radiation in winter can provide the heat required to maintain comfortable interior living conditions. South-east Queensland has, on average, seven hours of sunshine a day. If this solar gain is used with thermal mass, then effective passive heating can be provided and space heating can be avoided. The strategy is to selectively shade for solar access. Northerly windows receive indirect diffuse radiation to air charge thermal mass during the day. Thermal mass in the floors and reverse masonry veneer walls for internal thermal mass act as heat sinks. Curtains on windows and doors help to retain heat at night.

All units utilize this strategy. High levels of glazing are found on the north façade. The perimeter walls are made of reverse brick veneer that is lightweight cladding on the outside to protect the mass on the inside. In this way, the building can act as a thermal flywheel, reducing the need for space heating.

BUILDING HEALTH AND WELL-BEING

This material palette clearly demonstrates a life-cycle perspective through loose-fit long-life structures, integrating preventive maintenance strategies to

Source: Richard Hyde

7.30 External rooms are a feature of this development and are as important in a subtropical climate as the internal spaces for health and well-being

ensure durability. The main structure of the building comprises a masonry perimeter wall with suspended concrete floors, which represent long-life elements. The internal partitions are lightweight and short life, ensuring the flexibility to change the layout in the future. There are added benefits to this structural approach, where the mass of the building is increased, assisting the passive design. The perimeter wall, protected by lightweight cladding, provides a reverse brick veneer walling system, adding further to the mass effect.

Table 7.7 Design response to energy efficiency criteria

Energy efficiency and consumption criteria	Response
1 Evidence of energy audits, benchmarks, targets or thermal modelling, where undertaken	The building is based on previous prototype buildings, which had thermal performance audits
2 Measures to ensure minimal energy use for cooling in summer	Passive design strategies to maximize ventilation
3 Measures to ensure minimal energy use for heating in winter	Passive design strategies to maximize solar gain in winter
4 Measures to ensure minimal energy use for lighting and buildings ventilation	Fluorescent lighting used in public spaces. Efficient lighting inside building is at the discretion of the users and therefore beyond the scope of this study. Natural ventilation techniques are incorporated within the design
5 Use of energy efficient equipment, appliances and processes	No data available
6 Appropriate level of resource consumption (cost efficiency)	Eco-efficient construction and minimalist design
7 Exemplary design	High level of synthesis of technical aspects, poetics and place-making
8 Maximize use of renewable energy and low impact resources	Limited consideration of these issues

Source: Richard Hyde

CONTROLS

Overall, the units operate as passively controlled buildings, drawing power from natural renewable sources. The ingenuity of the design has enabled the building's operational energy to be reduced across the seasons, in comparison with other buildings of a similar typology. Many of the surrounding unit developments on the Sunshine Coast do not use microclimate control and have limited envelope control. As a result, mechanical cooling and ventilation is introduced to provide adequate thermal comfort. The Cotton Tree Housing project provides an alternative to this model.

PLANNING TOOLS APPLIED

The project demonstrates the application of a range of ESD strategies in the master planning, unit design and material specification, as shown in Table 7.7.

PRINCIPLES, CONCEPTS AND TERMS

The basic principles, concepts and terms for the project are outlined as follows:

- *Heat sink*: an area of the building or site that absorbs and stores solar heat through its mass and which continues to re-radiate to the immediate environment even after the solar load is removed.
- *Reverse brick veneer*: this form of construction was used in the walls and comprises lightweight cladding and insulation on the outside surface of the wall and masonry on the inside. It is ideal for skin-loaded buildings, such as those found in warm and hot climates where the dominant heat load is on the outside from solar gain and high external air temperatures. The advantages of thermal mass and insulation are combined. The lightweight cladding and insulation comprises the external layer, while the thermal mass forms the internal layer and protects from inward heat flux. This means that the thermal mass can stabilize internal temperatures more effectively.
- *Biodiversity*: the natural pre-existing flora and fauna of an area.
- *Cross- and stack ventilation*: in moderate warm climates, the purpose of ventilation is to provide fresh air to the occupants, to cool the building fabric and to cool the occupants through air movement and heat loss. Cross-ventilation is horizontal and is usually wind driven; stack ventilation is usually vertically driven by thermal difference.
- *Soft energy path*: this entails using energy much more efficiently than is common ('doing more with less'), obtaining energy from soft technologies and intelligently using fossil fuels for the transition.
- *Soft technologies*: these are renewable, running on sun, wind, water, and farm and forestry wastes.
- *Microclimate*: a local zone where the climate differs from the surrounding area.

ACKNOWLEDGEMENTS

Acknowledgements are due to John Byrne and Clare Design for assistance and advice, and for providing construction drawings. Acknowledgements are also due to D. Sanderson and J. Florence for provision of images.

INFORMATION

Byrne, J. (2000) 'An environmental design exhibition', Internal paper, Department of Housing and Local Government, Queensland State Government, Brisbane, Australia

Heshong, L. (1978) *Thermal Delight in Architecture*, MIT Press, Massachusetts, US

Hyde, R. A. (2000a) *Climate Responsive Design*, E&FN Spons, London

Hyde, R. A. (2000b) 'Cotton Tree Project', in *Environmental Design Guide*, *Case 21*, November, Royal Institute of Australian Architects, Canberra, Australia

Keniger, M. (1996) 'Cotton Tree Housing', *Architecture Australia*, vol 85, no. 4, July–August, pp. 58–63

Muschamp, H. (2000) 'Good buildings, and good for you', *The New York Times*, 16 April, p37

Olgyay, V. (1962) *Design with Climate*, University of Princetown Press, Princetown, US

Russell, J. (2000) 'Ten shades of green', *Architectural Record*, May, p3

SUMMARY

A changing context for housing in Brisbane's subtropical climate is occurring through the effects of urbanization on meso-climate conditions. The expansion and densification of the city is changing topographical features: reducing breezes, increasing hard surfaces, eliminating vegetation and elevating exterior temperatures. The changing needs of comfort, the desire for more stabilization of internal temperatures and reductions in humidity have resulted in the use of mechanical systems for climate control, such as air conditioning. The cultural imperative that this creates is to utilize less *interactive building solution sets* and to create the need for more *defensive solutions* that combine both passive and active systems, using mechanical systems to expel excess heat rather than the natural systems. Chapter 9 and 10 examine some of the solutions that facilitate this defensive approach.

Chapter 8
Kuala Lumpur:
A Hot Humid Climate

Sabarinah Sh. Ahmad

INTRODUCTION

The issues associated with hot, humid tropical climates and the appropriate architectural responses to the climate are discussed in this chapter. Typological solution sets for cost-effective design in hot humid climates are proposed for several types of developments. A few case studies are included to further demonstrate how architecture can respond positively to its environment.

Hot and humid regions are warm all year round, with daytime maximum temperature of 30–35°C. They extend up to 15–25 degrees latitudes on each side of the equator – for example, in Central and South America, Central Africa, Micronesia and South-East Asia. There is little seasonal variation in the climate, with a constant annual average temperature and humidity. The range of average monthly temperature is about 1–3°C; the average diurnal temperature variation is about 8°C; the annual mean temperature is about 27°C. Humidity and rainfall is high throughout the year; annual precipitation is greater than 1500mm. Coastal areas enjoy trade winds, while inland areas are windless, resulting in thermal stress during the day. Parts of the region are also subject to tropical storms, such as hurricanes and typhoons. Solar radiation intensity varies widely with cloud conditions. Further away from the equator, a wet season and a dry season occur.

The hot humid climate is uncomfortable and lacks variation when compared to climates that are further away from the equator. Architecturally, the climate could be a challenge when designing for passive cooling.

CLIMATE

Climate of Kuala Lumpur, Malaysia

Malaysia's weather is fairly hot and humid all year round (with the exception of the cooler climes of the central highlands). Average temperature is 26.7°C and average humidity is 83 per cent. Rainfall usually occurs in the form of thunderstorms. Malaysia's climate is dominated by the effect of two monsoons or 'rainy seasons', which affect different parts of Malaysia to varying degrees.

From November to February, the east coast is affected by the north-east monsoon. This brings heavy rainfall, strong winds and huge waves along the entire coast. From April to September, the west coast is affected by the south-west monsoon. It is weaker compared to the north-east monsoon. March and October are the transition months between the monsoons, characterized by light winds.

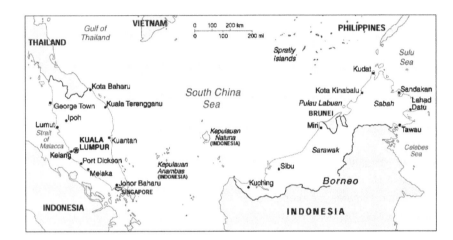

Source: Global Media (2007)

8.1 Map of Malaysia

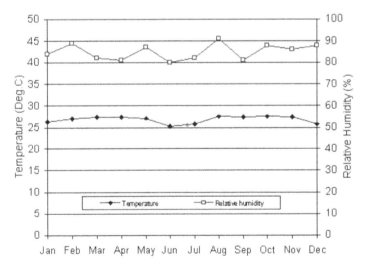

Source: Malaysian Meteorological Service (2002)

8.2 Annual variations of weather: temperature and relative humidity data

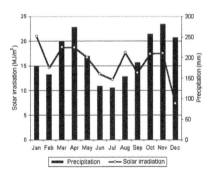

Source: Malaysian Meteorological Service (2002)

8.3 Annual variations of weather: solar irradiation and precipitation

The weather in Kuala Lumpur is hot and humid all year, with average temperatures of 23–32°C and average rainfall of 190mm. Showers occur almost daily, and downpours during the rainy season are not much worse than the rest of the year. Kuala Lumpur is affected by the south-west monsoon from April to September. The climograph (see Figure 8.5) of Kuala Lumpur shows that all temperatures fell outside the recommended American Society of

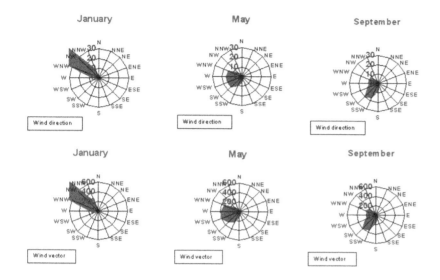

Source: Malaysian Meteorological Service (2002)

8.4 Wind roses for the months of January, May and September

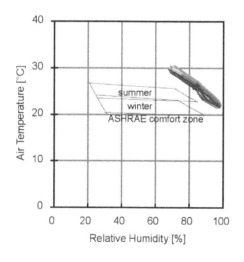

Source: Tamaki Fukuzawa (Climate Data: METEONORM Version 5.0 Edition 2003, Meteotest)

8.5 The climograph of Kuala Lumpur

Heating, Refrigerating and Air-Conditioning Engineers (ASHRAE) Standard 55 summer comfort zone. The cooling period is throughout the year.

The wind direction is mainly from the north-west to the south-west throughout the year, as shown in the wind roses in Figure 8.4.

SOLUTION SETS

Typological solution set for cost-effective design in hot humid climates

Hot humid regions are distinguished, from an architectural viewpoint, by two features. First, the climate is uncomfortable and is the most difficult to ameliorate by passive design. Second, many of the countries in the hot humid regions are developing countries. The latter fact has a direct impact upon the practicality of some modern concepts of urban and building design from a climatic viewpoint. The vast majority of people in this region cannot afford air conditioning. Therefore, thermal stress should be minimized primarily by environmentally friendly and low-technology design solutions (Chen, 1998). As a result, design principles for these climates should reflect these two features. Climate considerations such as maximum temperatures are not as high as in hot arid climates; but nights often remain above the comfort zone. Diurnal temperature variation is low, often less than 8°C, especially in the 'wet' season. Humidity is high, so skin evaporation is limited and evaporative cooling is not effective. Design aims are to reduce internal temperatures, maximize ventilation rates to increase the effectiveness of sweat evaporation, and provide protection from sun, rain and insects.

Appropriate strategies are as follows:

- Keep out direct sunshine and heat by:
  - using large overhangs to protect internal spaces from solar radiation;
  - ensuring that east and west elevations have few or no windows admitting low sun, and that walls on these elevations are reflective and well insulated;
  - using low thermal mass materials to minimize heat storage; and
  - using shading devices to minimize solar gain.

Source: Chen (1998)

8.6 Urban housing in a hot climate is dominated by the vernacular urban form comprising in shop house. Lessons learned from the shop house can be applied in new forms of bioclimatic housing

(a)

(b)

(c)

Source: Chen (1998)

8.7 (a) Shop house interior looking towards the street; (b) interior looking towards the air well; (c) internal floors and walls of shop houses are made of lightweight materials

(a)

Source: Chen (1998)

8.8 (a) Details of the air wells; (b) air well roof showing security devices

(b)

- Maximize natural ventilation by ensuring that:
  - north and south walls have large openings for ventilation;
  - double-banking of rooms is avoided, if possible;
  - rooms are arranged to aid cross-ventilation;
  - plans are open and free spaces between buildings are wide;
  - large volumetric ventilation is provided to remove internal heat;
  - spacing of buildings optimizes access to breezes; and
  - in free-standing houses, elevated construction is used, where possible, to improve wind exposure.

- Use orientation to best effect:
  - best orientation is for long façades to face north and south;
  - orientation of buildings should respond to available cooling winds, as well as to sun;
  - conflict between sun and breeze orientation should always be resolved to control sun, with the design of both building and landscaping modified to deflect available winds.

- Roofs should be pitched to facilitate water drainage.
- Mean radiant temperature should be kept as low as possible by using a reflective roof, ventilated air space and reflective foil above ceilings, as well as insulating ceilings.
- For row houses, courtyards and air wells on the ground floor, encourage cross-ventilation and daylight into the internal spaces.

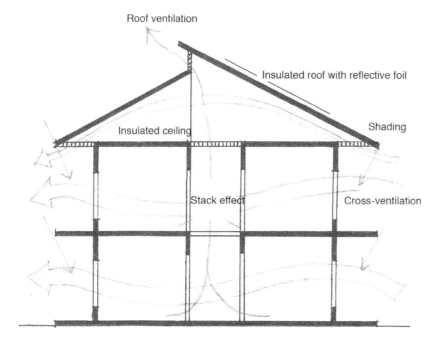

Roof ventilation

Insulated roof with reflective foil

Insulated ceiling

Shading

Stack effect

Cross-ventilation

Source: Sabarinah Sh. Ahmad

8.9 Solution set for a detached house: cross-section

6-10m

6-10m

10-20m

DRYING YARD

UTILITY BATH 5

WET KITCHEN

FAMILY

DRY KITCHEN

POWDER

DINING

COURTYARD

STORE

LIVING

FOYER

CAR PORCH

Ground floor

A/C   BATH 2   BATH 3   A/C

BEDROOM 2   BEDROOM 3

BATH 4

FAMILY

BEDROOM 4

MASTER BEDROOM

MASTER BATH

A/C

First floor

63' 9"

North

Source: Guthrie Property Development (2003)

8.10   Solution set for a row house unit: plans

(a)

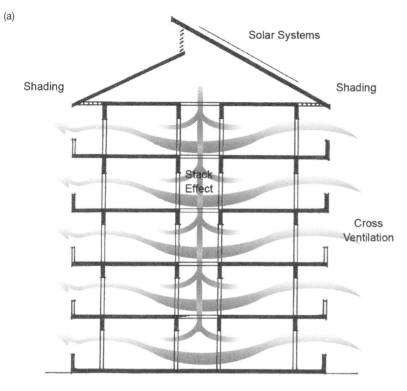

Source: Hicom Gamuda (2003)

8.11 Solution set for apartment units: (a) section; (b) plan

(b)

Table 8.1 Typological solution sets for a hot humid climate

Detached house	**Form:** compact for air conditioning to minimize surface area of envelope; spread-out building for natural ventilation
	Floors: two to three maximum
	Dimensional ratio (length/width): 1– 3 maximum
	Orientation (0° = south): 0° and 180°
	Roofing: pitched, ventilated attic, reflective foil under roof, separate and insulated ceiling
	Solar protection: façade-shadowing systems
	Active systems: photovoltaic (PV) collectors on roof
	Passive systems: cross-ventilation, shading, orientation
	Glazed/opaque surfaces ratio: south and north 30%
	Thermal time lag: >8 hours
	Ambient air exchange: 10 in summer (V x hour)
	Maximum yearly heating energy consumption: 0kWh/m²
	Reference U value: 0.3–0.6W/m²K
	Living-room orientation: south and north
Row house	**Form:** row of terraced houses
	Floors: one to three maximum
	Dimensional ratio (length/width): 0.3–1 maximum (single house cluster)
	Orientation (0° = south): 0° and 180°
	Roofing: pitched, ventilated attic, reflective foil under roof, separate and insulated ceiling
	Solar protection: façade-shadowing systems
	Active systems: PV collectors on roof
	Passive systems: internal courtyards
	Glazed/opaque surfaces ratio: south and north 30%
	Thermal time lag: >8 hours
	Ambient air exchange: 10 in summer (V x hour)
	Maximum yearly heating energy consumption: 0kWh/m²
	Reference U value: 0.3–0.6W/m²K
	Living-room orientation: south and north

Table 8.1 Typological solution sets for a hot humid climate (Cont'd)

Multi-family apartment house	**Form:** block
	Floors: four to five maximum
	Dimensional ratio (length/width): 1.6–2.5
	Orientation (0° = south): 0° and 180°
	Roofing: pitched, ventilated attic, reflective foil under roof, separate and insulated ceiling
	Solar protection: façade-shadowing systems
	Active systems: PV collectors on roofs
	Passive systems: 'double-skin' bioclimatic system
	Glazed/opaque surfaces ratio: south and north 30%
	Thermal time lag: >8 hours
	Ambient air exchange: 10 in summer (V x hour)
	Maximum yearly heating energy consumption: 0kWh/m²
	Reference U value: 0.3–0.6W/m²K
	Living-room orientation: south and north

Source: Sabarinah Sh. Ahmad

Simulation output for typological solution sets for a hot humid climate

Architecturally, the hot and humid region is one of the hardest climates to ameliorate through design. This is due to the high humidity and daytime temperatures that result in high indoor temperatures exceeding the ASHRAE summertime comfort upper limit of 26°C for most of the year. A cost-effective design for a hot and humid climate is one that uses the least amount of energy without sacrificing comfort.

Performance indicator I

Data for performance indicator I are as follows:

- days of discomfort for a detached house = 300–365 days per year;
- days of discomfort for a row house = 300–365 days per year;
- days of discomfort for a multi-family apartment house = 300–365 days per year.

Performance indicator II

Data for performance indicator II are as follows:

- predicted energy consumption for a detached house = 3600–6000kWh per year;
- predicted energy consumption for a row house = 3000–5000kWh per year;
- predicted energy consumption for a multi-family apartment house = 2500–5000kWh per year.

Case study 8.1 House at Kayangan Heights, Shah Alam, Malaysia

Sabarinah Sh. Ahmad

Source: Sabarinah Sh. Ahmad

8.12 House at Kayangan Heights, Shah Alam, Malaysia: view from the street

PROFILE

Table 8.2 Profile of a house at Kayangan Heights, Shah Alam, Malaysia

Country	Malaysia
City	Shah Alam, Selangor
Building type	Detached house
Year of construction	2003
Project name	A detached house at Kayangan Heights, Shah Alam, Malaysia
Architect	Rokiah Mohd Yusof

Source: Sabarinah Sh. Ahmad

PORTRAIT

Kayangan Heights is a gated residential enclave consisting of 40 detached houses with a clubhouse facility in the northern suburb of Shah Alam,

Malaysia. It is easily accessible via the Federal Highway, NKVE, through Sungai Buloh Highway. Each detached house sits on ample-sized lots of between 800 and 1200 square metres. Peace of mind is also assured as Kayangan Heights is guarded by a 24-hour security patrol. Other standard features are landscaped roadsides, concealed drainage and underground cabling.

CONTEXT

The freehold land was purchased through the developer, Arab Malaysian Properties. The owner of the land is then free to engage their own architect to build the house within a stipulated time frame.

ECONOMIC CONTEXT

This house is designed, built and furnished by the owner, architect Rokiah Mohd Yusof. The construction cost was kept relatively low at 1.2 million Malaysian ringgits because the owner sourced the building materials herself. The land cost was 27 Malaysian ringgits per square foot.

SITE DESCRIPTION

The house is situated in a cul-de-sac. The square-shaped lot size is 1000 square metres and the built-up area of the house is 570 square metres in extent. The land gently slopes up to the back of the lot.

BUILDING STRUCTURE

The house is a split-level two-storey-high building. The ground floor houses the living room, a dining area, a dry and wet kitchen, a study/guest room,

Source: Sabarinah Sh. Ahmad

8.13 House at Kayangan Heights: view of the porch from the entrance

Source: Sabarinah Sh. Ahmad

8.14 House at Kayangan Heights: view of the gazebo and swimming pool

Source: Sabarinah Sh. Ahmad

8.15 House at Kayangan Heights: view of the house from the gate

Source: Sabarinah Sh. Ahmad

8.16 House at Kayangan Heights: entrance
foyer to the house

Source: Sabarinah Sh. Ahmad

8.17 House at Kayangan Heights: view of the
entrance foyer from the first floor

Source: Rokiah Mohd Yusof

8.18 House at Kayangan Heights: view of the living room from the foyer

Source: Rokiah Mohd Yusof

8.19 House at Kayangan Heights:
        living room

Source: Sabarinah Sh. Ahmad

8.20 `House at Kayangan Heights:
        dining room

Source: Rokiah Mohd Yusof

8.21 House at Kayangan Heights: steps
        leading down to the dining room

Source: Rokiah Mohd Yusof

8.22 House at Kayangan Heights: master bedroom

a rumpus room, a store, a maid's room, two bathrooms and a generous porch. The wet kitchen area is for washing up and more vigorous cooking, e.g. Asian cooking. The dry kitchen area is more for the purpose of having an uncluttered space and for preparing simple snacks. The upper floor houses a master bedroom suite with an outdoor bath, two bedrooms, three bathrooms, a family/study room and balconies. The house is cooled by ceiling fans and natural ventilation, with the option of air conditioning during the hot season.

BUILDING CONSTRUCTION

The designer/owner of the house has largely used natural materials for the house finishing. The house is built using a post-and-beam reinforced concrete structure, with plastered clay-brick infill walls. Parts of the thick exterior walls are covered with sandstone and granite rubble stone. The granite rubble stones are quarried from Lembing River in Pahang and have a characteristic earthy colour due to the presence of iron oxides and other metals.

Some of the interior walls are also covered with sandstone. The roof is sloped with hardwood (*balau*) timber trusses and covered with clay roofing tiles. The ground floor is covered with white matt marble flooring that provides a cooling effect on the feet. The flooring material for the family and bedrooms comprises hardwood (*merbau*) timber strips. The paving material is a combination of slate, pebble wash and sandstone. Doors are made of timber with timber frames. Windows are timber framed with fixed plantation timber shutters. The natural building materials throughout the house give it an ageless look.

VENTILATION SYSTEM

The building uses cross-ventilation throughout the living, dining and bedroom areas, in addition to stack ventilation in the stairwell and entry foyer areas.

APPLIANCES

A solar water heater is used to reduce energy usage. This cuts the demand for electrical energy and reduces greenhouse gas emissions. The building is envelope sensitive and is lit by natural lighting during the day.

ENERGY SUPPLY SYSTEM

Electricity is directly supplied from the national grid.

SOLAR ENERGY UTILIZATION

Solar energy is used for daylighting and for a hot water system for the bathrooms and kitchen.

BUILDING HEALTH AND WELL-BEING

The housing estate is built 30 minutes from the Kuala Lumpur city centre and 10 minutes from Shah Alam city, and there is no industrial area in the immediate vicinity; therefore, it has low air, traffic and noise pollution. The

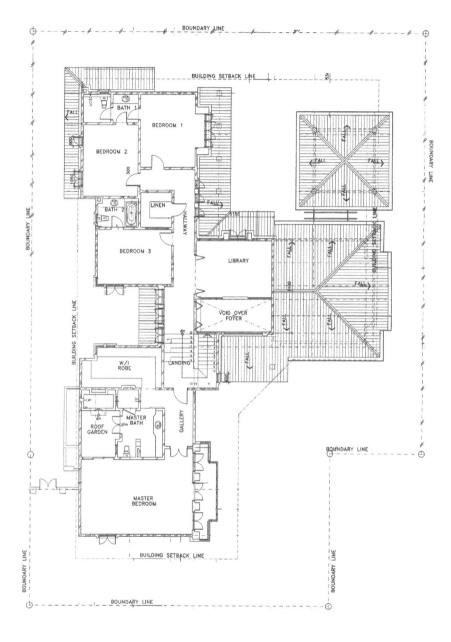

Source: Rokiah Mohd Yusof

8.23 House at Kayangan Heights: first-floor plan

surrounding hills, forest reserve, parks and tree-lined avenues provide green spaces and help to reduce air pollution in the housing enclave. The 'Balinese' tropical resort concept allows natural ventilation and daylighting to enter the house. The garden, koi pond, saltwater pool and courtyard gardens provide a visual and physical link to the interior spaces.

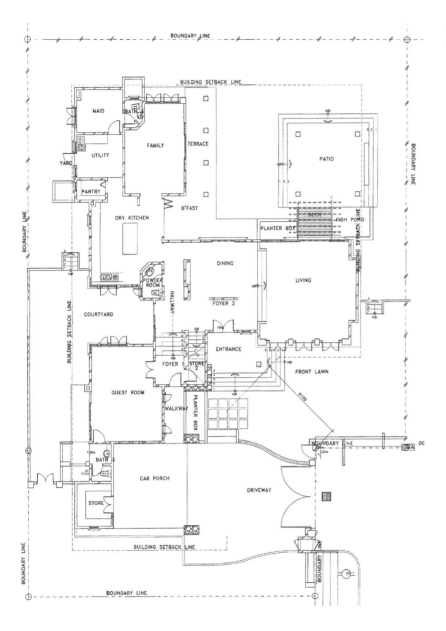

Source: Rokiah Mohd Yusof

8.24 House at Kayangan Heights: ground-floor plan

PRINCIPLES, CONCEPTS AND TERMS

The design concept is 'tropical resort' that provides a relaxing haven for the family. The F-shaped plan encourages cross-ventilation and daylighting. The lush and mature tropical garden, pool and koi pond create a cooling effect for the interior of the house.

Table 8.3 Strategies, measures and technologies: overview of the house at Kayangan Heights

Energy saving and rational use of energy		Solar and renewable energy use	
Passive mode	X	Passive solar heating	–
Active mode	–	Solar defence	–
Mixed mode	X	Solar thermal	–
Productive mode	–	Solar cooling	–
Building structure		Solar electric (water heating)	X
Sprawling design	X	**Building structure**	
Increased insulation	X	Improved daylighting	X
Reduced thermal bridges	–	Improved direct heat gain	–
Improved glazing (tinted)	X	Transparent insulation	–
Deep overhangs	X	Double height	X
Open section	X	Portal framing	
Cross- and stack ventilation	X	Recycled and plantation timber	X
Light colours to exterior	X	Sprawling building footprint	X
Selective shading	X	Low embodied-energy concrete	X
Technical service		Optimum orientation	X
Ventilation heat recovery	–	Microclimate control	X
Condensing boiler	–	**Technical service**	
Combined heat and power	–	Low PVC wiring system	–
Fuel cell	–	Solar air collectors	–
Heat pump	–	Earth-to-air heat exchanger	–
Advanced heat storage	–	Wood stoves	–
Energy efficient light sources	X	Backup fans for summer	X
Advanced controls	X	Thermal mass	X
Water storage and recycling		**Building health and well-being**	
Rainwater storage tanks	–	Low emission finishes	X
Rainwater purification	–	Increased air change rates	X
Grey water recycling for irrigation	–	Clean air source, but high on salt and humidity	–
Black water recycling	–		
		Reduced electromagnetic fields	–
		Lifestyle spaces	X

Source: Sabarinah Sh. Ahmad

Source: Sabarinah Sh. Ahmad

8.25 House at Kayangan Heights: outdoor bath and shower next to the master bedroom

Case study 8.2 Sunway Rahman Putra, Selangor, Malaysia

Sabarinah Sh. Ahmad

PROFILE

Table 8.4 Profile of Sunway Rahman Putra courtyard homes, Selangor, Malaysia

Country	Malaysia
City	Sungai Buloh, Selangor
Building type	Row house
Year of construction	2003
Project name	Superlink Courtyard Homes, Sunway Rahman Putra
Architect	Dr Ar. Tan Loke Mun
Developer	Sunway City Bhd

Source: Sabarinah Sh. Ahmad

Source: Sabarinah Sh. Ahmad

8.26 Sunway Rahman Putra courtyard homes: front elevation

PORTRAIT

The Sunway Rahman Putra is a residential enclave surrounding a 36-hole golf course. The first of its kind in Malaysia, this development encompasses 21.15 acres of land. It is easily accessible via the NKVE Highway, LDP Expressway, Middle Ring Road 2 and Sungai Buloh Highway. Nestled within this elite neighbourhood are 112 super-link courtyard homes. Each created as a tropical resort, these exclusive residences are embraced by a tropical landscaped garden designed by Made Wijaya. Safety is ensured through a 24-hour security patrol. Other standard features are concealed drainage and underground cabling.

CONTEXT

The super-link houses are built for property owners who prefer a landed property and are attracted to the facilities provided by this suburban residential enclave. The developer for the super-link houses is Sunway City Bhd, a privately owned company.

ECONOMIC CONTEXT

The super-link houses are built for the middle-class to upper middle-class set. The lot sizes are from 240 square metres (26 × 95 feet) to 265 square metres (28 × 95 feet) for intermediate lots, and up to 650 square metres for corner lots. The built-up areas are from 340 square metres to 450 square metres. The cost of the super-link house starts at 700,000 Malaysian ringgits for the intermediate unit.

SITE DESCRIPTION

The housing project is situated on a hilly site at Sungai Buluh, about 30km from the Kuala Lumpur city centre. The immediate surroundings are a 36-hole golf course, bungalow houses and double-storey terrace houses. The site was chosen for the views of the golf course and of the surrounding hills.

BUILDING STRUCTURE

Each link house is comprised of two storeys. The ground floor houses the living room, a dining area, a dry and wet kitchen, an indoor courtyard, a study/guest room, a store, a maid's room, two bathrooms and a generous porch that can accommodate four cars. The upper floor houses three bedrooms, three bathrooms, a family room and balconies. The house is cooled by ceiling fans and natural ventilation, with the option of air conditioners during the hot season.

BUILDING CONSTRUCTION

The houses are built using a post-and-beam reinforced concrete structure with plastered brick infill walls. The roof is sloped, with hardwood timber trusses, and is covered with clay roofing tiles. The ground floor has polished porcelain marble tiles that provide a cooling effect on the feet. The flooring material for the family room and bedrooms is timber parquetry. Doors are made of timber with timber frames. Windows are aluminium framed.

Table 8.5 Strategies, measures and technologies: overview for Sunway Rahman Putra

Energy saving and rational use of energy		Solar and renewable energy use	
Passive mode	X	Passive solar heating	–
Active mode	–	Solar defence	–
Mixed mode	X	Solar thermal	–
Productive mode	–	Solar cooling	
Building structure		Solar electric (water heating)	X
Compact design	X	**Building structure**	
Increased insulation	–	Improved daylighting	X
Reduced thermal bridges	–	Improved direct heat gain	–
Improved glazing and window technology	–	Transparent insulation	–
		Atrium, sunspace	X
Pavilion envelope sensitive design	–	Portal framing	
Open section	X	Recycled and plantation timber	X
Cross- and stack ventilation	X	Small building footprint	X
Light colours to exterior	X	Low embodied-energy concrete	X
Radiant barriers	–	Optimum orientation	X
Selective shading	X	Microclimate control	X
Technical service		**Technical service**	
Ventilation heat recovery	–	Low PVC wiring system	–
Condensing boiler	–	Solar air collectors	–
Combined heat and power	–	Earth-to-air heat exchanger	–
Fuel cell	–	Wood stoves	–
Heat pump	–	Backup fans for summer	X
Advanced heat storage	–	Thermal mass	X
Energy efficient light sources	X	**Building health and well-being**	
Advanced controls	–	Low emission finishes	X
Water storage and recycling	–	Increased air change rates	X
Rainwater storage tanks	–	Clean air source, but high on salt and humidity	–
Rainwater purification	–		
Grey water recycling for irrigation	–	Reduced electromagnetic fields	–
Black water recycling	–	Lifestyle spaces	X

Source: Sabarinah Sh. Ahmad

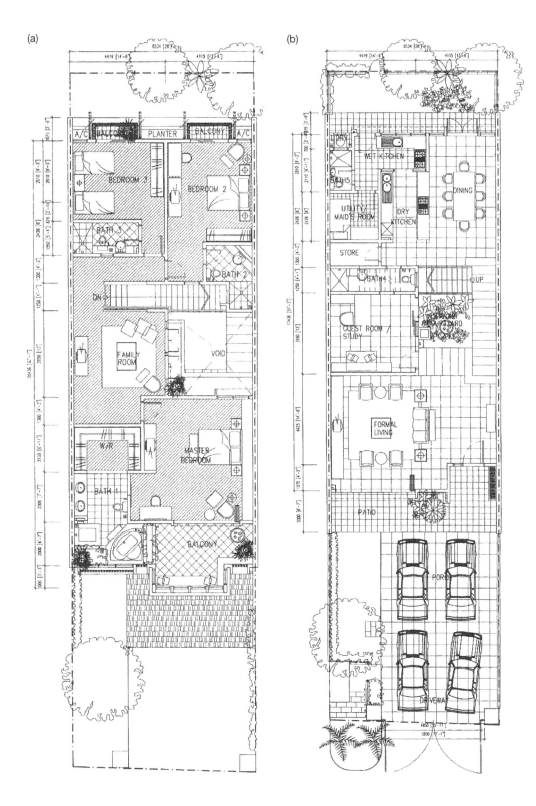

(a)

(b)

Source: Sunway City Berhad (2003)

8.27 Sunway Rahman Putra: (a) first-floor plan; (b) ground-floor plan

VENTILATION SYSTEM

The building uses cross-ventilation and stack ventilation (by the stairwell).

APPLIANCES

A solar water heater is used to reduce energy consumption. This cuts the demand for electrical energy and reduces greenhouse gas emissions.

ENERGY SUPPLY SYSTEM

Electricity is directly supplied from the national grid.

SOLAR ENERGY UTILIZATION

Solar energy is used for daylighting and for a hot water system for the bathrooms.

BUILDING HEALTH AND WELL-BEING

The housing estate is built 30 minutes away from the city centre and there is no industrial area in the immediate vicinity, so it has low air, traffic and noise pollution. The surrounding hills, parks, a golf course and tree-lined avenues provide green spaces and help to reduce air pollution in the estate. The courtyard concept allows natural ventilation and daylighting to enter the houses. The garden terraces and courtyards provide a visual and physical link to the interior spaces.

PRINCIPLES, CONCEPTS AND TERMS

Wider frontages (average row house frontage is between 6 and 7m) encourage more cross-ventilation and daylighting, and more comfortable internal spaces. The internal courtyard and stairwell benefit from stack ventilation and daylighting.

OTHER INFORMATION

Further information can be obtained from the developer's website at www.sunway.com.my/suncity. The sales and marketing agent can be contacted at sunprop@tm.net.my.

Source: Sunway City Berhad (2003)

8.28 Sunway Rahman Putra: view from the terrace towards the side garden of the corner lot

SUMMARY

Malaysia is situated in a maritime equatorial area, where the climate is generally the same throughout the year, with uniform temperatures, high humidity, light winds and heavy rainfall. Malaysia has a mean minimum temperature of around 22 to 24°C and a mean maximum temperature of 29 to 32°C, giving an annual mean of 26.75°C.

The very nature of the Malaysian climate may necessitate mechanically ventilated or air-conditioned interiors, especially in urban areas. However, poor design and indiscriminate use of air conditioning have resulted in huge increases in energy use. Passive and low energy design strategies are therefore better solutions for a sustainable future.

The comfort band for the Klang Valley area computed using Auliciems's equation for all building types is between 23.6°C and 28.6°C, with a neutrality temperature of 26.1°C (Auliciems, 1981). Since Malaysians, being acclimatized to hot and humid climates, are able to tolerate much higher temperatures, increasing the upper limit of the comfort range would result in greater energy savings.

These climatic consequences make passive design a challenging but feasible option for achieving commercial and residential building comfort standards. Thus, energy efficient design should be directed towards reducing energy demands through natural and fan-assisted ventilation, correct opening schedules, higher set-point temperatures (if air conditioning is being used), proper orientation and siting, reduction of solar gain (that is, shading devices), thermally efficient construction (that is, insulation and use of natural material with low U value), low energy equipment and plant, and improved use of daylight. Finally, trees and gardens that surround houses could also contribute to natural cooling.

REFERENCES

Auliciems, A. (1981) 'Towards a psycho-physiological model of thermal perception', *International Journal of Biometeorology*, vol 25, pp109–122

Chen, V. F. (ed) (1998) *The Encyclopedia of Malaysia: Architecture*, vol 5, Archipelago Press, Singapore

Guthrie Property Development Holding Sdn Bhd (2003) *The Lagenda Superlink Garden Homes*, Guthrie Property Development Holding Sdn Bhd, Shah Alam

Hicom Gamuda Sdn Bhd (2003) *Kota Kemuning Apartments*, Hicom Gamuda Sdn Bhd, Shah Alam

Malaysian Meteorological Service (2002) Weather website: www.kjc.gov.my

Sunway City Berhad (2003) *Sunway Rahman Putra 2003*, from www.sunway.com.my/suncity/projects/04/03rahmanputra.html

Global Media (2007) *Malaysia Country Map* from www.worldtravels.com/Travelguide/Countries/Malaysia/Map

Part III
Principles, Elements and Technologies

Redefining bioclimatic housing in Part I of this book involved a review of definitions, concepts and principles. New principles were developed specifically for housing, focusing on improving comfort and assessing the ecological footprint of both users and the building design. Part II focused on applying these principles to new housing projects in a range of locations and warm climates. From these projects, potential solution sets were identified for common housing types such as detached and row houses and apartments.

Part III derives some general recommendations from this work, providing a checklist of considerations for designing more climate-friendly housing. This chapter examines recommendations concerning the benefits of bioclimatic design. Areas that are examined involve the relationships between climate parameters, building elements and integration of passive and active systems, summarized as follows:

- *Better interpretation of location and climate parameters.* Climate interpretation involves examining the macro-conditions, particularly insolation, humidity, airflow and rainfall on the building typology. If it is to be bioclimatic, the layout of the building should be highly influenced by the macroclimatic conditions.
- *Greater synthesis between building elements and local climate conditions.* The influence of location involves interpreting meso- and microclimatic conditions. This has a profound influence on the extent to which bioclimatic strategies can be achieved.
- *Harmonizing passive and active cooling/heating strategies*. This may include sun shading, thermal insulation and cross-ventilation in combination with active systems, such as air conditioning. Harmonization through mixed-mode design allows for better use of energy. Higher thermal performance of the envelope can significantly reduce energy use, while the current practice of low

performance for the building envelope leads to high energy use and greenhouse gas emissions. The use of high efficiency equipment does not necessarily address this problem.

Chapter 10 examines a range of opportunities to improve the effectiveness of design outcomes, including:

- *Verification of predicted and operational performance.* Working from a performance model is crucial in 'tuning' the building over its life cycle. Bioclimatic buildings require verification performance through commissioning and post-occupancy study to assist with enhancing performance.
- *Integration of appropriate green technologies.* The use of a range of 'green technologies' that utilize sustainable materials and life systems creates substantial challenges of integration.
- *Engaging users in design and operation.* In order to achieve maximum performance, it is crucial that users understand the mechanism of the building, and control it according to the original design concept. Involving users in the design is central to making this concept work.

Chapter 9
Design, Elements and Strategies

Richard Hyde, Nobuyuki Sunaga, Veronica Soebarto, Marcia Agostini Ribeiro, Floriberta Binarti, Lars Junghars, Valario Calderaro, Indrika Rajapaksha and Upendra Rajapaksha

INTRODUCTION

The first part of this chapter examines an integrated design approach for housing. This can be understood as a set of principles, including bioclimatic principles. The second part discusses the building elements and design required to implement passive design principles. The third section examines passive cooling strategies, which include comfort ventilation, stack and wind-driven ventilation, night cooling through ventilation and radiation, direct and indirect evaporative cooling, and earth cooling. Related issues, such as daylighting and noise abatement, are also discussed. The final part presents a case study: Prosser House on the Gold Coast, south-east Queensland, Australia (see Case study 10.1).

INTEGRATED DESIGN PRINCIPLES

Richard Hyde and Nobuyuki Sunaga

The main principle of bioclimatic design for passive and low energy buildings is to provide a comfortable environment by virtue of the passive features of design. A second principle is to use the active systems (mechanical equipment, such as air conditioning) with the passive systems to create an integrated solution for climate control. This can produce an integrated approach to design.

This climate control concept requires a cross-disciplinary design method involving both architect and engineer to achieve an integrated solution. The spheres of influence are shown in Figure 9.1: one view of bioclimatic design is that the sphere of influence of architectural design lies in the use of passive design features, such as selecting the appropriate form and fabric of the building for the climatic conditions. The aim is to achieve an indoor condition as close as possible to the *comfort zone*. Hence, through building form and fabric, varying outdoor conditions are controlled in order to achieve comfort. There are limits to the effect of passive systems; hence, it is common to use active systems for times when comfort cannot be provided by passive systems.

Source: Richard Hyde

9.1 Elements from existing solutions can generate new strategies and also provide simple low technology solutions. Central atrium in Raffles Hotel, Singapore, and external courtyards provide elements that can be adapted to modern housing conditions

(a)

Source: The Ecovillage

9.2 Ecovillage, Currumbin, Gold Coast, Australia, aims to provide an integrated master plan, which is based, in part, on bioclimatic principles to achieve sustainable outcomes: (a) development codes for the site create a framework for using appropriate green technology, with verification and prediction through a monitoring system; (b) master planning of housing clusters; (c,d) bush house design; (e) EcoVision monitoring system to update users of resource usage and building management

(b)

(c)

(d)

(e)

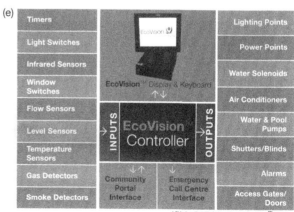

Source: Nobuyuki Sunaga

9.3   Jimmy Lim's house, Malaysia: (a) adaptation of the atrium for housing for cooling: the extensive use of plants is beneficial for improving air quality; (b) façade detail to facilitate safety and ventilation on the second floor; (c) roof ventilators which can be adjusted to control the ventilation to the atrium; (d) exterior view showing the atrium addition to the building

A more integrated view of both the passive and active systems is recommended for architectural and mechanical systems selection and design. Numerous opportunities exist for design improvements and innovative solutions:

- *Interaction of location and climate parameters.* Climate interpretation involves examining the macro-conditions, particularly insolation, humidity, airflow and rainfall on the building typology. If the building is to be bioclimatic, the layout should be highly influenced by the *macroclimatic* conditions. The use of psychometric information can assist in identifying the main bioclimatic strategies for building design.
- *Synthesis of location, climate and building elements.* The influence of location involves interpreting *meso- and microclimatic conditions.* One aspect of design is concerned with selecting elements that respond effectively to local conditions. New combinations of building elements help to address issues of thermal, acoustic and visual comfort. For example, microclimatic factors can be enhanced through courtyards to channel ventilation for improved thermal comfort, reduce glare and provide barriers to environmental noise.
- *Synergies between passive and active cooling/heating strategies.* The opportunities for linking passive and active systems to achieve greater synergies have significant benefits for energy reduction. Lack of synergy occurs when air conditioning is used as a Band-Aid to poor passive design. Opportunities for synergy include use of shading solutions, which reject heat but allow diffused light to reduce electric lighting.

CLIMATE AND LOCATION

Veronica Soebarto and Márcia Agostini Ribeiro

Macroclimate conditions

An important stage in design is gaining an understanding of the influence of local conditions on the climate. Szokolay (1991) defines three types of macroclimates for buildings:

1 hot temperatures that are mostly above the comfort zone;
2 moderate temperatures that are below and above the comfort zone;
3 cold temperatures that are mostly below the comfort zone.

Meso- and microclimate conditions

Meso-climate conditions will enhance or reduce macroclimate conditions. The effects of topography, vegetation and other physical characteristics create the meso-climate effect. Six main types of meso-climate can be found according to Goulding et al (1992):

1   coastal: influence of local sea/land breezes, higher winter temperatures than inland and lower summer temperatures;
2   flat open country: deserts and plains, with elevated wind speeds since there are usually few obstructions;
3   woodland and forests: more stable temperatures, higher humidity than open country and higher shade;
4   valleys: influenced by topographical shading, slope orientation and inclination (affected by solar access and local wind effects, such as funnelling);
5   mountains: reduced temperatures of 0.7°C for each 100m rise in altitude;
6   cities: 'heat island effect' and reduced airflow at ground level.

Quantification of meso-climate conditions is usually comparative and involves examining the relative difference between two areas. In examining the effects of meso-climate conditions, it is useful to develop zoning diagrams showing differences from macroclimate conditions, such as temperature variations, the effects of topography on wind flow and variations in solar radiation (see figure 9.4).

Understanding the site's microclimatic conditions is also a major priority. First, the amount and the direction of solar radiation are important factors to be considered in shading studies and when assessing passive solar heating. Maximizing shading of the site is also necessary. This is needed in particular to assist the cooling effects of cross-ventilation. This cooling effect can be negated if the air temperature rises as it moves across the site due to high incidences of solar radiation heating the surfaces which in turn heat the airflow. Plants frequently provide good shade onsite and in outdoor rooms. In hot and low latitude regions, evergreen trees are suitable for shading purposes.

Source: Richard Hyde

9.4   Sketch section through the hinterland of the Sunshine Coast, Australia, to show the changes to meso- and microclimate regarding the location of the Montville House

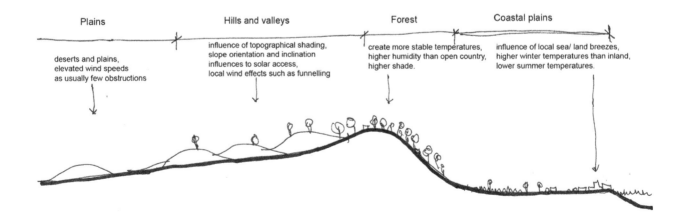

Plains

Hills and valleys

Forest

Coastal plains

deserts and plains, elevated wind speeds as usually few obstructions

influence of topographical shading, slope orientation and inclination influences to solar access, local wind effects such as funnelling

create more stable temperatures, higher humidity than open country, higher shade.

influence of local sea/ land breezes, higher winter temperatures than inland, lower summer temperatures.

(a)

Source: Richard Hyde

9.5 Use of meso- and microclimate enhancement of the Mapleton House site to provide passive heating and cooling. Mapleton House in Brisbane is located in a rainforest meso-climate. The microclimate enhancement involves placing the building on the southerly edge of the clearing in the rainforest. This allows summer shading from the forest in summer and winter heating from solar access

(b)

north (sunny) elevation

a kitchen
b dining room
c study
d bedroom
e terrace
f entrance hall
g workshop
h bathroom
i *tai chi*/bedroom

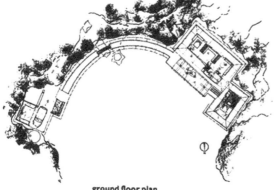

ground floor plan

In temperate and middle latitudes, deciduous trees are suitable in southern regions (northern regions in the southern hemisphere) for summer shading, while allowing in winter sunshine. Evergreen trees are used in the east and west for summer sun shading and, when located in the proper direction, for winter windbreaks. Furthermore, in urban areas, city planning that promotes heat-island phenomena should be avoided.

Second, mapping the primary wind direction in each season is a significant factor for site planning, building orientation and passive and low energy methods. The microclimate effects of wind exposure must then also be assessed for ventilation. Ventilation for passive cooling is important in summer, but can be a limitation in winter. For example, in a moderate climate where heating is needed in winter, if the site has high exposure to wind, a windbreak will be needed.

Climate interpretation

Comfort metrics and bioclimatic strategies

The potential influence on the comfort zone of strategies such as air movement, thermal mass, evaporative cooling, passive solar heating are plotted in Figure 9.6. The designer can assess the best possible building strategies by comparing the climate lines, which show the range of temperatures for each month with the extended comfort zone for the varying strategies. For example the hot humid climate (Figure 9.6a): the climate lines show high temperatures and humidity, so two options are to use thermal mass and air movement through ventilation to extend the comfort zone. This kind of approach should be carried out with site-specific data to support the strategies selected.

Hot humid climates demonstrate the problems associated with high humidity and temperatures above the comfort zone with little diurnal range. It is possible to extend the comfort zone by using bioclimatic strategies (see the section on 'Strategies and techniques'). Comfort ventilation can be achieved by utilizing airflow as seen in Figure 9.15. Strategies such as thermal mass are plotted, but do not overlap with the climate conditions; as a result, they may be less effective.

For hot arid climates, there are two main strategies for cooling: the use of evaporative systems and the use of thermal mass. Night ventilation is an important option to purge the thermal mass of heat gain during the day.

Moderate climates have periods of the year when both heating and cooling are required, and when the use of comfort ventilation and thermal mass are options. Extremes of temperatures in these climates must be considered.

Microclimate and ecological enhancement of the site layout

Conventionally, the microclimate of buildings has been largely ignored as an opportunity for advanced passive and active strategies. Many cities are dominated by hard and reflective surfaces, which create heat island effects, elevating local temperatures and creating a greater heat load for the air conditioning systems of buildings (Givoni, 1998).

(a)

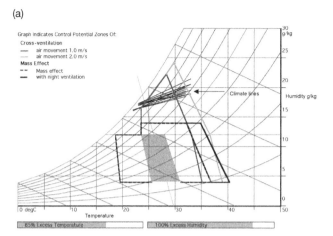

Source: Richard Hyde

9.6 A psychometric chart is a tool for checking design strategies. The comfort zone is shown as a shaded area in the centre with climate lines representing minimum and maximum temperatures for the year. In many cases the temperature lines are outside the comfort zone. The effect of building strategies is shown, which can extend the comfort zone so that all climate lines are included. Examples shown are: (a) Hot humid climate of Kuala Lumpur, Malaysia; (b) hot dry climate of Longreach, Australia, showing the potential effects of passive solar and mass to extend the comfort zone; (c) hot dry climate of Longreach, Australia, showing the potential use of evaporative cooling and passive solar to extend the comfort zone

(b)

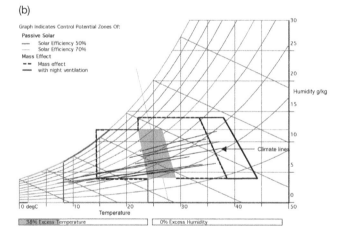

(c)

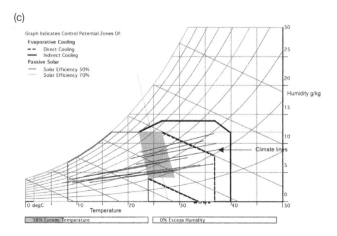

(a)

(b)

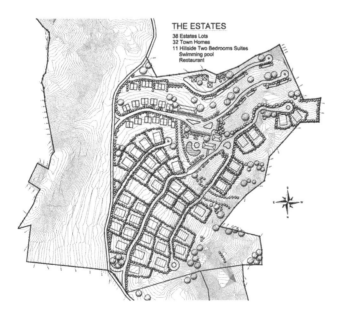

THE ESTATES

38 Estates Lots
32 Town Homes
11 Hillside Two Bedrooms Suites
Swimming pool
Restaurant

Source: Pt Wisata Bahagia Indonesia

9.7   Sundancer Spa and Beach Resort Stage 2, Lombok, Indonesia: (a) master planning to include the provision of bioclimatic principles, such as orientation of the lots for solar control and wind access in this hot dry climate; (b) zones for habitat conservation also facilitate improvement in ecological value

Yet, a key aspect of the designer's knowledge base is to know the site's biophysical preconditions prior to development. This includes climatic conditions and other influences such as topography, flora and fauna. *Site analysis* is carried out to determine the biophysical preconditions. At the briefing and feasibility stage of the design, the microclimate's potential should be taken into account. At the sketch design stage, it is possible to plan the layout of the building and development to achieve *microclimate enhancement*. For example, if the summer sun is problematic from the west, landscaping can be used to protect the building.

Sun shading of the site is necessary to enhance the effects of cross-ventilation. Lack of shading has been found to create a localized heat island on the site, increasing the temperatures of the air used to cool the buildings. Selection of appropriate plant species is a major element of providing shade onsite. Some proven examples have been found. In hot climates, evergreen trees provide permanent shading. In temperate and middle latitude climates, deciduous plants are suitable for the equator-facing façade. When heating is needed, plants loose their leaves, and when shading is necessary, they grow their leaves. Questions have arisen whether enhanced microclimates create indirect benefits for the health and well-being of a building's users and occupants.

ELEMENTS

Veronica Soebarto, Richard Hyde, Márcia Agostini Ribeiro and Floriberta Binarti

The main elements of a building comprise the physical aspects of the building, the building envelope and external environment, and the external rooms and spaces. The principles of climate responsive design create opportunities for better comfort and energy consumption through the use of microclimate control, the selection of appropriate building forms and fabric, and minimizing mechanical plant and equipment (which will be covered in Chapter 10).

In order to minimize cooling needs in a building located in warm and hot regions, the main aim is to reduce solar heat gain through the microclimate surrounding the building and through the building envelope. Bioclimatic design strategies to minimize cooling should include climate responsive layout of the area and orientation of the building façade's windows; adequate size and details of window openings; the use of shading devices for fenestration; the use of insulation; the inclusion of vegetation in the building's surroundings; and the use of courtyards. While the optimum bioclimatic design of elements is strongly dependent upon climate, additional factors often drive the design to the least climate-optimum conditions.

The challenge is to use appropriate elements that mitigate negative effects in order to create positive outcome. In view of the new principles of bioclimatic design, additional features, such as the need to preserve ecological value, the use of public open spaces and low impact mobility, should be incorporated as elements of housing design.

Building layout and site planning

The master planning of the site should result in a building layout that responds to bioclimatic principles. This approach is theoretically possible, but rarely happens – that is, while there can be a number of principles for the layout and subdivision of land and the creation of road systems that reflect climate, these factors are rarely given a priority. Hence, there is a wasted opportunity for establishing energy efficient measures at a macro-scale (Hanisch, 1998).

Box 9.1 Measures for location and site planning

Indicator measures for establishing building layout and site planning include the following:

- precinct location ratio: transport mobility impact reduction;
- habitat conservation ratio: habitat conservation measures;
- public open space ratio: public space and amenities;
- building footprint ratio: footprint minimization;
- environmental landscaping rating: landscaping for resource conservation;
- site disturbance ratio: minimizing the effects of building onsite.

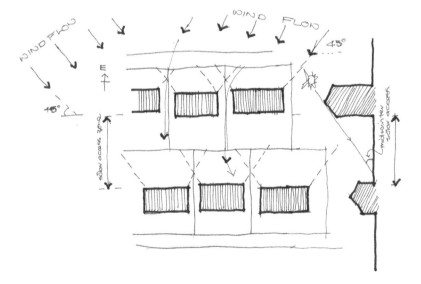

Source: Richard Hyde

9.8 Typical layouts can be developed to optimize orientation for solar and favourable wind vectors. This layout works well for north-east and south-east cooling breezes. The building's solar footprint and wind flow can be mapped. For moderate winter climates, the sun angle midwinter can be used to size the solar access zone. Permeability between buildings is necessary for wind. Good ventilation can be achieved with wind flow from 90° to 45° perpendicular to the building façade

Lot size, density and layout for energy efficiency

There is concern about the impacts of land-use patterns on energy use and comfort. One of the principles of building design is that energy use for transport increases with physical separation. Yet, the argument for relating energy use to site layout is not as straightforward due to the lifestyle of the occupants. Nevertheless, from the viewpoint of transport energy, higher density solutions are more efficient.

Another important principle relates to the orientation of streets and allotments since this can have a significant effect on the potential for energy efficient housing. Codes are needed to help with ensuring that this principle is addressed because the market is not sensitive to this issue. Developers do not see this as a marketing issue for their projects.

For energy efficiency in a warm climate, solar access and access to wind flow is strongly correlated with a dwelling's orientation and the location of living rooms, which are in turn influenced by lot orientation and road layout.

Therefore, as density increases to achieve better affordability and create more efficient transport, lot sizes become smaller. These consequences mean an integrated solution for house orientation, room layout and lot geometry. Road layout is critical if a bioclimatic response to building is to prevail. Furthermore, densities above 40 houses per hectare create additional concerns, such as protection of neighbours' interests and privacy issues, leading to a greater need for an integrated bioclimatic solution.

Building suggestions that have arisen from this scenario have included an east–west street pattern, skewed to incorporate connections. Planners have recommended that this pattern could be integrated with other elements, such as topography.

Zero lot line development (the wall of the house is built on the lot boundary) can help with improving energy efficiency by allowing more privacy and better solar access. For tropical climates, a westerly zero lot line using a buffer wall is deemed effective at cutting low-angle solar access and at providing privacy.

Plan form and orientation

Thermal zoning functional needs and energy efficiency

If the functional plan of a building is related to the site, then one primary consideration is the thermal zones created. Rooms facing the polar area will receive less heat than those facing the equator. A general recommendation would therefore be to place air-conditioned rooms on the southern edge to avoid environmental heat loads.

Rooms will also receive differing solar radiation at different times of the day. Bedrooms are usually placed on the east side to avoid the evening heat gain. Examining thermal zoning options can thus lead to energy efficiency through appropriate functional planning. For a compact highly cellular building this approach can be effective.

Wind vectors, breezeways and personal comfort

If the building is to profit from comfort ventilation, the layout will vary depending upon the climate, especially if it is dry or humid. For warm and dry

(a)

(b)

Equator

(c)

Equator

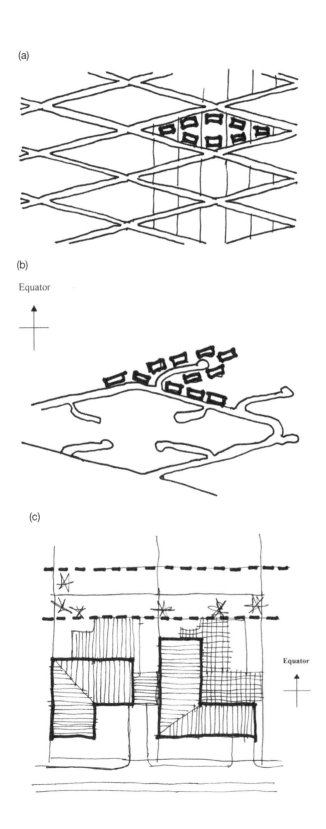

Source: Richard Hyde

9.9 Subdivision planning considers road pattern and hierarchy, topography, views, gradients, drainage, physical obstructions, density, block size and geometry, but has little regard for the orientation of solar access or wind flow: (a) the skewed grid may assist with addressing this principle; (b) with densities above 40 houses per hectare, an integrated solution is needed so that all factors are considered; (c) adaptation house forms, such as zero lot line development, assist with this principle

climates, it is necessary to keep the inner temperature below outer temperature levels. Therefore, it is important to reduce outer heat gains during the day: the volume must be compact (reduce envelope area) and the ventilation rate must be kept at a minimum (air changes per hour (ACH) = 0.5). At night, since the outer temperature usually drops drastically, it is important to increase ventilation in order to cool any internal thermal mass of the building.

In warm and humid climates, the house can be viewed as a spatial volume rather than a set of rooms. Utilizing this volume and keeping it open encourages ventilation during the day and night. The goal here is to reduce humidity and therefore enhance the psychological effect of cooling through comfort ventilation.

Orientation is important: the main rooms and windows should be located according to the sun's position and the direction of prevailing winds. In a hot dry climate, whenever possible the longer side of the building should be oriented towards the sun (in the southern hemisphere, this means orienting the building to the north). In this way, shading of the northern elevations can be done with relative ease (as the sun angle in summer is high); this will also allow as much solar penetration as possible in winter (required for passive heating). West- and east-facing elevations should be kept to a minimum since the summer morning and afternoon sun can be quite intense while the sun angles are quite low, making it more difficult to shade walls or windows. Rooms such as the laundry and bathroom can be located on the east or west side of the house to protect the living areas from summer sun.

In a hot humid climate, ventilation is the most important aspect to be taken into account. Therefore, wind direction will determine the orientation of the building and windows. Of course, solar heat gain should also be avoided or reduced; but this can be achieved through the correct use of sun shading devices, vegetation, etc. It is important that a building has its longest façade oriented between 30° and 120° to the prevailing wind direction to be effective for comfort ventilation (the optimum angle is 90°).

Influences of air conditioning on plan form

In warm climates, the layout of a building, its form and the composition of its rooms can depend upon climate control assumptions. Is the building to be *free running* – that is, will it use comfort ventilation – or will it will use mechanical air-conditioning devices that provide *mechanical cooling*?

In the second case, the building should be planned as a compact volume in both dry and hot humid climates. The area-to-volume ratio should be as small as possible so that the heat gain can be improved in winter and reduced in summer. With free-running buildings, the opposite layout requirements are needed: a more linear plan, aligned perpendicularly to the prevailing cooling breeze, is favoured.

This advocacy belies a more complex planning process, which focuses on issues of how to make a compact plan more climate interactive so that if the mechanical system fails there are additional climate control systems in place. Similarly, with the linear plan, when free-running conditions are extreme, providing comfort is frequently challenging. The planning process for development, therefore, hinges on reconciling some of these basic relationships.

Building envelope

With regard to reducing environmental impacts, there is an increasing call to improve standards for the building envelope. A first cost view of the building envelope has resulted in the adoption of minimum standards for performance.

An important principle for bioclimatic design is to increase envelope performance through four measures:

1   integrated climate responsive design;
2   setting targets for performance improvements – for example, a 30 per cent improvement in the performance of the building envelope is recommended by the Building Research Establishment (BRE) Eco Homes rating tool;
3   improved thermal performance of individual elements of buildings, such as roofs, windows, walls and floors; and
4   integration of resource-producing technologies (see Chapter 10).

The building envelope comprises elements of the building from the building's line of enclosure, marking the inside from the outside. The building envelope can be thought of as primarily comprising opaque and semi-opaque elements, such as walls, floors, roofs and shading devices, as well as transparent elements, such as windows and skylights. The primary function of the building envelope is to filter the external environment in such a way as to facilitate passive and, possibly, active systems of cooling (and, in some cases, also heating). The design challenge is complex, involving a balance of often contradicting requirements. Invariably, however, a poor-quality building envelope results in less comfort for the occupants. The lack of comfort means a reliance on active systems such air conditioning, which uses more energy, as well as creating the need for larger plant and equipment.

The building envelope in a solar sustainable house should be of the highest quality possible. This does not mean using exotic materials or highly expensive materials; rather, it involves simple and effective design that responds to the local environment, including the thermal, lighting and acoustic environment and specific conditions in which the building is located. These specific conditions include environmental hazards, such as rain, storms and fire.

(a)

(b)

Source: Richard Hyde

9.10 Functions of the building envelope: (a) a filter for rain screening, solar shading, air infiltration, moderation of light, moisture, heat and noise; (b) the design consequences are often ignored, compromising the building appearance

(a)

(b)

Source: Richard Hyde

9.11 (a) Queenslander: quick response buildings;
     (b) Mediterranean house: slow response

First, the opaque and transparent elements are examined with respect to thermal environmental conditions. In moderate and hot climates, these elements are highly defensive, rejecting heat flux and creating shelter from wind and rain. Second, fenestration is examined, which is a complex element since the design and position of window controls provide a large range of functions, controlling the flux of heat, light and sound, as well as providing ventilation, viewpoints and a sense of the external environment.

Opaque elements

Conventionally, building construction and materials have been classified into two types: heavyweight systems that are primarily made from masonry, such as brick, concrete or earth, and lightweight systems, such as timber, steel and fibre cement sheets. There will be significant differences in the performance of these systems in warm climate regions, especially with regard to the kind of building construction and materials used. The performance of these types of building construction systems is generally examined in terms of their influence on internal temperatures and the time delay in responding to changes in external temperatures. The thermal inertia of materials is an important aspect in the response of opaque systems.

Lightweight approach – quick-response buildings

Lightweight buildings (without insulation) are called quick-response buildings since they will generally allow quick response to changes between external and external conditions. This type of building is suited to climate conditions that are moderate, where microclimate control can be used to assist in moderating temperatures and where adaptive measures can be used by the occupants in extreme conditions.

The Queenslander is an example of a lightweight building. The construction comprises suspended timber floors; single and/or double-skin external walls, single-skin internal partitions and a metal roof on timber purlins. The form and fabric have been adapted to minimize the limitations of this building system.

Improvements to lightweight systems can be achieved through a number of measures (see Case study 9.1):

- increased insulation in the roof, floors and unshaded walls, particularly reflective foil insulation to reduce reflective heat gain;
- generous roof overhangs to shade walls and windows to exclude sun;
- use of microclimate enhancement for shade; and
- solar-access direct gain heating.

Generally, the peak temperatures inside these buildings will occur two hours after the external peak temperatures. This means that the buildings will cool down quickly at night.

Heavyweight construction – slow-response buildings

Heavyweight construction normally comprises thick masonry external walls, internal masonry walls and concrete floors and roofs. The terms are often confusing as designers commonly use roofs and internal walls of lightweight materials (such as timber) for cost and practical structural reasons and still call the construction heavyweight as the predominant material for thermal purposes is heavyweight. Heavyweight construction creates a slow response to external temperature variations due to the high thermal capacity of the materials. To achieve good performance, sufficient material is needed to prevent heat flux from reaching the interior over the daily cycle. The consequence will be small changes to internal temperatures, with a peak temperature approximately seven hours after the external peak temperature. Traditional construction in many warm climate areas where there is abundant stone, features walls up to 1m thick in order to achieve sufficient thermal inertia.

There are general improvements to the use of thermal mass that can be achieved, as well as several climate-specific measures. Improvements include:

- use of light colours and increased insulation on the outside of the thermal mass to reduce heat flux;
- improved ventilation during the night to coincide with peak temperatures inside the building; and
- closing up the building during the day to provide maximum cooling for occupants.

Climate-specific measures comprise the following:

- *Climates with good diurnal temperature variations*: in hot dry regions, there is usually a substantial difference between day and night temperatures, as well as variations between winter and summer temperatures. Although temperatures during the day are high, at night they drop drastically. Comfort ventilation must be well planned so that hot and mainly dusty air does not affect the building. The use of heavyweight construction (for example, double brick, aerated concrete) in the walls and ceilings can help to reduce external heat gain during the day. At night, due to the delay of heat transmission caused by the heavy mass of the building, the internal space of the building tends to be warmer than outside. Night ventilation is recommended for periods when night temperatures are high.
- *Climates with high humidity and little diurnal range*: in hot humid climates with little diurnal range, the performance of the thermal mass can be improved with the use of good cross-ventilation at night (Givoni, 1994b).

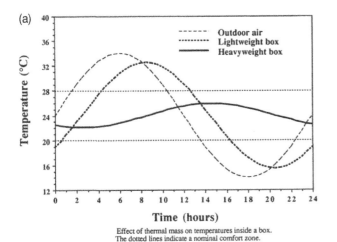

(a)

Effect of thermal mass on temperatures inside a box.
The dotted lines indicate a nominal comfort zone.

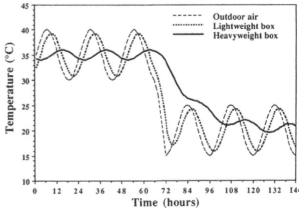

Effect of sudden temperature drop on temperatures
inside a lightweight or heavyweight box.

(b)

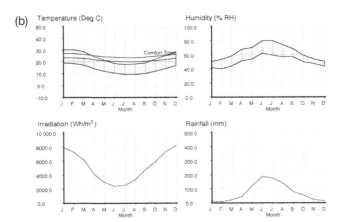

Source: Delsante (1998)

9.12 (a) Perth House demonstrates the use of thermal inertia as found in heavyweight construction; (b) the Perth climate has hot dry summers with a diurnal range of 10°C; (c) the building envelope has 400mm thick rammed limestone walls, concrete slab and concrete floors; (d) the building has external walls of heavyweight materials to utilize their thermal mass but internal walls are lightweight for practical reasons

(c)

(d)

- *Climates with variations in temperature/humidity in their daily and annual profiles*: in moderate climates, the summer is hot and humid, while the winter presents low temperatures with dry air. There is no great daily or annual temperature span. Use of heavyweight construction in walls and ceilings is recommended, as well as day and night ventilation. Buildings should also have adequate amounts of internal thermal mass. Uncovered concrete floors, reverse brick veneer and internal brick walls are some examples of thermal mass. This mass will help to moderate the internal temperatures by averaging the diurnal extremes. During summer days, the mass absorbs the heat, keeping the internal space cooler than the temperature outside. To avoid overheating, however, the mass must be shaded from direct solar radiation. In winter, the mass stores the heat from either solar radiation or heaters and releases the heat at night to help warm the internal space. Notice, however, that in winter all openings should be closed and the external walls must be insulated to minimize heat loss.

Solar integrated cool roof

In the tropics, the roof is a part of the building that receives the highest amount of solar radiation. High solar radiation can be a cooling problem if it accumulates and is then transferred to the indoor space. The amount of solar heat gain depends upon the roof form, the surface material and the cavity. Although the role of the roof's form in a cooling strategy is of less importance than other factors, especially in warm humid regions, roof form should be designed to minimize solar heat gain, promote ventilation to the indoor space and protect the indoor space from high precipitation and wind forces.

Two main types of roof form exist:

1   attic types with a large cavity between the cladding and ceiling; and
2   slab types with a small or no cavity between the cladding and ceiling.

The design of roof types has reconciled a number of parameters, such as the colour of the material, the location and type of insulation and the ventilation of unwanted heat from the roof. Additional parameters include the use of the roof for storage and the use of the roof for collecting water and solar radiation (see Chapter 10 for further details).

Improved performance can be achieved in the following climates by adhering to the following measures:

- moderate climates: cool and warm temperate:
    - a light colour to reflect solar radiation;
    - reflective foil and bulk insulation reduce downward heat flow;

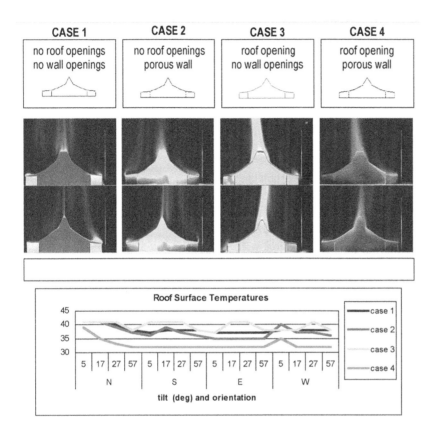

Source: Floriberta Binarti

9.13 Computational fluid dynamics simulation results – roof surface temperatures and temperature fields of Joglo type, with variations in roof and wall opening with solar altitude 47° and the azimuth 315° or north

- bulk insulation to the ceiling to prevent heat loss in winter;
- roof ventilation to allow temperature equalization during the summer;

- hot arid and hot humid:
  - a light colour to reflect solar radiation;
  - reflective foil insulation to reduce downward heat flow;
  - fly or parasol roof to allow temperature equalization.

In cases where these conditions vary (for example, where dark-coloured roofs are used for aesthetic and glare reduction), compensation occurs through the other measures. For instance, the use of dark-coloured roofs requires additional insulation and ventilation to mitigate the effects of higher surface temperatures caused by the lack of heat reflection. In moderate climates, the lack of roof ventilation in attic spaces can mean more ceiling insulation and reflective foil barriers to encourage heat to flow downwards.

(a)

(b)

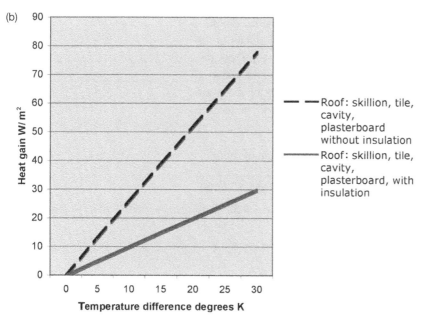

Temperature difference degrees K

— — Roof: skillion, tile, cavity, plasterboard without insulation

—— Roof: skillion, tile, cavity, plasterboard, with insulation

Source: Richard Hyde

9.14 (a) Construction of this house with a black roof is likely to mean increased insulation and ventilation to reduce heat gain; (b) black colours absorb heat and increase back radiation into the artic space and into the house; surface temperatures of the roof can be as high as 50–60°C; while the neighbours receive less glare, passive measures are needed to ensure that the house is comfortable. The graph shows the effects of heat gain during an increase in surface temperatures; roofs without insulation are likely to be less comfortable than those with insulation

Insulation

In hot climates, the use of insulation in the building envelope is primarily to reduce heat in order to lower solar heat gains. Three types of insulation are common:

1   reflective insulation: use of foils that reflect radiant heat;
2   resistive insulation: bulk insulation uses air gaps in the fabric material to resist conduction and convection through the material; examples are cellulose, wool and fibreglass;
3   capacitive insulation: use of thermal mass to increase the time lag of heat flow through walls acts as a form of insulation provided that walls and roofs have sufficient mass; a time lag of eight hours or more is needed.

Insulation is rated according to its R value or U value. A high R value is good for controlling heat flow; the converse is true if a U value is used.

As a result, with roof design the inherent problem is that the roof cladding materials (tiles, metal sheets, fibre cement and concrete) are poor insulators with high conductance. It is therefore essential to place insulation under the roof with a minimum 25mm air gap beneath the cladding, as seen in Table 9.1. Increasing standards of insulation is an important principle for bioclimatic design and is a central part of new standards for the thermal performance of a building (see Chapter 1 for a discussion of the BRE EcoHomes rating tool).

An integrated design approach to using insulation is needed through careful consideration of the environmental conditions surrounding the building. For example, is the roof or are walls shaded or unshaded; is the colour of the surfaces dark or light; do the surroundings reflect heat from other buildings?

This kind of analysis will determine the heat gains to the surfaces of the envelope and, hence, the level of insulation needed (see Table 9.2). From this analysis, it is seen that the difference between internal and external temperatures drives the need for insulation. So, whether roofs or walls are shaded or unshaded will determine the temperature difference and, hence, the

Table 9.1 The effects of insulation on the thermal performance of roofs

Materials	U value (W/m²K)	R value (W/mK)
Metal roof	2.69	0.37
Metal roof with 50mm bulk insulation	0.56	1.79
Metal roof with 100mm bulk insulation/and or reflective foil	0.31	3.23

Source: Zold and Szokolay (1997)

Table 9.2 Environmental loads and heat gains through elements

	Heat gain (W/m²)	Admittance (W/m²/K)	Temperature difference (K)
Environmental loads			
Solar radiation	1000		
Building envelope			
Roof (skillion): tile, cavity, plasterboard, without insulation	65	2.6	25
Roof (skillion): tile, cavity, plasterboard, with insulation	25	1	25
External walls: unshaded, solid brick	105	4.2	25
External walls: unshaded, brick veneer: single brick,100mm, cavity 50mm, stud frame and plasterboard.	55	2.2	25
Windows: unshaded, single glazed, clear glass, aluminium frames	150	6	25
Windows: unshaded, single glazed, heat reflecting glass, aluminium frames	130	5.2	25
Roof lights: unshaded clear glass	142.5	5.7	25
Roof lights: plastic dome, double skin	80	3.2	25

Source: Richard Hyde

heat gains through the element. Unshaded windows, even with solar glass, still provide significant heat gain.

The effects of environmental conditions on external surface temperatures can be significant, such as using dark-coloured materials; this can raise surface temperatures by heat absorption. Additional insulation and ventilation are needed to reduce the heat gain.

Fenestration

Fenestration is concerned with the design and placement of transparent elements in the building envelope.

Size and type of window glazing

Windows have a number of functions in any climate: providing maximum view to the outside, increasing solar heat gain in winter and ensuring comfort ventilation.

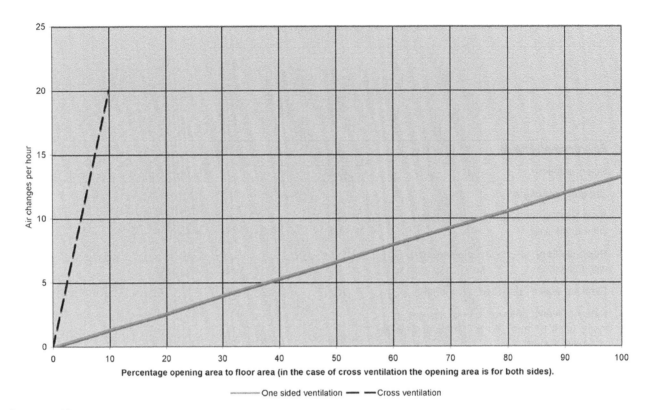

Source: Lim (1998)

9.15 The effectiveness of cross-ventilation compared to single-sided ventilation

Having large windows, therefore, is often desirable; but without good design, large windows can be detrimental both in summer and winter. In summer, if not properly shaded, large windows become the major source of solar heat gain. On the other hand, during winter nights, a substantial amount of internal heat will escape to the outside through large windows if the glass is not treated properly.

Again, there will be differences between a hot dry and a hot humid climate. For a hot dry climate, windows must be very well insulated and shaded. Outside temperatures during the day must be excluded, along with wind, which carries a lot of dust. A good solution is to plan a compact building with an atrium where the windows can be larger and protected from the sun and wind. Openings to the outside should be small and always shaded.

In a hot humid climate, it is important to ensure comfort ventilation throughout the day and occasionally during the night. Windows should be placed according to the wind direction and be very well protected from the sun. This climate mainly indicates protection against rainfall, especially in summer. In some regions, in order to protect from heat losses in winter, the use of double glazing is recommended. This will minimize heat losses in winter and reduce solar heat gain during summer.

Based on monitoring studies of the performance of several houses in the Adelaide region in South Australia, a window area of approximately 30 per cent of the floor area will minimize summer heat gains, while still maintaining access to views and providing contact with winter sun.

(a)

(b)

Source: Szokolay (1991)

9.16 Design methodology for shading depends upon the climate; for moderate climates where heating is needed in winter, selective shading is required – that is, equator-facing sun shading devices should admit solar radiation for the underheated part of the year and exclude sun for the overheated part of the year: (a) horizontal sun shading at Couran Cove, Gold Coast, Queensland, Australia, provides selective shading to the north façade; (b) vertical shades provide shading to the east and west

Shading

In warm climates, the control and reduction of heat gains through windows is extremely important. However, this should not disturb the view or the use of daylight in the building. It is also important to ensure enough heat gain in winter.

There are a number of shading strategies that can be fixed or adjustable, placed outside or inside the building envelope. When placed outside the window (for example, roof eaves and awnings, pergolas with or without deciduous plants), these devices block sunlight before it reaches the glass and therefore prevent heat gains inside the building. If they are placed on the inside of the building (for example, curtains and louvres), part the heat will be trapped between the shading device and the window, and part of it will be conducted to the room.

The art and form of shading devices depend strongly upon the location of the windows, as in the north, south, east and west façades. The east and west

(a)

(b)

(c)

Source: Richard Hyde

9.17 Fundamental shading geometry can be used in a creative way to reduce solar gain, but also to create a variety of spaces to engage occupants with the view: (a) equator-facing shading stems; (b) easterly vertical shades; (c) integrated bay window and shading system façades will be the most difficult to protect, due to the low sun angles in the early mornings and late afternoons. For the north and south façades, generally, fixed or adjustable shading devices placed horizontally above the north-/south-facing windows is suggested. For the east and west façades, adjustable vertical shading is advised (such as external blinds and timber louvres) or fixed horizontal and vertical shading devices. The overhang of these devices will depend upon the latitude of the building's location. Preliminary studies can be easily conducted with simulation tools, such as Sun Tool for Ecotect (see Chapter 2; demo software is available at www.squ1.com).

In many cases, adjustable devices are effective in all positions as long as the occupants are willing to operate them to control the levels of heat and daylight. A good example is shown in Figure 9.17, where sliding timber screens, which are applied outside the windows, primarily shade the equator-facing windows. During summer days, these screens are closed to reduce solar heat gains while still allowing enough daylight penetration into the building. Late in the afternoon, the screens are opened (as well as the windows) to allow the warm air inside to escape outdoors. During winter days, these screens are usually opened to admit solar heat, but are closed at night to reduce heat loss.

Solar chimney

Lars Junghans

Concept and benefits

A solar chimney is a channel that is glazed on the sunny side and is backed with a collector wall. It is situated at the top of the room, which should be ventilated. The added height and temperature difference combine to significantly increase airflow, if designed correctly. If there is an opening high in the building and another low in the building, a natural flow will be induced; this is called the stack effect. The sun-warmed air in the duct increases the natural stack effect. Therefore, the solar collector is useful in extracting warm room air, which ideally is replaced by cooler air from the shady side of the room. During a sunny day, a solar chimney is efficient; in overcast weather, it is less effective but still helps to discharge room air due to the height difference and stack effect. The influence of the chimney's height and the effect of the vertical extent of the chimney above the roof is shown in Figure 9.18. It should be noted that during the design of these elements, a common concern is that the chimney exhaust is too close to the roof, causing problems with the system's effectiveness.

The benefits of solar chimneys are as follows:

- Electricity consumption, otherwise consumed by a fan mounted in the roof, can be saved.
- Airflow rates of between 0.25m per second and 0.4m per second can be achieved.

- The solar collector induces higher flow rates than a conventional chimney.
- Night ventilation is also possible.

Limitations

The limitations of solar chimneys are as follows:

- Occupants must tolerate higher indoor temperatures and more variable ventilation rates than with an electric fan system.
- During winter conditions, building heat losses may be substantial. Insulated dampers are expensive and are difficult to make airtight.
- In warm humid tropics, because of large fluctuations in solar radiation (from strong sun to overcast conditions), airflow rates can vary greatly.
- The system cannot be combined with mechanical air conditioning since it leaks humid warm air back into the conditioned space.

Examples of solar chimneys are:

- *Trombe wall*: the first generation trombe walls included dampers at the top and bottom of the air gap, which allowed the system to function as a solar chimney. According to needs of individual occupants, the wall could be simply vented to the exterior, or the lower flap could be opened to the room, the lower flap to the exterior could be closed and the trombe wall could be used to draw cool outside air from the north side of the room through the room, to be then exhausted up and out of the top exterior opening of the trombe wall air gap.
- *Thermal chimney*: a vertical shaft works like a solar chimney, but without the solar assistance to accelerate the upward flow of air.
- *Rooms with high ceilings*: these rooms can benefit from airflow derived through stack effects. Good effects can be achieved if the rooms are at least two storeys high. Alternatively, the rooms can be coupled with an open stairwell or tall atria.

Synopsis

The design of solar chimneys should ensure that the extract stack is as high as possible above the roof. The exterior of the chimney should be dark to increase the air temperature within. Side openings in the chimney should be avoided in order to prevent potential backward flow. Insulation of the chimney reduces its effectiveness since heat from solar gain can assist in driving the thermal movement air. Solar access can be provided, and in such cases the interior surface of a transparent chimney facing the sun acts as an absorber and should be painted black. Rotating metal scoops at the top of the chimney preserve the airflow, even if there is wind.

The connection of the chimney to the ventilated room is important. The path that the air takes through a room and out the chimney should allow for easy airflow. Inlet openings should be placed to let cold air inside the house, and should be located on a room's windward side to obtain more ventilation pressure due to the prevailing wind.

Solar chimneys are useful in low-rise buildings in climates with a high average solar radiation. The channel should be as high as possible and the collector area should be large in order to maximize airflow. Figure 9.18a shows the temperature difference caused by the chimney. In Figure 9.18b the resulting air flow is explained. Calculations are made in *Tryflow* Trnsys 15 based star network fluid calculation program.

Courtyards and external rooms

Indrika Rajapaksha and Upendra Rajapaksha

Buildings with courtyards can offer a substantial potential for utilizing natural ventilation for indoor thermal comfort. As an open space within a building (or building cluster), courtyards are a design element in most vernacular buildings and were originally used in the Mediterranean, Middle Eastern and tropical regions.

A number of design variables are identifiable in most traditional building solutions that incorporate a courtyard. Agreement between building geometry, enclosure, orientation, density of the building context and access to wind flow can carry considerable architectural implications in modifying the microclimate of courtyards. In order to optimize the climate response of courtyards and to increase their thermal buffering characteristics, design variables have responded to the local climatic condition. This has resulted in subtle regional variations to the form of courtyards.

The climate response of the courtyard form in hot dry climates is based on the ability of the courtyard to act as a cool sink for the surrounding built spaces (see Figure 9.19). Examples of building solutions include compact building forms with less surface area to volume; flat roofs for radiant cooling; high thermal mass in the walls and roof for nocturnal cooling; vertically deep courtyards for daytime internal shading; and windowless external fabric to avoid internal breezes, sandstorms, etc.

In warm humid regions, particularly in the tropics, courtyards demonstrate a different response to the climate. The primary strategy is defensive in order to obtain cooling inside buildings involving shading and ventilation. The layout of the typical courtyard building acts as an air funnel (see Figure 9.19). In most cases, openings in the external fabric are used to bring wind-induced ventilation to the interior. Other examples include large roof overhangs and spread-out plan forms for shelter from sun and rain, as well as better exposure to outside breezes.

Air funnel effect

Courtyard buildings in warm humid regions have often been adapted for wind-induced cross-ventilation (see Figure 9.19). The sectional geometry creates different pressure fields along the wind-flow axis, thus making the building an

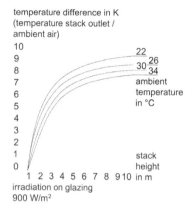

(a)

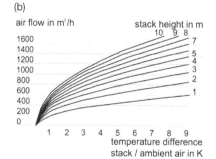

(b)

Source: Lars Junghans

9.18 (a) Temperature difference stack outlet and ambient air. Stack properties: width 1m, depth 0.3m, single glazing with different heat gains, 900W and 600W; (b) air flow of a solar chimney

air funnel (Rajapaksha, 2003). In most cases, openings occupying the entire external envelope are used to bring wind-induced ventilation to the interior.

The entrance veranda on the windward side acts as a wind tunnel, focusing the incident wind into the courtyard that lies on this air funnel, which in turn ventilates the indoor spaces around the courtyard. Examples of this type of courtyard building are found in Sri Lanka (see Lewcock et al, 1998) and South India. To avoid the heat of the wind-induced ventilation from reaching the interior, the perimeter openings are protected with heavy shade, in most cases achieved using wide roof eaves and other intermediate spaces, such as verandas.

Passive zones

The incorporation of a courtyard into a building form offers a microclimate (buffer zone) between the outdoor and indoor environments of the building. From the climate design viewpoint, a courtyard building presents a greater flexibility in promoting larger areas of internal passive zones, which can benefit from natural ventilation and daylight. With a courtyard, the functional depth of the plan can be kept to a minimum for better cross-ventilation potential from the wind than a deep-plan solid building of the same floor area provided that the spacing and density of nearby buildings, internal layouts, openings and orientation of the building to the prevailing wind are optimized.

A 'selective mode' building (Hawkes, 1996, p38), where the admission of substantial elements of the outdoor environment into the building is allowed, would benefit from a smaller volume-to-surface area ratio. Such buildings with

(a)

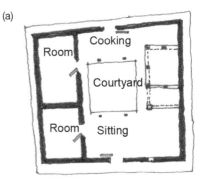

(b)

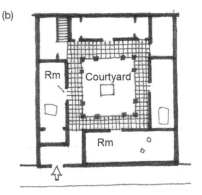

(c)

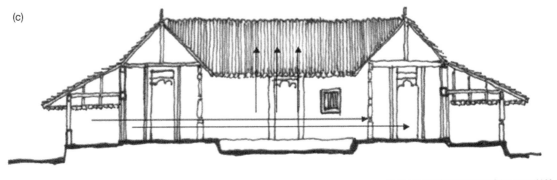

(d)

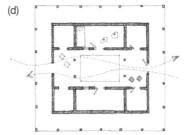

Source: (a) after Galotti (1925) in Schoenauer (2003, p155); (b) after Coomaraswamy (1956, p41)

9.19 An inward-looking courtyard in hot, dry and warm, humid climates: (a) Moorish dar, a typical North African urban courtyard house of the medieval period, Morocco; (b) small yeoman's house in Kandy, Sri Lanka; (c) tropical courtyard building as an air funnel, monk's house in Kandy (1771 AD); (d) plan of a typical courtyard dwelling in Sri Lanka

shallow plans would have a higher potential for employing passive climate-control measures because of the higher proportion of the passive zones to the total floor area of the building. A courtyard building, in this respect, presents a greater flexibility in promoting larger areas of passive zones and, therefore, natural ventilation and daylighting (see Figure 9.20).

The courtyard as a service space can potentially bring environmental benefits if this space and the surrounding servant spaces maintain favourable environmental conditions for thermal comfort. This objective depends upon the appropriate control of heat transfer between the:

- courtyard and outdoor microclimate; and
- courtyard and spaces served by it (see Figure 9.20).

In the tropics, the courtyard can overheat through solar gain and transmit this heat to adjacent occupied spaces served by the courtyard, reducing comfort. Avoidance of this problem can be promoted by the airflow effect, roofing, shading and thermal mass.

The heat transfer between the courtyard, its adjacent spaces and the outdoor environment is regulated by the effects of passive strategies – the

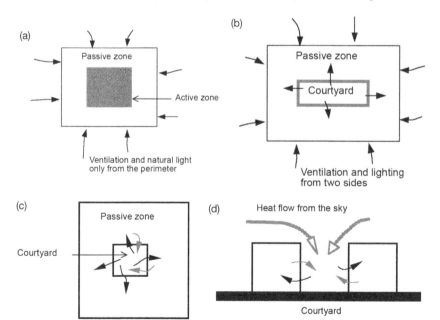

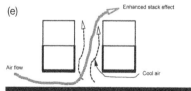

Source: Upendra Rajapaksha

9.20 Comparing the area of passive zones: (a) deep plan building with active systems in the centre; (b) courtyard building – the passive zone area is relatively greater in a courtyard building than in a deep plan solid building of the same floor area; (c) heat transfer in a courtyard building; (d) heat flow in section; (e) enhancing stack effect in a deep courtyard through use of massing

airflow, thermal mass and passive solar energy. A general guidance of these effects can be found in Rajapaksha (2003). In the tropics, the courtyard can overheat and transmit solar heat to adjacent occupied 'servant' spaces, creating overheating. The avoidance of this problem can be promoted by the airflow effect, roofing, shading and thermal mass.

Airflow effect

Airflow is a primary effect that dictates the thermal environment inside the courtyard. The effects of airflow promote comfort cooling of occupants and structural cooling, control overheating and remove solar heat out of the building's interior. This is desirable when the incoming air is at a lower temperature than the indoor air. Airflow is moved through buildings either by the wind pressure effect or stack effect, which can be regulated by the following:

- wind permeability of the geometry; and
- wind permeability of the enclosure.

Sectional geometry and building enclosure can regulate the airflow pattern inside courtyards. Fully enclosed courtyards receive wind-generated ventilation from their sky roof opening involving two distinct airflow patterns in the exchange of air within the courtyard. One pattern comprises a circulation at the top level of this space, 'top vortex', with stagnant air below, and in the other, a vortex occupies the full width and depth of the courtyard, known as 'full vortex' (see Figure 9.21).

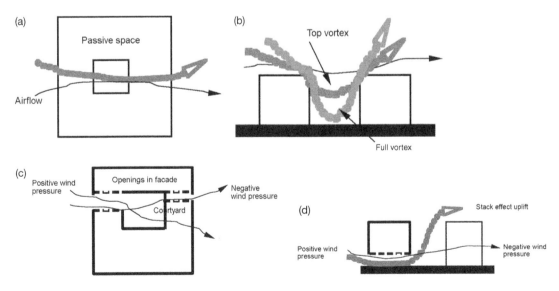

Source: Upendra Rajapaksha

9.21 Airflow behaviour in a fully enclosed courtyard: (a) plan; (b) section – airflow patterns show a top vortex and full vortex in the exchange of air within the courtyard; (c) semi-enclosed courtyard as an air funnel in plan; (d) section showing wind permeability of plan enclosure and sectional geometry affecting upwind airflow

These courtyards provide the least potential for preventing indoor overheating and promoting passive cooling since the top and full vortexes bring heat from direct and diffused solar radiation into the courtyard. Therefore, a solution for these courtyards is to increase wind-driven ventilation, ideally through an upward air movement generated horizontally from openings found in the building's wall enclosure (envelope). In addition, shading of the courtyard through landscape and/or roofing systems can be designed to facilitate this improvement. Semi-enclosed courtyard buildings can promote upward air movement if the sectional geometry and plan enclosure are suitable. Figure 9.21 explains how the wind permeability of the geometry and enclosure promote upwind airflow through the courtyard.

Results of a computational fluid dynamic (CFD) analysis reveal a reduction of indoor air temperature below ambient levels when the courtyard of a semi-enclosed building functioned as an air funnel, discharging indoor air into the sky. The analysis was carried out using measured climate values of wind direction and speed in the tropical city of Colombo. A thermal investigation was carried out for the same building in 2000. The results revealed a convective heat transfer from air to mass in the wall surfaces, with a significant correlation between the wall surface temperature, the indoor air temperature and the indoor airflow pattern. A quantification of this can be found in the Chapter 2 in the case study of Bandaragama House (see Case study 2.1).

Stack effect and shading

Stack ventilation will always takes place from a lower opening to an upper opening through the buoyancy action of heated air. Moreover, relatively taller courtyards – for example, courtyards with a smaller aspect ratio (degree of openness to the sky) tend to promote ventilation due to stack effect to the extent that the immediate external environments remain below the courtyard temperature. In deeper courtyards, the lower occupied zone shows greater shading; thus, thermal stratification could be used to promote stack ventilation. However, if courtyards receive ventilation from a sky roof opening, the wind pressure effect is stronger than the pressure forces generated by the stack effect. Hence, the wind can override and negate stack effect. This problem can be avoided by incorporating a windward opening at the lower section of the envelope, which can promote upwind air movement inside the courtyard (see Figure 9.21).

Synopsis

The use of a courtyard as a vertical air funnel in increasing Venturi effects for the natural ventilation system of a building is a significant design improvement. Manipulating the building cross-section and layout to create a negative pressure zone above the courtyard's roof opening and a positive pressure zone at the bottom of the building section facilitates the upwind flow of air inside the courtyard. The upwind flow of the air funnel promotes cross-ventilation, assists stack effect by thermal buoyancy and Venturi effects, and removes warm air away from the courtyard. More analytical design phase assessment and monitoring of airflow behaviours in real buildings may be appropriate to ensure the efficiency of this system.

STRATEGIES AND TECHNIQUES

Veronica Soebarto and Márcia Agostini Ribeiro

Research concerning passive cooling strategies is relative young, although many cooling strategies have been used for centuries. With the oil crisis during the 1970s came substantial interest in renewable energy sources; but this interest focused on heating and not cooling. Only at the beginning of the 1980s did the interest in cooling increase; but the use of air-conditioning systems had already become a standard procedure in most warm climate countries. The actual interest in natural cooling systems has many different possible reasons; certainly, the increase in energy prices as much as the environmental damage to ecological cycles play a significant role in refocusing attention in this area.

Designing passive cooling systems requires an understanding of various definitions, an understanding of environmental science principles of heat flow pertaining to cooling and an appreciation of the limitations of bioclimatic strategies.

Passive cooling comprises any cooling strategy that may sink the inner temperature of a building through natural energy sources *under* the limit of the annual average outer temperature. In warm climates, it is important to note that the use of such cooling strategies must be preceded by the discussion on minimizing cooling needs by building design and site planning. This approach is bioclimatic when linked to the needs of the occupants for thermal comfort. Heat gains must be reduced, in the first instance, through the use of bioclimatic strategies in order to sink the inner temperature of the building until it reaches values *near* the annual average outer temperature. Therefore, a bioclimatic architectural design is a prerequisite to using passive cooling strategies. It is also important to note that most warm climates also experience low temperatures during the night or winter, which should not be forgotten during the planning process.

The first part of this section, therefore, investigates strategies for passive cooling, as well as other environmental factors pertaining to the utilization of these strategies. Importantly, the need for daylighting and noise abatement can have a bearing on the use of passive cooling systems. For example, the use of highly effective shading can reduce daylight, while the extensive use of roof lights can admit unwanted heat flux. Furthermore, the external acoustic environment may require ventilation if there are high levels of unwanted intrusive noise. This explains the need to investigate strategies for noise abatement.

Passive cooling

There are many passive cooling strategies that can be applied in a warm climate, using different heat sinks: air, water, earth and high atmosphere. Each

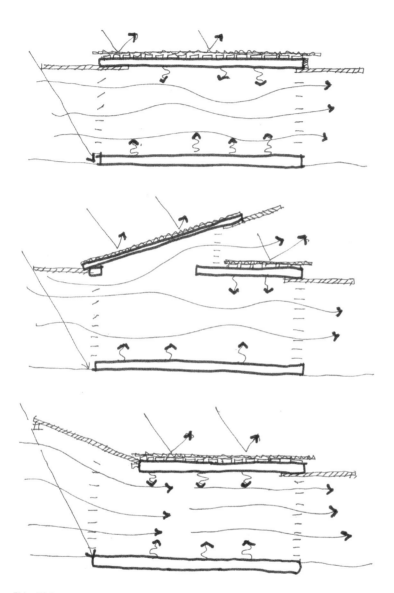

Source: Richard Hyde

9.22 Comfort ventilation – operated best with external air temperatures at the comfort zone. The greater the number of openings, the better, and the least number of internal obstructions, the better

of these heat sinks can be used in different strategies. Six passive cooling strategies are commonly used for passive cooling and their application depends upon the local climate and upon the specific comfort limits of local users:

1 Comfort ventilation uses air as the heat sink.
2 Night cooling ventilation uses air as the sink.
3 Radiant cooling uses the upper atmosphere as the sink.

Table 9.3 Strategies for passive cooling

| Climate | Passive solar systems | Solar protection | | Passive cooling | | | | | Dehumidification | Air heat exchangers (reduction power equipment) | Daylighting |
| | | Transparent element | Opaque elements | Ventilation (day and night) | Nocturnal radiation | Evaporation | | Earth cooling | | | |
						Direct	Indirect				
Moderate	X	XX	XX	XX	X	X	X	XX	X	X	XX
Hot dry		XX	XX	XX	XX	XX	XX	XX		X	XX
Hot humid		XX	XX	XX	XX			XX	XX	X	XX

Source: Valerio Calderaro

4   Direct evaporative cooling uses the phase change of water as the heat sink.

5   Indirect evaporative cooling uses the phase change of water as the heat sink.

6   Earth cooling uses the earth as the heat sink.

Comfort ventilation

Functions of ventilation

Ventilation can be defined as the effects of air movement in the building. There are three main functions of ventilation. The first is the provision of a sufficient quality and quantity of air for people's life processes and activities. This is the provision of healthy ventilation – that is, air that is free from pollutants or other harmful effects. Furthermore, in the case of urban sites, where buildings are built next to areas of high levels of acoustic and atmospheric pollution such as around freeways, the location of fresh air input to buildings has to be carefully sited to minimize these effects.

The second main function of ventilation is to provide occupants with personal cooling; occasionally this is called comfort ventilation, created by the passage of air across people's bodies. This is related to the availability and velocity of cooling air from outside the building. The velocity of air that can achieve thermal comfort increases with temperature until the skin surface temperature is reached at approximately 35°C. This is problematic in hot dry climates, where the summer air temperatures can peak at 42°C. In this case, the air is dry and dusty and is above the comfort zone – bringing this air into the building is therefore not beneficial for personal cooling. Other forms of cooling thus are required, such as evaporative or mass cooling. Buildings designed for comfort ventilation can operate in extreme conditions (35–45°C) provided that:

- The building has a well-performing envelope (such as good insulation, good shading or thermal curtains for the windows);
- There is a method of heat exhaust through a thermal chimney, ventilator or roof ventilator.
- Windows and doors can be closed up to prevent the ingress of external air during the day.
- Comfort ventilation is used at night.

Ventilation's third main function is to cool the fabric of the building, commonly called structural cooling. The amount of cooling provided by the air will depend upon the relative temperatures of the materials and the air. Air entering the building that is cooler will absorb heat from the materials and vice versa.

The design consequences of these ventilation functions suggest that it is highly context specific; general levels of airflow from mean wind speeds are modified by the site and through the building skin through the disposition of elements. In addition, occupants compete for personal cooling with the need for structural cooling of the building. Thus, the lack of openings and the volume of airflow through the building, as well as temperature differences, are crucial parameters if ventilation is to be effective. One of the first tasks when designing for ventilation is to assess the exposure of the building to the microclimate and prevailing breezes (check wind rose data.) In particular, it is necessary to carefully assess the upwind microclimate to check for good air pathways to the building and the influence of meso-climatic conditions, such as topography and the presence of trees and other building obstructions.

Another important task is to assess how ventilation will provide comfort. Natural ventilation of this type is highly variable and dynamic. As a result, backup systems such as mechanical ventilation are used for practical reasons since these systems provide more consistency in the provision of comfort. The variable and dynamic conditions of ventilation are apparent if the forces that drive these systems are considered.

Natural ventilation is generated by pressure differences in and around the building. These pressure differences come from air movement generated by air temperature and by wind. Temperature-driven ventilation is usually the lesser pressure of the two. Thus, when both are expected to occur at the same time, the wind pressure will prevail. Temperature-driven ventilation is called stack effect because it uses the natural buoyancy of the hotter air to rise and displace cooler air. As a result, stratification will occur in a space with hotter air at the top and cooler at the bottom. An advantage occurs in buildings when the external temperature is lower than the internal temperature. The internal air will rise up and exhaust from the building, bringing in fresh air. In warm climates, the effectiveness of the stack effect is questionable since the outside and inside temperature differences are small. Therefore, since stack is driven by temperature difference, the pressures are also minimal. Wind-driven ventilation is therefore commonly used in warm climates.

The third task is to decide how the building might attenuate the effects of airflow. The design factors affecting ventilation are as follows:

- Reduce plan depth and increase the openness of the section to facilitate cross-flow and the vertical flow of air.
- Provide optimum orientation of rooms to the prevailing breeze and a link between the leeward and windward side to make use of pressure differences.
- Maximize the skin opacity through the number and size of openings (for example, single-, double- or three-sided openings to rooms; horizontal versus vertical stacking of openings);
- Reduce internal obstructions.
- Select the building site to increase exposure to airflow effects.

In principle, it is best to conceive of rooms and spaces as large ducts that can moderate and direct airflow. Any obstruction will direct or block the airflow and provide friction, reducing its energy. Ideally, wind flows from the windward side to the leeward side by pressure difference through the linking of internal spaces. The deeper the building plan and the greater the distance from the envelope to the interior of the building (and the more cellular its internal form), the greater the friction will be. As a result, less airflow will reach the interior.

In situations where, for functional reasons, such as where there are high casual gains from occupants and the need for cellular spaces, natural ventilation may not be feasible and mechanical assistance for natural ventilation is required. This can be provided in a number of forms:

- Floor or desk mounted fans can be used to bring in external air if placed next to the window. This is useful for providing individual cooling due to local air movement.
- Ceiling mounted fans, which recycle air and give air movement, cannot lower the air temperature in the space but provide cooling due to the movement of air across the skin. Subjectively, the internal air temperature is reduced by up to 3°C.
- Exhaust fans extract air from the space and draw in external air to provide cooling.
- Ducted air using input fans bring external air into the space.

The efficiency of this system is improved if air can be brought in from a cool location. This type of system is useful for moderating peak loads from environmental or high casual gains. A computer simulation study of school classrooms in Queensland, Australia, illustrates the application of this system.

On days where there was little breeze, peak internal temperature were found to exceed the comfort zone by 5°C to 10°C. The use of the ducted air system was found to reduce peak temperatures to that of 1°C to 2°C above ambient temperature. The use of ducted external air is also a more energy efficient form of conditioning than full air conditioning. This is because with full air conditioning, energy is required for air movement and for temperature reduction. In ducted external air, energy is only required for air movement. At best, these systems can only bring the external temperature to the ambient air temperature; yet they do provide this temperature consistently. Thus, in design conditions where behaviour modification cannot take place – that is, in classrooms – this consistency of temperature is a crucial factor. In this way natural ventilation can be supplemented by mechanical ventilation with a reduced energy component rather than full air conditioning.

Comfort ventilation principles
This kind of cooling strategy aims to provide comfort to users, mainly during the day. It is achieved through the control of airflow inside the building and is based on the psychological effect of wind speed on users. It is especially useful for hot humid climatic conditions, where the airflow will reduce the humidity near users and therefore enhance their heat resistance. The general principles are as follows:

- *Spatial planning*: to ensure ventilation inside the building, it is recommended that the geometry should be open – that is, the section is open to allow airflow through the roof ventilators. The plan is also open to allow air to move through from the windward side to the leeward side, creating cross-ventilation. A dispersed plan form is ideal. Not only will there be more window area, but there will also be different wind pressures on the external walls. It is also possible to increase the wind pressure by placing the windows at different heights.
- *Position of windows*: the optimal solution to achieve cross-ventilation is that each room should have at least two windows placed on different external walls; when this is a possibility, it is important to note the effect of different window sizes and positions on the airflow rate.
- *Ceiling height*: even with only one window (single-sided ventilation), it is also possible to achieve good ventilation provided that air can flow vertically through the bottom of the window and can diffuse through the room and out through the top of the window. With high ceilings (a floor-to-ceiling height of approximately 3m) it is important to achieve this kind of airflow pattern. Hot air trapped near the ceiling can be evacuated by using this method.

- *Size of windows*: varying the relative size of windows has important consequences for ventilation:
  - If the opening at the high pressure side is larger than the one at the low pressure side, there will be a better distribution of the airflow inside the room.
  - If the openings are equal, the higher speed of the airflow will be achieved.
  - If the opening at the low pressure side is bigger than the one at the high pressure side, the average airflow speed will be smaller.
- *Types of window:* different types of windows at the high pressure side will have different effects on the airflow pattern inside the building; the following are some experiences with using windows for ventilation:
  - Louvred windows can provide a 100 per cent opening area, can be used to finely tune the ventilation flow and are flexible in term of materials and opacity. Leakage problems and security problems have also been solved with new designs.
  - Horizontal sliding windows are poor performers, often reducing the flow rate to 50 per cent of the window area.
  - Sash windows are good for single-sided ventilation with high ceilings as the windows can be slid up at the bottom and down at the top, controlling vertical airflow.
  - Awning windows require casement stays to control the position of the window pane and should by opened to the windward side to act as wind catchers.

Overall, the increase of the airflow inside a building can be achieved either mechanically or naturally. In both cases, it is important to ensure a good airflow distribution inside the building or room, the right airflow speed (2 to 4m per second) and that the building construction will not act as a heat store at night.

Specific limits are as follows:

- climate: warm and humid;
- wind speed: 1.5 to 2m per second;
- maximum temperature: 28 to 32°C;
- maximum relative humidity: 70 per cent; and
- daily amplitude: <10°C.

Mechanical comfort ventilation

This kind of ventilation will be achieved with the use of fans inside the room in order to increase the airflow speed. It is recommended that the air reaches the room through the windows and that the fans are used as exhaust fans. It is mostly used when the wind speed from the local climate is not enough to ensure the minimum airflow speed inside the room.

Airflow dynamics

In order that comfort ventilation can be achieved, the wind speed at the location should be sufficient to provide an airflow speed inside the room of 1 to 4m per second. Furthermore the building's orientation, geometry and massing are also important. These elements control the aerodynamic flow and allow the wind to reach the inside of the building and to flow through the rooms. Pressure differences between windward and leeward sides control this flow of air and can help direct airflow into the building. High pressure at the windward side and low pressure at the leeward side can drive air through the building.

Some building characteristics must also be taken into account when designing for comfort ventilation since these can impede airflow. These characteristics are the spatial planning, ceiling height, the position of the windows, the window type, the location of the layout of rooms inside the building and the location/tasks of occupants (see the earlier section on 'Fenestration').

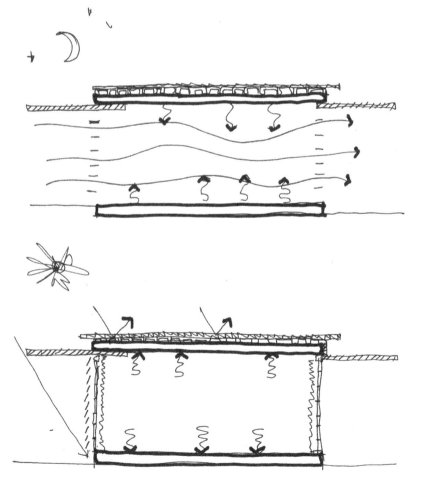

Source: Richard Hyde

9.23 Night ventilation: when conditions during the day are outside the comfort zone, it is often preferable to close the building up during the day. Thermal mass can absorb heat gain from the occupants during the day. This system operates best with a well-performing envelope and ventilation of the thermal mass during the night

Night ventilation

The ventilation of a building during the night can cool down its mass so that it can be used as a heat sink during the day. The cooler mass of the construction absorbs the heat during the day and therefore sinks the inner temperature of the building. The effectiveness of this strategy depends highly upon the daily temperature amplitude. Specific limits to this system are as follows:

- climate: warm and dry;
- night temperature: under 20°C;
- daily temperature: 30°C to 36°C;
- wind speed: 1.5 to 2m per second;
- maximum relative humidity: 15 per cent; and
- daily amplitude: 12 to 15°C.

The building should be closed during the day and opened at night when the outdoor temperature drops. Therefore, the use of mechanical fans during the day is recommended to create air movement inside the building. It is also important that the building envelope combines both a high level of insulation and thermal mass. Insulation is recommended with R values of 3 or higher with a number of radiant barriers and should be placed external to the thermal mass (that is, high capacity materials of concrete, stone or dense bricks).

In order to improve the efficiency of this strategy, the construction must be planned according to the wind direction during the night and should provide either cross-ventilation in the rooms or inside the building structure. There are three sorts of night ventilation cooling strategies: cooling rooms, cooling the building structure (hypocaust) and cooling through a special storage system.

Cooling rooms and hypocaust

In this case, night ventilation of the building is achieved as comfort ventilation. Some precautions must be taken into account regarding security and comfort. In a residential building, it is important to note which rooms are occupied during the night. In occupied rooms airflow should be kept at a comfort level by natural ventilation. When rooms are not used at night the airflow speed can be higher and mechanical ventilation can therefore also be used if necessary. Hypocaust systems allow air to pass through in the building structure to effect changes in temperatures (along walls, ceilings or between the floors). The structure must be designed to allow the wind to pass through it. This system allows for better contact between the heat sink (mass) and the cool air and it also eliminates problems such as security and privacy. The air is blown, at first, through a heat sink storage system such as a cold water tank or even the attic, and afterwards through the building. The airflow in the building is achieved through either a hypocaust or directly to the rooms.

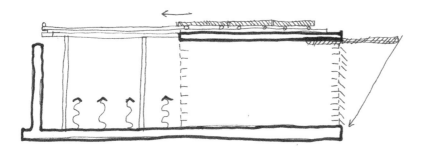

Source: Richard Hyde

9.24 Radiant cooling uses the atmosphere as a heat sink. The design principles involve utilizing cooling surfaces that radiate to the sky, such as roof ponds, courtyards and swimming pools. The spaces either act as a refuge, as with the swimming pool, provide direct cooling, as with the roof pond, or provide a space to draw cool air into the building, as with the courtyard

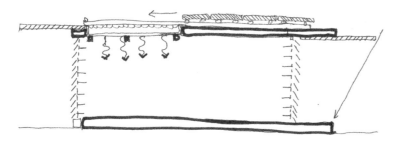

Radiant cooling

This passive cooling strategy uses the roof as a heat radiator based on the fact that any area radiates long wave to the sky and that the radiance balance will only be detrimental at night. The specific limits are as follows:

- climate: almost all regions, clear sky;
- building height: one floor; and
- mechanical ventilation: used to draw in the cool air from the roof.

There are many sorts of radiant cooling, of which three are described below.

Massive roof with mobile insulation

This is the easiest variant of radiant cooling. The roof, built with a massive and highly conductive material (for example, concrete) is insulated externally. During summer days the insulation helps to minimize solar heat gain and the roof acts as a heat sink, absorbing the internal heat and making the inner space cooler. During summer nights, the insulation is removed and the roof radiates the absorbed heat to the sky.

Skytherm system

This radiant cooling system consists of a series of 'roof pond' water elements and insulation panels. In winter, solar heat is gathered in the roof's water bags during the day and it is conducted through a metal ceiling, providing a warm radiant panel in the spaces below. Movable insulation panels above the water bags control the amount of heat retained. For summer cooling, the water bags are exposed to the cold night sky to reject heat and are covered during the day. The cool ceiling absorbs heat from the occupied space.

Long-wave radiators

In this case, the system uses conventional insulation and over it (5 to 10cm air space) metal plates. The cool air produced over the insulation is blown inside the building and its cooler mass acts as heat sink during the day.

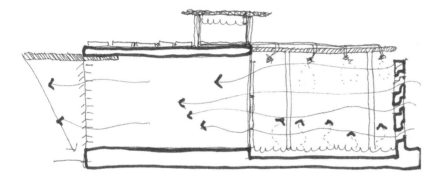

Source: Richard Hyde

9.25 Evaporative cooling can be provided directly using semi-enclosed spaces adjacent to the building. In arid areas, local cooling of these spaces by a misting system, which sprays water into the air, can raise humidity and reduce temperatures, contributing to improved comfort inside the building

Evaporative cooling

Evaporative cooling systems take advantage of the fact that it takes energy to change water from liquid to gas. This phase change energy comes from the sensible heat in the air, therefore lowering the temperature of the air. The specific limits are as follows:

- climate: warm and dry;
- maximum wet bulb temperature (WBT) in summer: 22 to 24°C; and
- maximum dry bulb temperature (DBT) in summer: 42 to 44°C.

Evaporative cooling techniques fall into two categories: direct evaporation and indirect evaporation.

Direct evaporation

In direct evaporation systems, water is evaporated directly into the air, which is circulated to the space being cooled. This raises the relative humidity of the air and, consequently, of the space. These systems include using evaporative pads in windows, building a pond outside the windows on the windward side of the building, and using a cooling tower. The last one brings outside air inside the building through a tower where the air is passed through wetted pads.

Without any mechanical fan to draw in the outside air, however, this system must rely on adequate wind speeds. This means that the tower has to be oriented in the right direction and have the right height to capture the wind, which could be a problem for a house in an urban setting. A good example of a downdraught passive direct evaporative-cooling tower is the one developed by Cunningham and Thompson. This employed four wetted pads for outdoor air to flow through so that, as the water evaporated, it provided a cooling of the air in the house, quite close to the outdoor temperature. A thermal chimney also attached to the home provided additional pressure to drive the system, pulling warm air out of the home as the cooling tower dropped cool air in. Misting systems can also be used to spray water into the air, which evaporates.

Indirect evaporation

Indirect evaporative approaches use evaporation to lower the temperature of a medium, which is separated from the building air, therefore not adding moisture to the air. When cooled, the roof or wall becomes a heat sink and absorbs the heat from the internal space. An example of applying this system is a shaded roof water pond over a non-insulated roof. By shading the roof water pond, the effect of radiant cooling is eliminated and the only cooling process is evaporation. Evaporative roof ponds have been applied in Sacramento, California; Colima, Mexico; and Israel (Givoni, 1994b).

Another variant of these systems uses a flat plate air-to-air heat exchanger. The water is added to the exhaust air, just before it enters the heat exchanger, lowering its temperature. In the heat exchanger core, heat is transferred from the supply air to the cooled exhaust air, thereby cooling the supply air. Indirect machines typically are larger and use more power than direct evaporative machines. Multiple-effect machines are possible, in which the first stage is indirect and the second stage is direct. Each stage adds power and complexity to the machine.

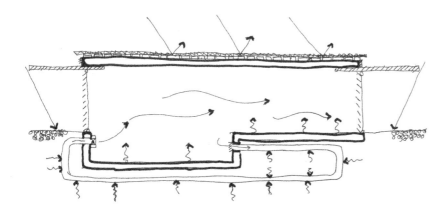

Source: Richard Hyde

9.26 Earth effects rely on the earth as a heat sink provided that earth temperatures are sufficiently different from air temperatures

Earth effects

In this cooling system, the earth is cooled to below its natural temperature and, afterwards, air is blown into the building. The temperature of the earth 2 or 3m under the surface is slightly above the average annual air temperature. This temperature can be further lowered through these techniques:

- mulching the surface of the soil with pea stone or wood chips and irrigating, if necessary, to provide moisture for evaporation (the mulch must be vapour permeable);
- shading the earth's surface – for example, by raising the building above the ground on posts.

Specific limits include the following:

- climate: moderate or warm;
- necessary underground area: 1 square metre of underground area for 5 square metres of occupied area (Zimmermann, 1996); and
- surface of the earth: for the installation of tubes, it is better if the soil is sand or gravel.

The surface temperature of the ground is dependent upon colour, vegetation and soil moisture, as well as soil conductivity and soil diffusivity. The net radiant flow into the soil is usually positive in summer, though shading can reverse this effect; it is negative during winter. The net convective heat flow is usually from the earth to the air; but this also reverses with shading, so it makes sense to minimize convection with a vapour permeable covering, such as the mulch described above. Dense low vegetation actually maintains air near saturation at the ground, which reduces evaporation from the soil. Water that percolates into the soil can raise or lower the soil temperature. Rain runoff from unshaded soil can raise the temperature of the shaded soil as it sinks in. If the soil is covered with pea stone, watering the stone at the end of the night can capture some of the radiant cooling stored by the stones.

Earth-coupled cooling approaches include partial or full earth sheltering of the building, and the use of earth tubes in either closed or open loops. It is important to design the system so as to minimize energy used by the fans used to blow the air through the tubes. This leads to designing tubes to minimize friction. Longer tubes of a smaller diameter dehumidify better. A US Department of Energy (USDOE) report on earth tubes contains the following conclusions:

- Small diameter tubes are more effective per unit area than large tubes.
- Long tubes are unnecessary.

- Tubes should be placed as deeply as possible.
- Closed-loop systems are more effective than open-loop systems.
- Tube thermal resistance is immaterial as the ground thermal resistance dominates.

The basement walls of a building can also be used as an earth–air heat exchanger, with the understanding that condensation and biological growth could be present.

Synopsis

In principle, types of climate and other conditions that form the building context – such as daylight and noise – determine the selection of a passive cooling system. For example, comfort ventilation may not be appropriate in a noisy urban macroclimate, where the airflow is restricted. Careful site analysis is needed to determine the viability of the use of these strategies or the selection of backup active systems to operate with a mixed mode operation – that is, passive and active systems.

Daylighting

Lars Junghans

Daylighting in residential houses located in hot climates requires consideration of both the need for daylight and the heat gain associated with sunlight. The main principle is to use diffuse light in housing to avoid the issues of heat gain. The most important properties of an effective day-lit space are:

- adequate and uniform light distribution in the space;
- natural shadow projections and colour rendition; and
- avoidance of disabling glare and minimization of contrasting glare.

Upon entering a room, the quality of natural lighting strongly affects the first impression of the space. A well-lit, bright space without glare gives a good impression. Dark corners or a strong decrease of brightness in the room depth leave a negative impression. Small windows protect a room from overheating solar gains, but leave a room dim. As a result the designer is confronted with a dilemma.

In climates with a large number of annual sunshine hours, people more readily accept dark rooms (which are cooler than sunlit rooms) than is the case, for example, in cold countries with less sun.

Strong direct light and severe contrasts in brightness are uncomfortable. There is a fundamental difference between the hot arid and the warm humid tropics. In arid regions, glare is usually caused by reflection of irradiated surfaces of the ground or other buildings. In warm humid regions with more diffuse light, glare occurs, especially, from the sky.

Sky conditions

This section considers three sky conditions:

- *Overcast sky*: the light coming from the zenith is triple that coming from the horizon.
- *Clear sky without sun*: the light coming from the horizon is brighter than from the zenith.
- *Clear sky with sun*: the intense light from the sun dominates the sky globe.

Light distribution in a room differs according to sky conditions. With an overcast sky, light from the zenith is approximately three times brighter than from the horizon. A clear sky without the sun appears dark blue because of the lack of dust. The light at the horizon is then brighter than from the zenith. Near a window, a space receives more light from an overcast sky than from a clear sky without direct sun. This is explained by the difference of the light distribution of the sky's hemisphere. The illuminance values by direct sunlight are so high that the difference from design variations is obscured in the absolute values:

- Warm, humid (equatorial), cloudy and hazy during the entire year and bright when cloud cover diminishes and there are clear sky conditions: dull grey sky conditions are found when cloud cover is 60 to 90 per cent.
- Warm humid climates have more overcast sky conditions than sunny conditions; accordingly, there is less direct sun in the room.
- Hot dry climates, usually clear sky with extremely high light intensity, periodically somewhat reduced by dust particles in the air, with minimal clouding: light from the sky can be obscured by sand and dust storms.
- Subtropical climate, varied with season, clear blue sky shortly after the rainy season: increasing glare occurs due to the concentration of dust in the air.

The rule that the window must be as high as possible to maximize the depth of light penetration into a room is mainly valid in climates with prevailing overcast sky conditions.

Visual comfort: Sky illuminance

Sky illuminance varies considerable within the range of warm and hot climates. In tropical regions such as Malaysia, overcast conditions can form the dominant sky condition, producing 80,000 Lux. In moderate climates such as Brisbane, there are seven hours of sunshine, on average, per day, giving clear sky conditions and available illuminance of up to 200,000 Lux. This is complicated by the fact that in moderate and hot humid climates, both overcast and clear sky conditions can prevail for long periods. In summer in Brisbane, monsoonal rain conditions provide overcast conditions, while the tropics can have clear sky conditions for extended periods. The impact of these widely different sky illuminance conditions has an important consequence for visual comfort and the design of the building envelope, creating the need for an interactive envelope that can be adapted to these light conditions. During recent years, it has been the trend to design for one set of conditions and then use electric lighting when lighting levels need to change with sky conditions.

Glare and perception

Glare is a human perceptual condition, which occurs when there is a high contrast in the visual field. Hence, glare is a condition of vision where there is discomfort or reduction in the ability to see objects. The most dangerous type is disabling glare, where the sudden occurrence of a bright light can impair the visual observer. This is uncommon in buildings but can occur in climates with the movement from very bright external sky conditions to dark interior spaces. More common is discomfort glare, where the field of vision is too bright for a task (such as reading on the beach). The final more common type of glare is caused by reflection – that is, reflection of bright sources from surfaces, such as the sun reflected from mirror glass façades or light reflecting in computer screens.

Qualitative and quantitative standards

Planning issues often result in internal rooms with no daylight. This should be avoided since it increases energy use in a building as permanent supplementary lighting is needed. Quantitative standards have been developed to assist with ensuring adequate daylight to rooms. Daylight factors (ratio of the illuminance due to daylight at a point on the indoor working plane to the simultaneous outdoor illuminance on a horizontal plane from an unobstructed sky) provide one standard for measuring daylight. The following rules of thumb can be used:

- Determine the ratio of daylight to non-daylight rooms:
  - percentage of rooms without daylight should be zero.
- Evaluate the room, as well as window area and position on the wall:
  - ceiling height/room depth method: effective daylight penetration is 2.5 times floor height;
  - window height/room depth method: effective daylight penetration is 2.5 times window height; and
  - glazing ratio method: average wall glazing ratio (transparent to opaque area); common target is 15 to 20 per cent.

Room design

Shape and depth

If daylight is used to generate room design, then the shape and depth of a room is controlled by the use of side, ceiling height and top lighting. The rules of

(a)

(b)

(c)

Source: Richard Hyde

9.27 Glare is a condition of vision where there is discomfort or a reduction in the ability to see objects; types of glare include: (a) disability – inability to distinguish detail (for example, reading in a highly contrasting visual field or dark threshold spaces); (b) discomfort – field of vision is too bright for the task (for example, reading on a balcony); (c) reflected (for example, reflections from surfaces such as roofs)

thumb show that the use of side lighting with high ceilings (3m) can achieve room depths of 7m. Introducing top light through the roof increases this depth. In climates with high sky illuminances, the effective room depth should be considered where verandas and external rooms are used. Thus, a 3m veranda attached to a room will reduce the daylight penetration by 3m.

Surfaces

The level of reflectance in the room is important in controlling glare and the light distribution in the room. A light-coloured floor makes the space appear brighter. The real effectiveness in illuminance by changing floor reflectivity is comparable to changing wall reflectivity. However, a dark floor appears solid and stable. To choose a floor colour slightly darker than the wall is a good compromise. Ceiling colour has a smaller influence. Warm colours such as red and yellow are preferred in cold climates. Cold colours such as blue are preferred in hot regions.

Window design

Window size

Window size strongly affects the first impression of a space. Increasing the window area by up to 50 per cent increases the light level, while also reducing the luminance contrast and improving visual comfort. Both the increased absolute amount of light and reduced luminance contrast give a better first impression upon entering such a space.

Light diffusers

There are several types of light diffusers:

- *Venetian blinds*: conventional Venetian blinds allow the occupant to adjust the mix of sunlight and daylight. A simple Venetian blind, even with white slats, obstructs much daylight, even if the slats are set horizontally. Only a minimal amount of light is reflected to the ceiling. The room impression is usually dim.
- *Fixed architectural shading*: features such as roof overhangs or *brise soleil* are often used in hot climates. Such a system protects the window from direct sunlight; but in a diffuse sky the room may become too dark. It can also be frustrating that the direct view to the sky zenith is permanently blocked. On west or east façades, a *brise soleil* must have an egg-crate geometry. Vertical elements are essential to block lower sun angles in early mornings and late afternoons.
- *Mashrabia:* this turned wood spool work is traditionally made of beech and mahogany wood, and is used to provide privacy, shade, light and ventilation. In residential buildings located in dry climates with many

annual sun hours a traditional *mashrabia* can be a good shading device. The *mashrabias* deliver pleasant light into the room, the carving creates a silhouette and occupants are able to look outside. The illuminance values are very low, but are sufficient for rooms in private homes. *Mashrabias* are ineffective in climates with frequently overcast sky because illuminance values are then too low, creating gloomy space.

- *Light funnel or light tube:* day lighting is directed from outside to inside. As the skylight admits light, it is directed through the ceiling by a reflecting tube and then dispersed to the interior by a ceiling diffuser. The benefit of this system is it that it can bring light into areas that are normally lit by electric lighting during the day. The light tube can also be used to bring light into areas of the room that are not effectively lit. In this way daylighting can be used to save energy and also improves the light quality. The effect of the sky tube is to make the room appear brighter by reducing the contrast between the window wall and the interior walls, creating a more uniform illuminance.

Window location and orientation

The daylight qualities of a reference room were analysed so that different proportions of a window opposite the entry to the room were considered – in this case, a window in the south wall. The window area is kept constant in all variations:

- A wide window with short height results in large differences between luminance values at the window and the wall; this is uncomfortable for occupants.
- A window as a door offers occupants a large vertical view angle from the sky to the ground; the illuminance near the window is good, but decreases rapidly with room depth, and the first impression upon entering the room is positive.
- The function of a high window is to achieve good daylight penetration, while a low window offers a view to the outside. With a high window, the light distribution in the room depth is slightly improved, but the human impression is unfavourable because of the limited view outside and the dark corner.
- A window in one corner strongly illuminates the adjacent wall and corner, while the opposite corner is quite dark so that the furniture arrangement must take account of this asymmetry.
- A room with a window in the side wall has less glare than a room with a window in the wall opposite from the point of entry; hence, the room seems to be brighter and more pleasant.

(b)

(c)

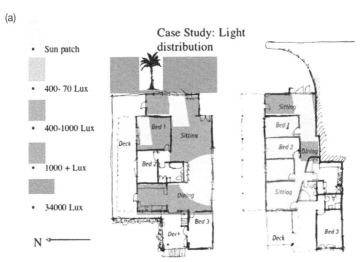

(a)

Case Study: Light distribution

- Sun patch
- 400- 70 Lux
- 400-1000 Lux
- 1000 + Lux
- 34000 Lux

N

Source: Richard Hyde

9.28 Lighting studies are useful to identify shading and glare: (a) sun patches in the morning cause problems for the home office space to the east; (b) studies of the easterly façade showing the large area of glass, which causes lighting and overheating problems in summer; (c) glare from morning sun in the home office; (d) a better location for the office is on the south side of the building; (e) roof lights should avoid admitting direct beam radiation: more diffusers are needed – in this case, below this roof light

(d)

(e)

External conditions to the window

The external reflectance of areas adjacent to the window will affect the internal illuminance. Assessing the quality of the external environment is an important condition of window design.

Synopsis

For a given window size, it is possible to maximize daylight effectiveness in a number of ways:

- Strategies that increase light penetration while also reducing heat into a room are most effective in creating a balanced, visually bright atmosphere, even if the absolute level of light is reduced and a thermally neutral internal environment is created.
- A light funnel cross-section of the wall opening allows a small opening to give a bright impression.
- Light levels are markedly higher in a space with light-coloured surfaces; but not all surfaces are equally important. In order of importance are side walls adjacent to a window, the floor and the ceiling. The wall containing the window is least critical for illumination; but a light colour minimizes the luminance ratio and improves visual comfort.
- Movable shading devices should be mounted so that they can be moved fully out of the aperture area of the window.
- Fixed architectural shading devices block valuable zenith sky light, which is the best light source by overcast skies.
- Diffusing external conditions prevent glare from the ground, and tall plant growth prevents glare from the sky. If plants are not possible, the surfaces of the ground should not be too light. Surfaces reflecting light directly into the window of a house should be dark.

Noise abatement

Valerio Calderaro

Hearing is one of man's most important communication channels, perhaps only second to vision. But while the eyes can be shut when there is too much light or there is an unwanted scene, the ears are open throughout life to unwanted noises, as well as to wanted sounds. Protection, if necessary, will have to be provided in the environment. Noise is the term used for any unwanted sound; thus, the definition of noise is subjective. One man's sound is another man's noise. The science of sound – acoustics – can be broadly divided into two major areas:

1 designing for *wanted sound* (creating the most favourable conditions for listening to a sound we want to hear: room acoustics);
2 designing for *unwanted sound* (controlling noise).

The purpose of this section is to examine the application of these principles in residential buildings in warm and hot climates. A number of issues are particularly important. From the viewpoint of a residential building, which is to be designed, it is useful to distinguish between external and internal noises.

External noise

Strategies to mitigate external noise include the following:

- distance;
- avoiding zones of directional sound;
- screening;
- planning: using parts of the building that are not noise sensitive as barriers;
- positioning openings away from the noise source; and
- noise-insulating building envelope.

Distance and zoning

If a site is given on which the positioning of a building is subject to the designer's choice and there is a noise source to one side of the site (for example, a busy road), it is advisable to place the building as far from the noise source as possible.

Some sources are strongly directional. There may be a band of maximum noise across the site (particularly if it is a large site), either due to such directional sources or to the funnelling effect of local topography. Its existence can be discovered by a 'noise climate' survey of the site. The building(s) should then be placed away from such noise bands.

Screens and barriers

Screening or barriers in the path of sound can create an 'acoustic shadow' – if the sound is of a high frequency. At low frequencies, diffraction will occur at the edge of the barrier; thus, the 'shadow' effect will be blurred. If the dimension of the barrier (in a direction perpendicular to the sound path) is less than the wavelength of sound, the shadow effect disappears. As at 30Hz the wavelength is over 10m, any barrier less than 10m will be ineffective for such low frequency sounds.

The screening effect of walls, fences, plantation belts, etc. can be used to reduce the noise reaching the building. These should be positioned in such a way as to fit in with any advantageous effects of local topography. As a general rule, it can be established that a given barrier will be most effective when it is as near to the source as possible. The second best position would be near the building that is to be protected: it would be least effective halfway between the source and the building.

Screening can rarely be relied upon as a positive means of noise protection, but it will help to ameliorate an otherwise critical situation.

Planning, openings and insulation

Planning of the building will obviously be governed by a whole series of factors other than noise; but noise protection should be included among the factors taken into account. The relative importance or weighting of the noise control aspect will depend upon the particular design task: it may be dominant in the case of a school classroom near a motorway, or it may be quite subordinate. External noises can be controlled at the planning stage in two ways:

1   Separate areas that are not noise sensitive, where noise would not cause disturbance, and place them on the side of the building (possibly in a separate block or wing) nearest to the noise source. Thus, the areas or block would provide screening and protection to the more critical areas.

2   Position or orientate the major openings away from the noise source. Usually in the external envelope of a building the openings (doors and particularly windows) are the weakest points for noise penetration, so it is logical to place them in the least exposed positions. Furthermore, the plan shape can be adjusted so as to provide protection or screening from the sides. Special elements (wing walls and screens) can also be introduced to provide additional sideways protection.

Noises within the building

When designing for noises generated within the building, the designer can take the following measures:

- Reduce at source.
- Enclose and isolate the source, or use absorbent screens.
- Planning: separate noisy spaces from quiet ones, placing indifferent areas in between.
- Place noisy equipment in the most massive part of the building (such as in a basement).
- Reduce impact noises by covering surfaces with resilient materials.
- Reduce noise in the space where it is generated by including absorbent surfaces.
- Reduce airborne sound transmission through airtight and noise-insulating construction.
- Reduce structure-borne sound transmission through discontinuity.

Noise implications for climate types

This section examines how typical building forms in warm climates respond to the problem of noise.

In hot dry climates, walls and roofs are usually of massive construction. Windows and openings are typically small and often face an enclosed courtyard. All of these are favourable from the viewpoint of external noise exclusion. Such a building would only be vulnerable to noise generated within the courtyard or by overhead noise sources (for example, overflying aircraft). Since the occurrence of the latter is infrequent (except near airports) and the former are 'familiar' noises, there will be no serious noise problems.

The very large thermal capacity of the building may necessitate the erection of a lightweight shelter on the roof, to be used for sleeping – at least during the first half of the night. Noise protection for this shelter will be practically impossible, but at least the users will have the option to choose between a thermally comfortable but noisy space, or a quiet but somewhat overheated one. The only special measure that could be taken to reduce external noise penetration is the protection of the small windows or ventilators in the external walls for the period when they are open (when closed, the heavy shutters, used for thermal reasons, would provide adequate protection).

Since the only potential source of discomfort are the noises inside the courtyard, it may be helpful if soft surfaces are used as far as practicable – for instance, lawn instead of paving, or absorbent materials on the soffit of the veranda roof around the courtyard.

In warm humid climates, the buildings would typically be of a lightweight construction, with very large openings exposed to wind and air movement. The building envelope will not be able to control noise. At the very best, it can reduce the penetration of outside noise by skilful use of absorbent surfaces. From the viewpoint of inside noises, the situation may be somewhat better than in the hot dry climate courtyard house: inside sounds are free to escape, will not be reflected from bounding surfaces and there will be no build-up of a reverberant sound.

Planning controls, such as distance, positioning or various forms of barriers will have to be relied upon to a great extent. Fortunately, there are two areas of concurrence between thermal and sonic requirements:

1   Densities in this climate are and should be much less than in other climate zones; distances between buildings must be kept greater to allow air movement – this would also help the noise problem.
2   Since the positive control of humidity is only possible with air conditioning, the use of such an installation is much more warranted here than in any other climate. Air conditioning implies a sealed envelope, which in turn

makes positive noise control feasible. Noise control requirements, especially in the case of highly noise-sensitive buildings, would assist or reinforce the case for installing air conditioning.

In temperate climates, buildings are likely to be of massive construction. Windows and openings would probably be reasonably large to provide air movement in the warm humid season, but with provision for closure in the cold season and during the day in the hot dry season. Generally, the building would be closer in character to buildings in hot dry climates; thus, the noise problems would also be similar – not very serious. Window shutters and doors should be massive, both for thermal reasons and for noise reduction.

Problems may arise in the warm humid season, when windows are open for ventilation. In such a situation, it would be futile to attempt noise insulation. Absorption could be used to reduce the noise, as in warm humid climates; but the benefit of this would not normally justify the cost for three reasons. First, the warm humid season, when such ventilation is necessary, is usually short, up to about three months in duration. Second, absorption is not very effective at reducing noise penetration: absorptive materials are mostly vulnerable and, exposed to changing climatic conditions, they may rapidly deteriorate. Third, in critical situations (such as in lecture rooms), a suitable aural environment can only be ensured with a sealed envelope, which would demand air conditioning. In fact, aural requirements may dictate the use of air conditioning, even if its use would not be fully justified for thermal reasons.

Synopsis

Controlling noise in the external environment is an important part of the passive cooling of buildings and is often ignored at the planning stage. Since it is difficult to control noise through the use of distance, principles of controlling unwanted sound at source have been developed. One of the most sensitive conditions for comfort in warm and hot climates is sleeping. Intrusive noise and elevated temperatures often make sleeping environments unfeasible. Careful planning, envelope design and a passive cooling system should be selected to meet site conditions.

Case study 9.1 Prosser House (Healthy Home Project), Queensland, Australia

Richard Hyde

This project is a demonstration project for lightweight houses in moderate climates (Watson and Hyde, 2000).

PROFILE

Table 9.4 Profile of Prosser House, Queensland, Australia

Country	Australia
City	Surfers Paradise
Building Type	Detached
Year of construction	1999
Architects	Richard Hyde and Upendra Rajapaksha

Source: Richard Hyde

PORTRAIT

The project demonstrates how a new suburban house can achieve a high level of environmental sustainability by employing, among other environmental strategies, improved air quality, passive solar design, active solar systems and

Source: Richard Hyde

9.29 Prosser House (Healthy Home Project): front elevation. The house is constructed of low embodied-energy materials, timber and fibre cement. The steel roof is designed to collect rainwater. Also installed on the roof is a solar hot water heater and a photovoltaic system. A grey water system and water storage tank were installed to cut water consumption by up to 80 per cent of a normal house's consumption

low embodied-energy materials. It challenges existing norms of house design and legislation concerning the onsite reuse of resources such as water and collection, purification and the use of rainwater.

This has been achieved in the form of an environmental prototype: a single residential dwelling constructed on the Gold Coast in Queensland, Australia. In addition, the project brings together cutting-edge environmental technologies through partnerships with over 30 industry organizations and other collaborating groups. These groups have supported the project through an extensive range of products, services and contributions. An innovative feature of the project is the environmental assessment of the house, supported by grants from the Australian Research Council and the Forest and Wood Products Research and Development Corporation.

During 2000, the house won a national and South Queensland Excellence Award in Energy Aware Housing. The project has been cited extensively in the press and in national publications for good residential design.

CONTEXT

Location

The home is a privately built, new detached home located on an existing block. The original house was a beach house, which became obsolete over time with the extensive modernization and commercialization of the area. The existing timber building was removed and a new building was built that characterized the past, but also addressed contemporary issues of sustainability. Hence, the style and construction of the building follow the lightweight tradition of Queensland housing, but is built to accommodate site, climate and urban issues.

Economic

The building is over 200 square metres in area and was built at a cost of Aus$800 per square metre, excluding the environmental technologies. A sponsorship programme in which approximately 30 companies participated supported the financing for the project's materials and systems. The environmental technologies, such as the photovoltaic electric systems and solar hot water, were purchased through conventional means. The technologies for water recycling were made available through government funding. At present, legislation in Queensland prohibits onsite water recycling; hence, a virtual grey water system was devised to demonstrate the potential for water recycling.

Site description

The building is located four blocks from the Pacific Ocean. A mixture of dwelling types, from high-rise to low-rise apartments, dominates the area. The zoning for the area allowed up to five storeys in height. Overshadowing is not problematic, although available breezes for cooling are reduced due to the shelterbelt effects of adjacent buildings. Hard surfaces and the absence of vegetation produce a heat island effect, which creates heat sinks onsite. The site chosen has many of the contextual problems found in urban areas and, hence, presents significant challenges for designing a passive house.

(a)

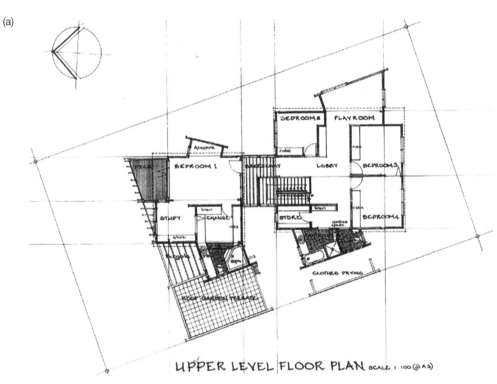

UPPER LEVEL FLOOR PLAN SCALE 1:100 (@ A3)

(b)

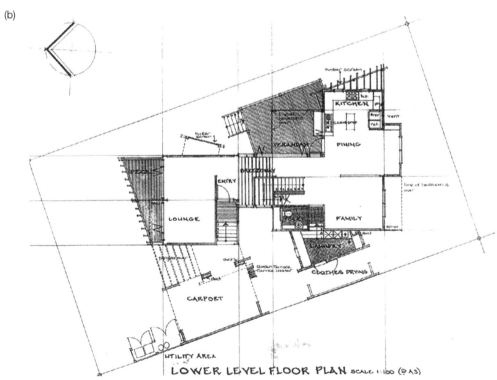

LOWER LEVEL FLOOR PLAN SCALE 1:100 (@ A3)

Source: Richard Hyde

9.30 Prosser House: (a) upper plan; (b) ground floor plan; (c) section

(c)

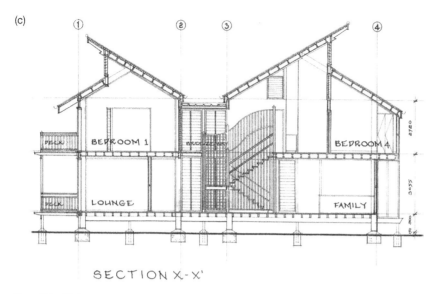

SECTION X-X'

Source: Richard Hyde

9.30 Prosser House: (a) upper plan; (b) ground floor plan; (c) section *(Cont'd)*

PLANNING TOOLS APPLIED

A design phase assessment was carried out as follows:

- water flume tests, used to demonstrate the effectiveness of the ventilation system;
- lighting simulations using radiance;
- embodied and operational energy calculations;
- shading calculations;
- post-occupancy details;
- questionnaires;
- lighting, thermal and ventilation monitoring;
- air quality assessment;
- energy audit;
- water audits; and
- life-cycle costing.

CONCEPTS

The brief for the house stipulated that the building was to be 'healthy for the occupants and for the environment'. This was achieved through addressing the main eco-home principles, respecting users, working with the climate, respecting the site and conserving resources such as energy and water. The building uses comfort ventilation and passive solar heating for climate control. Green materials are employed in the form of plantation timber, recycled timber and natural paints and finishes.

Source: Richard Hyde

9.31 Prosser House: water flume tests to assess ventilation

(a)

(b)

Source: Richard Hyde

9.32 Prosser House: construction systems use a timber framing system, which was prefabricated to reduce construction impacts and to lower embodied energy

BUILDING FABRIC

Building construction and structure

The building is two storeys high and has an open plan and section. The building is split into two pavilions connected by a breezeway. It has an interconnected kitchen, living and dining area, with a separate living room.

An alfresco shaded deck area leads from the kitchen and has a northeast aspect. Four bedrooms and a family area are placed on the first floor. The house is passively heated and cooled with backup systems, fans for space cooling and a radiant heater in the living/dining area. The coastal location of the building provides a microclimate that modifies temperatures.

A portal frame structural system is used for its low embodied energy, speed of construction and flexibility. The frame forms a structural armature so that the internal rooms can be changed easily, without structural consequences. In addition, large voids can be created to maximize ventilation. Lightweight construction is used (corrugated iron roof), as well as suspended timber floors with pine wall framing. Fibre cement cladding is used on the exterior, with plasterboard lining. The portal frame is prefabricated to give top-down construction. Early completion of the roof allowed the remainder of the building to be built onsite, offering workers shelter from the environment.

PASSIVE SYSTEMS

Ventilation system

The building uses cross- and stack ventilation, and is designed to increase internal wind speeds through pressure effects created by the building mass and geometry, juxtaposed to the prevailing breezes. The climate produces strong cooling breezes in summer, which can be used for cross-ventilation.

There are also periods of calm; in this situation, stack ventilation is used. Air rises into the atrium and syphons through pop-up roof monitors. To facilitate this type of ventilation system, a pavilion form is used with the building split into two sections and connected by a breezeway. The portal frame structure allows the building to be opened inside to facilitate air movement without the need for bracing walls.

Direct solar gain

The site enjoys, on average, seven hours of sunshine a day. The solar loads can be as high as 1000W/m². Illuminance values from clear sky conditions prevail, with levels in the range of 100,000 Lux to 150,000 Lux. Passive solar heating is used in winter through direct gain and through air charging. Solar penetration is permitted in winter to provide midday air temperatures in the range of 15 to 20°C.

Solar shading

Solar shading of windows is used during summer with a highly defensive envelope. Light colours and radiant barriers are needed to maintain this defence. For summer conditions, north-facing windows are shaded with horizontal awnings, while east and west window areas are reduced and vertical shades are used. Radiant barrier insulation is used in the east and west walls to accommodate solar loads. Non-habitable spaces are zoned to the west in order to avoid heat gain to bedrooms and other spaces. Landscaping is an important feature in summer, providing cool pools of air that can be used to ventilate the building.

Source: Richard Hyde

9.33 Prosser House: atrium space is used for vertical circulation and airflow; the bedrooms are linked to the atrium to provide additional stack ventilation

(a)

(b)

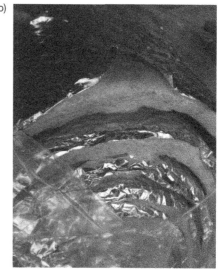

Source: Richard Hyde

9.34 Prosser House: the two-storey form of the building with a large wall surface area to the east and west; high levels of radiant insulation were used throughout the envelope, particularly on the unshaded walls and roof, to eliminate heat gain

Insulation

The walls utilize resistive insulation comprising foil membranes and foil batts. These provide an R value of 2.1. This system is also used in combination with bulk insulation in the roof, giving an R value of 3.5. The use of light-coloured materials with high emissivity reduces heat gain.

Windows

Plantation pine louvred and casement windows were used. The louvres are either body tinted glass or timber blades. The glass used in the windows aims to reduce heat gain, but also allow passive solar heating in winter. This gives a U value of 2.2, which is high, so the building was designed with no windows facing west to avoid the high heat gain from this aspect.

The louvred windows were found to have an advantage in that the material could be changed to either glass or timber, depending upon orientation and lighting requirements. Therefore, clear glass was used in the top light in the atrium and body-tinted glass to the north and east. Timber was used where privacy was needed. In this way the owner could customize the window system.

Ground effects

The ground floor utilizes a suspended timber floor raised 600mm above the ground. To achieve ground cooling effects in summer and ground heating effects in winter, a skirt was placed to block cross-ventilation of the under-floor space. The benefit of this strategy was to assist in stabilizing the internal temperatures when external conditions are extreme.

ACTIVE SYSTEMS

Solar-generated electricity and solar-heated hot water are used. Backup heating and cooling systems are employed, particularly during periods of high humidity. Ceiling fans give 2°C of effective cooling and extend the comfort zone in these conditions. In very hot weather, the house is sealed and stratification occurs, with hot air rising in the atrium and with cooler air remaining at the ground level, cooling the living spaces.

APPLIANCES

A range of energy saving appliances are used. Furthermore, 40 per cent of household energy is used in the hot water supply, which is normally heated by electricity generated from fossil fuels. Hence, a solar hot water system is used with a gas booster, cutting the demand for electrical energy and reducing greenhouse gas emissions. The building can be lit through natural lighting during the day, thereby reducing the need for electric lighting. Energy efficient light fittings and controls are used.

Energy supply system

A grid-tied photovoltaic system is used with an inverter to generate 240 V electricity. This is fed directly into the grid so that no power is diverted for house consumption. A meter is used to monitor the power provided to the grid. The energy carrier pays about Aus$0.24 per kWh for the energy, up to a maximum of 1000kWh.

Electricity consumed in the house is fed directly into the home and is separately metered and purchased at an average cost of $0.12 per kWh. The system design is integrated with the roof to an optimum angle of 35 degrees to maximize power generation. The system is devised to make the building grid neutral – that is, it uses as much as it generates, although the economics of the buy-back scheme do not seem to be passing the savings back to the owner. No energy storage system is used, nor is electrical hot water heating. A 240V PVC reduced wiring system is used to minimize the environmental impact of the materials in the energy system.

Water recycling and conservation

A 22,000 litre rainwater tank is provided to collect rainwater. This is used primarily in summer when there is plentiful rainfall, which tops up the tank. The tank can provide 75 per cent of the water consumption of the house. The water from the tank is purified through two methods. A first flush device sends to waste the first quantity of rain from the roof in order to avoid dirt that has collected on its surface. An ultraviolet pasteurization system removes bacteria from the rainwater, making it safe to drink.

Grey water is recycled using a three-chamber bio-system comprising a sand filter. The water from the grey water can be used for irrigation. At present, state legislation prevents this practice; but the system is being monitored to provide evidence for a change in policy.

The pumps used in the water treatment contribute approximately 1kW to the energy budget of the building, which is, in turn, provided by renewable sources.

BUILDING HEALTH AND WELL-BEING

Materials

The selection of materials for the interior reflected the need to reduce off-gassing. The paints have low levels of volatile organic chlorides. An organic floor finish was used to reduce gas emissions. The floor finish enables floors to be vacuumed and disinfected with a water and mentholated spirits solution. Rugs used on the floor can be removed easily for cleaning. Solar radiation is used to control insect infestation in the rugs, avoiding insect sprays.

(a)

(b)

Source: Richard Hyde

Figure 9.35 Prosser House: grey water system

Source: Richard Hyde

9.36 Prosser House: recycled materials and materials with a low environmental impact were chosen

The increased ventilation in the building also ensures that the build-up of emissions from materials is removed. Potential sources of pollution, such as the garage, are located at a distance from the building. Pollution from the kitchen, which can leak into the bedrooms, is avoided. Lifestyle spaces, such as the deck, are created to provide functional and amenity spaces outside of the building.

INFORMATION
Further information on the project can be found at www.healthyhomeproject.com/ (accessed 6 June 2005).

SUMMARY

This chapter provides a synopsis of the main design methods for a bioclimatic approach to housing in warm climates.

REFERENCES AND OTHER INFORMATION

AboulNaga, M. M. and Abdrabboh, S. N. (2000) 'Improving night ventilation into low-rise buildings in hot-arid climates exploring a combined wall-roof solar chimney', *Renewable Energy*, vol 19, pp 47–54

Adam, S., Yamanaka T. and Kotani, H. (2002) *Mathematical Model and Experimental Study of Airflow in Solar Chimneys*, Department of Architectural Engineering, Osaka University, Osaka, Japan. Available online at www.arch.eng.osaka-u.ac.jp/~kotani/rv2002-3.pdf

Commonwealth of Australia (2003) *Your Home: Design for the Lifestyle and the Future*, on CD and at www.yourhome.gov.au/index.htm

Coomaraswamy, A. K. (1956 [1908]) *Mediaeval Sinhalese Art*, 2nd edition, Broad Campton, New York

Delsante, A. (1998) 'Building materials: massive versus heavy weight', in *Design and Building of Energy Efficient Houses,* Queensland Department of Resource Industries, Queensland, Australia

Givoni, B. (1994a) *Basic Study of Ventilation Problems in Housing in Hot Countries*, Building Research Station Report, Technion, Israel Institute of Technology, Israel

Givoni, B. (1994b) *Passive and Low Energy Cooling of Buildings*, Van Nostrand Reinhold, New York

Givoni, B. (1998), *Climate Considerations in Building and Urban Design*, Van Nostrand Reinhold, New York

Goulding, J. R., Owen, J. L. and Steemers, T. C. (eds) (1992) *Energy Conscious Design: A Primer for Architects*, Batsford for the Commission of the European Communities, London

Hanisch, J. (1998) 'Land sub-division for energy efficiency', in *Design and Building of Energy Efficient Houses,* Queensland Department of Resource Industries, Queensland, Australia

Hastings, S. R. (ed) (1994), *Passive Solar Commercial and Institutional Buildings: A Sourcebook of Examples and Design Insights,* Paris, France International Energy Agency. Source http://www.ae198.dial.pipex.com/literaturereferences.html

Hawkes, D. (1996) *The Environmental Tradition: Studies in the Architecture of Environment*, E. & F. N. Spon, London

Hyde, R. A. (1999) 'Prototype environmental/work home infrastructure building for warm climates', in the *Proceedings of PLEA 1999 Conference* Brisbane, 22–24 September, pp289–293

Kumar, S., Sinha, S. and Kumar, N. (1998) 'Experimental investigation of solar chimney assisted bioclimatic architecture', *Energy Conservation and Management*, vol 39, nos 5–6, pp 441–444

Lewcock, R., Sansoni, B. and Senanayake, L. (1998) *The Architecture of an Island: The Living Legacy of Sri Lanka*, Barefoot Pvt Ltd, Colombo

Lim, B. (1998) 'Ventilation with sun protection', in *Design and Building of Energy Efficient Houses,* Queensland Department of Resource Industries, Queensland, Australia

Lippsmeier, G. (1969) *Tropenbau = Building in the Tropics*, Callwey, Munchen

Rajapaksha, I. (2002) *Passive Cooling Design Strategies for Tropical Courtyard Buildings, Onsite Thermal Investigation and a Computational Analysis*, PhD thesis, Nagoya University, Japan

Rajapaksha, U. (2003) *An Exploration of Courtyards for Passive Climate Control in Non-domestic Buildings in Moderate Climates*, PhD thesis, University of Queensland, Australia

Romero, M. A .B. (1988) *Princípios Bioclimáticos para o Desenho Urbano*, Projeto, São Paulo

Schoenauer, N. (2003) *6000 Years of Housing*, W. W. Norton and Company Inc, New York

Szokolay, S. V. (1991) *Climate, Comfort and Energy: Design of Houses for Queensland Climates,* Architectural Science Unit, University of Queensland, St Lucia, Queensland

Van Lengen, J. (1997) *Manual do Arquiteto Descalço,* Instituto Tiba, Rio de Janeiro

Watson, S. and Hyde, R. A. (2000) 'An environmental prototype house: A case study of holistic environmental assessment', in *PLEA 2000*, Cambridge University, Cambridge, pp170–175

Zimmermann, M. (1996) *Bericht EMPA-KWH*, IEA-BCS, Switzerland

Zold, A. and Szokolay, S. V. (1997) *Thermal Insulation*, University of Queensland, Brisbane, Queensland

Chapter 10
Green Technologies, Performance and Integration

Nathan Groenhout, Richard Hyde, Deo Prasad,
Shailja Chandra, Yoshinori Saeki and Lim Chin Haw

INTRODUCTION

The aim of this chapter is to provide an overview of the green technologies that can be used in designing solar sustainable homes and to review the issues that arise.

The first part of the chapter describes the functions and purposes of green technologies, grouped into life systems and material systems. Life systems are the bioclimatic systems that operate within the ecosphere and interact with it. These are mainly active systems that include resource-producing systems, such as solar hot water, and climate control systems, such as air conditioning. Material systems are derived from the lithosphere and concern those materials that have environmentally friendly qualities. There is a strong political imperative to utilize green technologies to improve environmental performance of houses (AGO, 1999a, b; USDOE, 2003).

This information covers a large area, so the scope of this part of the chapter has been limited to providing a framework for evaluating these systems in terms of their potential use for a project.

The second section of the chapter, 'Integration', is concerned with the integration of these technologies with the building form and fabric. Synergies exist at a number of levels:

- conceptual;
- technical;
- spatial; and
- elemental.

Examples of such synergies are given to provide some solution sets for design.

Source: SIRM Bhd

10.1 Prototype Solar House, Malaysia, has been developed to test the application of green technologies

Finally, a number of case studies are presented to demonstrate the performance advantages of integration. They include the following buildings:

- Mobbs House, Sydney, Australia;
- Prototype Solar House, Kuala Lumpur, Malaysia;
- Prasad Residence, Sydney, Australia; and
- Sunny Eco-house, Japan.

LIFE SYSTEMS

Nathan Groenhout and Shailja Chandra

Life systems provide the necessary resources to support human existence in houses, including systems for energy and water supply and for waste disposal. Solar electric and thermal systems aim to harvest solar energy available onsite for use in the home. These are important technological approaches to creating a closed-cycle system for energy onsite, with the advantage that they avoid increased reliance on the electrical grid for energy, particularly at times of peak demand. The technology also relies on renewable energy as a fuel source and therefore avoids pollution by greenhouse gases.

Photovoltaic energy generation

Photovoltaic (PV) cells are made of a semiconductor material, frequently crystalline silicon. In PV cells, electrons are freed by the interaction of sunlight with this semiconductor material, generating a current in the form of direct current (DC) electricity. This DC electricity can then be converted to alternating current (AC) electricity for various end uses through the help of inverters and step-up transformers. Depending upon the area of the solar panels and the energy efficiency of the house, solar panels can produce more electricity than required by a standard household during a typical day. The surplus electricity can then either be fed back into the utility grid, for grid-connected systems, or stored in battery banks for stand-alone systems. In grid-connected systems, electricity may be drawn from the grid to cater for night-time loads or during periods of low insolation. In stand-alone systems, electricity for these purposes is drawn from the battery banks.

Generation of electricity through the use of solar panels is one of the most benign ways of producing energy. Despite this, the proportion of renewable energy (not including hydraulic sources) remains at the disturbingly low figure of 1.7 per cent of the total global electricity generation capacity (*Photovoltaics Bulletin*, 2003). Out of this small share, photovoltaics contribute the highest proportion, greater than thermal solar water heating.

While the field of photovoltaics has seen some remarkable technological enhancements in the past few years, the low demand for PV is still a concern. PV installations in the residential sector can act as a significant stimulant in boosting the demand for PV. To make solar energy suitable for residential applications, a number of companies such as BP, Sanyo and Philips are developing cutting-edge, ready-to-install and attractive forms of solar panels (see the web links at the end of this chapter for more information).

PV basics

A PV system connected to the utility grid generally has the following components:

- PV modules or array, connected to an inverter;
- an inverter, which converts the system's DC electricity to AC;
- batteries to provide energy storage or backup power in case of a power interruption or outage on the grid (this is optional for grid-tied arrays and a requirement for a decentralized or stand-alone system); and

- reversible metering to provide an indication of system performance and also to monitor electricity flow to and from the utility grid.

PV modules produce energy intermittently, according to the time of day and sky conditions; so a decentralized energy generator may need a storage device – batteries are often used for this. An alternative system that is gaining popularity for decentralized power generation is the use of an uninterruptible power supply (UPS), such as a fuel cell.

PV performance factors

Orientation

The orientation and tilt angle of solar panels is instrumental in maximizing total energy output. In Australia and other southern hemisphere locations, building surfaces that have a northerly aspect will be well exposed to solar irradiation. As a rule of thumb, the optimum angle for a solar panel is true north orientation, tilted at the angle of latitude of its location. For example, Sydney's latitude is 34° south, so a panel oriented true north at 34° to the horizontal would achieve maximum solar exposure.

PV panel shading

Good PV power performance requires adequate direct solar irradiation. Sites should be selected that avoid unnecessary shading, both from obstructions on the building itself and from surrounding objects.

Soiling of panels

Soiling of PV panels can reduce power output; but significant losses can be avoided by sensible site selection, and if the panels are suitably inclined they can be self-cleaning when it rains. PV module manufacturers normally provide advice on the minimum angle for self-cleaning.

Value adding of PV systems

House owners have various reasons for installing PV systems. Apart from a decision of conscience to be environmentally responsible, solar energy is also considered by many to be the energy choice for the future (USDOE, 2003), so for many prudent owners it makes sense to invest in a type of energy that provides a more secure supply, independent of the utility grid.

The added values of the use of building integrated photovoltaics (BIPV) include:

- enhancement of the building value by the presence of BIPV elements;
- synergies with other forms of energy, such as heating and cooling, where the BIPV element adds to insulation value or to shading;
- prestige and marketing opportunities;

- architectural merits and aesthetics;
- satisfaction of the consumer's desire for environmentally sound products; and
- satisfaction of the desire for increased energy autonomy (see www.sanyo. com/industrial/solar/, www.bpsolar.com, www.pvdatabase.com).

Domestic solar water heating

The concept of using solar energy to produce hot water is not new; domestic solar water heaters in a variety of forms have been in production for well over a century. These systems range from the bush shower, a simple black canvas bag full of water hanging from a tree, to complex systems that integrate state-of-the-art technologies into rooftop systems.

In Australia and many other parts of the world, the predominant system for domestic solar hot water systems is the flat plate collector coupled with a storage tank. These systems come in a number of configurations, both active and passive. Modern flat plate collectors used for domestic solar water heaters may be categorized into a number of basic types. These are:

- thermosyphon or passive systems, in which circulation of the working fluid is by natural convection (buoyancy driven) within the system (a common example is shown in Figure 10.2a);
- forced circulation or active systems, in which circulation is by a pump (see Figures 10.2c, 10.2d and 10.2e);
- open-loop or direct systems, where the potable water is circulated through the collector directly, either by thermosyphon action or by a pump (see Figures 10.2a and 10.2c); and
- closed-loop or indirect systems, where a separate working fluid, usually a glycol mix, is circulated through the collector, and then a heat exchanger is used to transfer the heat to the potable water for use (again, these may be active or passive systems; see Figures 10.2b, 10.2d and 10.2e)

The most common type of system in use in Australia is the flat plate thermosyphon collector as shown in Figures 10.2a and 10.2b. This type of system has been manufactured in a variety of designs by a number of local companies for over 50 years and Solahart is credited with developing a new style of close-coupled collector and storage tank system (Butti and Perlin, 1980). This design generally consists of one or more absorber plates connected in parallel to a horizontal hot water storage tank with the plates mounted flush on the roof and the tank located above the absorber plates.

The absorber plate consists of a series of riser tubes connected to an upper and lower header tube, operating on the thermosyphon principle. The upper header connects to the inlet of the storage tank and the lower to the outlet. The thermosyphon principle works because of the change in density

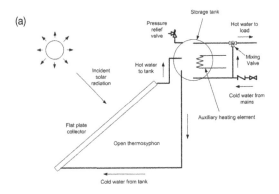

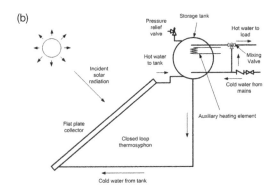

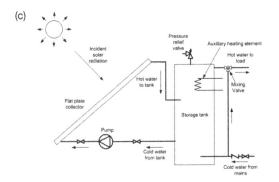

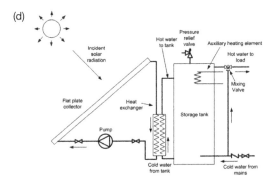

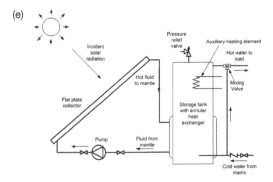

Source: Nathan Groenhout

10.2 Domestic hot water systems: (a) close-coupled system using an open-loop thermosyphon for direct solar water heating with an auxiliary heating element in the storage tank; (b) close-coupled system utilizing a closed-loop thermosyphon with a mantle heat exchanger tank and an auxiliary heating element; (c) separate collector and tank in an open system with forced circulation and an auxiliary backup element in the storage tank; (d) separate closed-loop system with external heat exchanger – system has forced circulation on the collector side and natural circulation on the storage tank side; (e) separate closed-loop system with forced circulation and annular heat exchanger integral to the storage tank

that occurs in fluids when they are differentially heated, so that the warmer fluid, being less dense, rises while the cooler fluid falls. Incident radiation on the absorber plate causes the fluid inside the plate to heat up and flow up the risers into the header and the storage tank. The cooler fluid is forced out of the tank and back down to the lower header.

The system may work as a closed-loop or open-loop system. Modern closed-loop systems typically use a mantle-type heat exchanger such as that shown in Figure 10.2b, which is essentially a narrow annulus between the two shells of the storage tank. A heat transfer fluid such as polypropylene glycol (antifreeze) is used as the working fluid, circulating through the collector and into the narrow annulus, transferring heat to the water inside the inner shell. The use of the heat exchanger makes these collectors particularly suitable for use in cold climates or where water quality is a problem.

Flat plate systems have a number of drawbacks. Stationary flat-plate collectors mounted flush on a building roof tend to have excess output during the summer months when the sun is high; however, during winter months performance drops due to the solar radiation striking the absorber at a more acute angle. Maximum energy collection occurs when the incoming solar radiation is normal to the collector surface. For stationary collectors, it is therefore recommended that the collector be mounted facing due north (in the southern hemisphere), at an angle of inclination equal to the latitude at that location. For most installations this would require mounting the collector on a stand, rather than flush on the roof, to increase the angle of inclination of the collector, since the roof pitch rarely coincides with the latitude. This arrangement is not considered aesthetic and so, in the vast majority of installations, the collector is mounted flush on the roof at the expense of optimal energy collection. Performance is further reduced by the fact that the majority of houses do not have a roof facing due north. The drop in performance occurs when demand tends to be highest and so these systems require gas or electric boosters to guarantee a constant supply of hot water.

Other developing technologies include evacuated tube collectors, double-sided flat plate collectors (as shown in Figures 10.3 and 10.4) and combined heat and power systems. Most of these systems are still expensive, require further improvements in quality or have not yet been commercialized.

Climate control systems

Mechanical ventilation

The aim of mechanical ventilation is to improve the quality of the indoor environment by providing fresh air, and removing stale air, odours and moisture from occupied spaces. In typical building applications, fans for mechanical ventilation are heavy consumers of energy, and if fossil fuels are used for supply this can lead to high greenhouse gas emissions. Typically, mechanical ventilation is used for kitchen and bathroom exhausts; however, it may also be used for the whole house.

In the kitchen, the use of exhaust fans and filters can reduce the levels of moisture and odours, as well as helping to prevent the build-up of grime, mould and bacteria on surfaces. In the bathroom, the primary aim is to remove moisture and prevent the growth of mould and bacteria on the walls and ceiling. The growth of mould and fungi in damp bathrooms is linked with

Source: Adsten (2002)

10.3 MaReCo design for a roof aligned in the east–west direction

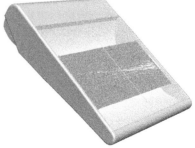

Source: Groenhout (2004)

10.4 Image showing the proposed 2000B collector as a packaged unit increased respiratory problems in children and adults, while excess moisture can lead to structural damage of timber and timber products.

Due to the high energy consumption of mechanical ventilation fans, the use of timers, occupant sensors and renewable energy sources such as photovoltaic panels should be considered.

For more general applications, mechanical ventilation of roof and/or sub-floor spaces can relieve heat build-up and prevent excess heat gain to occupied spaces, while also preventing the build-up of moisture. This is particularly important in humid locations or in regions where termites are particularly active.

Mechanical ventilation may also be used to provide ventilation for an entire house through the use of a centrally located ceiling exhaust fan. In addition to providing ventilation, these systems provide evaporative cooling since moisture evaporates from a person's skin as the air moves across it, providing a cooling effect. The fan draws air in through open windows to a central location, such as a hallway, and then exhausts it into the roof space, typically providing over 30 air changes an hour throughout the house. A damper in front of the fan closes when the fan is not operating to prevent warm air from being lost in winter from the occupied regions – for example, to the roof space. This type of system requires high flow rates, which may result in unacceptably high air speeds in some spaces; they also need adequate gaps below the doors to allow airflow when they are closed, and acoustic issues need to be considered as well.

The specific limits of mechanical ventilation include the following:

- power: up to 3kW;
- climate: all;
- air speed: 1.5 to 3m per second;
- maximum temperature: provide physiological cooling through evaporation of moisture from the body (physiological effects become negligible if air temperature is greater than 30°C); and
- maximum relative humidity up to 100 per cent (the fans cannot reduce external humidity, but aid the body's cooling by increasing the air speed around the body and, hence, improving evaporation).

Fans

Fans are a form of mechanical ventilation and are typically used for providing air movement rather than for direct cooling, although air movement can also provide indirect cooling through evaporative processes. Fans are the most cost-effective form of cooling and typically cost less than a few cents a day to run.

Fans can also be effectively used in winter for air mixing. For instance, in houses with interconnected floors, the warm air tends to accumulate on the upper floors and cold air on the lower floors. The use of a fan in the ceiling of the connecting void or stairwell between the floors can be used to mix and share the warm air between the two floors, providing better thermal comfort.

The specific limits of fans are:

- power: up to 0.6kW;
- climate: all;
- air speed: 1.5 to 3m per second;
- maximum temperature: can provide up to 4° to 5° of cooling through evaporation; fans become ineffective above 27 to 30°C; and
- maximum relative humidity: up to 100 per cent.

Evaporative coolers

Evaporative cooler systems, also known as 'swamp coolers' or 'swamp boxes', operate on the physical phenomenon that occurs as water evaporates in air, causing a reduction in the air temperature. Evaporative cooling is one way of converting hot and dry air into a cool breeze, using the process of evaporating water. Evaporative cooling is one of the oldest methods used by humans to make living spaces more comfortable in hot conditions.

Evaporative coolers utilize the natural process of water evaporation along with an air-moving system to create effective cooling. Fresh outside air is pulled through wetted pads, which also act as filters and cool the air through water evaporation. A small pump recirculates the water from a sump to continuously wet the pads. The pads used in the evaporative cooling system require regular replacement, and the replacement period depends upon hours of usage and outdoor air conditions. Another form of pad is used in rigid-sheet pad coolers, which use corrugated sheets that allow air to flow between them at higher velocities. A fan blower then circulates the cool air throughout the conditioned space.

There are two types of coolers: direct and indirect systems. Direct types commonly use a downdraught ventilation system to bring air from the roof, through a gauze filter that is kept moist by sprayed water, delivering it into the room below. Indirect types link the evaporative process to an air-conditioning system. A combination of refrigeration and/or evaporative cooling is built into the equipment.

There are two basic types of evaporative coolers, known as single stage or double stage, which utilize direct or indirect cooling or a combination of both:

- Single-stage coolers: there are two types of single-stage evaporative coolers. In direct coolers, water evaporates directly into the air, reducing its dry-bulb temperature and increasing its humidity; in indirect evaporative coolers, heat is transferred between two airstreams via a heat exchanger without contacting each other.
- Two-stage or indirect/direct evaporative coolers: these coolers use an air-to-air heat exchanger, which reduces the temperature of the incoming air

without raising the relative humidity, by using indirect evaporative cooling. The incoming air then flows through a direct evaporation stage, further reducing its temperature and increasing its humidity.

Since evaporative coolers use water to reduce air temperature and increase humidity, they are ideal in environments with abundant water and low humidity, such as in arid climates.

Evaporation creates cooling through a phase change process; when water turns to vapour, heat is absorbed, lowering the temperature of the air. Thus, in an arid climate the relative humidity in the air can be around 40 per cent and the air temperature 35°C. Adding moisture to the air increases its relative humidity and reduces temperature. This can create moist air at a more comfortable temperature in dry conditions.

Evaporative coolers can provide cooling, using less energy than refrigerated air-conditioning systems because of their simple mechanism; but in recent years the consumption of water has become problematic. These systems can use in the order of 100 litres per day and represent 2.5 to 66 per cent of summer water use (*Journal of the American Water Works Association*, 1998).

The specific limits of evaporative coolers are:

- power: < 1kW;
- climate: warm temperate and hot arid;
- air speed: 1.5 to 3m per second;
- maximum temperature: 1 to 3 degrees of cooling, < 35°C; and
- maximum relative humidity: < 30 per cent at maximum temperature.

Air-conditioning/heat pump systems

Air-conditioning systems represent an enormous burden on energy supply utilities due to their high energy consumption and their effect on peak demand. The boom in poorly designed mass-produced housing, with lightweight construction and large areas of glazing, along with changing expectations of thermal comfort by occupants, has led to a jump in the number of air-conditioning systems installed. Systems are typically large reverse-cycle ducted systems designed to cool large spaces.

Domestic air-conditioning systems come in a range of configurations. The simplest are the wall-mounted 'window rattler' systems, which can typically provide cooling for spaces up to about 50 square metres. Small units can be plugged into a 15 amp wall outlet, while larger units are hard-wired like a kitchen stove.

Another simple system is the portable split system, with a separate condenser and evaporator. The fan unit is located inside the house and the heat rejection coil is located outside; flexible hosing connects the two sections. The

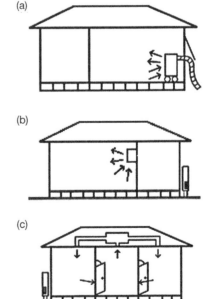

(a)

(b)

(c)

Source: adapted from Sustainable Victoria (2007)

10.5 Air-conditioning system: (a) portable split system; (b) split system; (c) ducted system

fan coil unit is often mounted on wheels and can be moved around the room as required. Units typically have a single-phase motor running off a standard 10 amp wall outlet and can cool up to around 30 square metres.

Split systems are similar to the portable system except that the internal fan coil unit is mounted on the wall, floor or ceiling. These systems can achieve higher efficiencies than simple wall-mounted types and can provide cooling for up to 50 to 60 square metres. Single-split systems have a single fan coil unit per condenser, while multi-split systems have multiple fan coil units per condenser.

Ducted systems are similar to split systems in that the fan coil unit is separate from the heat rejection equipment; however, the air is distributed throughout the space and other rooms through ducting. This allows a greater number of spaces to be cooled by the same system, as opposed to a conventional single-split system, which can only cool the space in which the fan coil unit is installed.

Ducted systems are typically larger and more expensive to install due to the additional cost of ducting. Ducted systems are also generally slightly less efficient than conventional split systems. Ducted systems allow zone control of houses provided that sufficient controls and dampers are installed.

The use of inverters, whereby adjusting the power supply frequency varies the compressor motor speed, can provide energy savings and improve performance. The use of inverters increases the installed cost of the system but saves on operation costs.

Specific limits of air-conditioning systems are:

- power consumption < 1kW to 10kW;
- climate: warm and humid;
- maximum temperature: up to 43°C;
- minimum relative humidity: 70 per cent daily amplitude;
- maximum relative humidity: 100 per cent daily amplitude.

The main limit to these systems is normally set by the sink conditions, the temperature or the humidity in the external environment; as the temperature and humidity differentials across the systems widen, the efficiency decreases rapidly. Therefore, cooling to a dry-bulb internal temperature of 24°C with 60 per cent humidity while the external temperature is 45°C with 100 per cent humidity requires greater capacity and energy use than would be the case with a narrower differential.

This points to the importance of control systems, such as thermostats, that can adjust the running of the system to suit external and internal conditions. Using demand control – for example, by raising an AC unit's set point to 27°C in high outdoor temperature conditions – is likely to deliver significant savings in energy and cost while still providing sufficient comfort. The technology of control systems in many products still requires development and lacks the sensitivity to match these kinds of requirements; therefore, there is a large reliance on users, rather than mechanisms, to manage energy efficiency.

Control technologies

Control technologies can play a significant role in energy management for sustainable houses. Automated buildings reduce energy consumption by managing lighting and ventilation to suit the needs of the occupants efficiently. Inappropriate selection or use of control technologies can, however, lead to increased energy use. Control systems can be broadly categorized into demand control, regulating the need for active systems; and supply control, regulating the operation and efficiency of active systems.

Demand-control technologies

These technologies are designed to provide demand-side management of energy use by controlling energy efficiency and peak energy consumption to avoid the need for additional energy production by suppliers. Some technologies that can be incorporated within demand-side energy management include:

- the use of stand-alone or grid-connected renewable energy systems that reduce the use of non-renewable fossil fuel energy during peak energy usage periods;
- thermal storage and thermal mass to shift the period of peak energy use to times when off-peak energy tariffs may be charged;
- use of variable speed drives or inverters on air-conditioning systems to reduce energy consumption when reduced cooling or heating is required; and
- the use of high efficiency electric motors in appliances such as washing machines, dishwashers and air conditioners.

Supply-control technologies

These types of technologies control appliances and equipment in a more efficient manner to reduce energy consumption. Such technologies include:

- Occupant movement sensors: these are infrared sensors that detect motion and body heat and are used to switch lights and ventilation systems on automatically when people enter a space, and to switch them off when no motion is detected for a set period of time. Occupant sensors typically require a certain level of movement to occur continuously within a space, so these systems are commonly used for outdoor security lighting. They can also be used in hallways and indoor and outdoor transit spaces that are not occupied for long periods.
- CO_2 sensors: these measure the amount of carbon dioxide within a space and control ventilation systems to ensure that acceptable fresh air levels are

(a)

Source: Pt. Wisata Bahagia Indonesia

10.6 Mixed-mode system used in housing units in Indonesia: (a) windows have sensors, which shut off the air-conditioning system if the building is operated in passive mode; a combination of fans, air conditioning and passive systems, such as cross-ventilation, can be optimized using this control system to achieve comfort and significant increases in energy efficiency; (b) natural ventilation can be used, but this needs to override the control system for the air conditioning so that the benefits of energy savings can be achieved

(b)

maintained. When a space is only partly occupied, the CO_2 level will typically be low, so the ventilation system can reduce airflow and cooling or heating.

- Programmable timers and thermostat controls allow more flexible control of systems – for example, an air-conditioning system may be set at 24°C and programmed to turn on half an hour or an hour before the occupants return home from work, allowing the room to come down to the desired temperature slowly and, hence, more energy efficiently, rather than switching on the air-conditioning system when the occupant enters the room and trying to bring the air temperature down quickly by setting the thermostat to 18 or 19°C, well below the final desired temperature. These systems can also be 'set back' to provide less cooling or less airflow during the night.

- Thermostats can be used for controlling energy needs by mounting them away from windows where the temperature variations are wider, and also by setting them a degree or so higher than common practice – for instance, between 25 and 27°C. Every 1°C lower can increase running costs by up to 15 per cent (see www.sustainability.vic.gov.au/).
- Mixed-mode controls can be used to allow buildings to operate with either passive or active systems. A switching system allows users to choose which mode is appropriate for the climate conditions inside and outside the house.

Water supply and recycling

Richard Hyde

Supply of water is central to life and traditionally the collection, storage, processing and distribution of water has been centralized within communities as a 'mains' service. Two principal systems are used: the potable water supply is piped into houses, and wastewater is taken from the site through the sewer. Rainwater is normally not collected but taken away separately via the storm water drains.

In countries such as Australia, which have a growing demand for water and a variable natural supply, it is argued that there will be benefits from a hybrid system with both centralized and decentralized components. In theory, collecting, processing and treating of wastewater makes sense. Research from the Healthy Home Project (Hyde, 1999) has shown that a hybrid system where black water is sent to the sewer, rainwater is collected, grey water is reused and water efficient fittings are installed can save 60 to 70 per cent of water consumption. If this technology were to be used extensively in housing on the Gold Coast in Australia, where this project is located, it would defer construction of dam infrastructure, giving a saving of Aus$85 million, and retain water in the local ecosystems. The current plans for this region are to implement desalination systems to provide water; this would require energy derived from fossil fuels, imposing a large environmental penalty. Such a desalination strategy is simply transferring the problem from one area of environmental impact to another.

Decentralized systems, which reuse rainwater and treat wastewater onsite, are proven technologies and are worth considering as an alternative viable strategy.

Water technologies and reuse

The current use of water technologies is based on a premise of unlimited supply and large centralized processing systems to maintain the health and welfare of consumers. Water is collected in centralized catchment areas, is processed, and then is distributed through large pipes to the houses where it is used for personal consumption and transportation of solid waste from the site. The disposal of solid waste from the site again requires large pipes and needs central treatment so that the water can be safely returned to the natural ecosystem. During recent years, systems have been proposed and built that provide alternatives to this current approach; these are called 'green water systems'.

To understand the rationale behind these green water systems it is necessary to understand shifts in thinking with regard to urban development. If we take a more sustainable approach, we begin to question whether 'we want health, security, freedom and pleasure for ourselves and our families. For distant generations, we wish the same, but not at any great personal cost' (Edwards, 1999). So, we need to consider the selection of technologies on the basis of different criteria, not simply our current personal needs, but also our future needs and the needs of the environment. The concept of *sustainable cities* engenders a set of criteria for green technologies; a city can become more sustainable if it reduces its input of resources (land, energy, water and materials) and its waste outputs (gases, liquids and solid wastes), while simultaneously increasing its liveability (health, employment, community activities, leisure activities, public spaces, land use and pedestrian accessibility) (Newman and Kenworthy, 1999).

In this respect, green water systems are, first, about land use; second, about conserving resources; and, third, about generation of resources through waste recycling. A site can therefore be more intensively used than at present, helping to reduce the footprint.

Green water systems are about collecting, storing and processing water onsite. There are different types of water:

- rainwater – precipitation that falls on the site;
- grey water – water used on the site that has a low solid content (that is, from laundries and showers); and
- black water – water used on the site for kitchens and toilet flushing that has a high solids content and a high health risk.

There are very good reasons why systems exist in their current form – four major issues concern green water systems:

1 *Aesthetics*: how comfortable are we about drinking water from the skies of our cities or using water that has passed through our bodies?
2 *Health*: will the system create more environmental impacts (health problems) than it eliminates? The risks from failure of the system are high since it could spread disease. In some countries there is legislation that forbids the installation of these systems in urban areas for this reason.

3   *Cost*: what is the additional capital cost of installing these systems? With the present low cost of water to the consumer, owners are unlikely to recoup an immediate return on investment. The advantage of these systems is that they gives owners a degree of autonomy, as well as providing a reduction in the use of the resource because of inherent supply restrictions since the owners or users can only utilize what is collected, unlike the mains system where use is unlimited and control of demand is by price or legislation.

4   Maintenance: is additional maintenance needed to regulate the performance of the grey water systems? It is wise to analyse the output from the system periodically to check bacterial counts. Solid waste also needs to be removed at regular intervals. Rainwater systems are less problematic; maintenance needed is similar to that for the normal system, although tank sludge should be removed regularly. The systems must be designed, and checked, to prevent insects (such as mosquitoes) from breeding because of diseases related to these carriers.

An inherent problem is that the current economic paradigms applied to housing do not take into account the externalities of the system; the capital cost, health and maintenance responsibilities work against the owners' immediate interests, while the benefits are largely experienced at a city scale because capital expenditure on large infrastructure projects, such as dams and pipe work, can be deferred or limited. For these reasons, some countries are beginning to operate rebate schemes or give subsidies to owners who install green water systems.

Onsite water collection and recycling

The main aims are to collect the precipitation falling on the site and to recycle both black and grey water-borne waste onsite. Green water systems usually comprise the following elements, which fall into three groups of technology. First, there is the *plant* and equipment used for collecting, storing and processing the water; second, there are *distribution* elements (pipes); and, finally, there are the *diffusion* elements (taps, showers and irrigation components).

Plant consists of the following elements:

- rainwater collector, usually a roofing system that is benign, having no free chemicals to pollute the water;
- *rainwater tank* for storage;
- *diverter* to minimize pollution of the collected water; the initial runoff from the roof is diverted from going directly into the rainwater tank;
- *micron filter* to filter the water for drinking;
- *bio-system tank* to process the grey and/or black water by bacterial action on the solid waste;

- *wastewater utilization equipment*, which uses processed grey and black water for irrigation or other purposes (processed grey water can be collected in a tank to be reused for toilet flushing).

Distribution is usually through *pipes*; the rainwater is usually gravity fed from the roof, via the diverter, to the rainwater tank, through 100mm pipes. From the tank, an electric pump is used to distribute water through the house; these pipes are about 35mm in diameter. Mains water can be supplied to the tank as a backup (a back-flow valve is needed to avoid contamination). The bio-water system uses a similar system of pipes, 100mm for the gravity system and 35mm for grey water recycling. There is generally a secondary piping system to transfer the grey water from the tank to the toilets for flushing purposes.

Diffusion involves items such as *taps* and *showerheads*; the use of *low-flow fittings* is recommended to cut the demand for water use. Drip irrigation systems can be fed from the processed black water tank by gravity or pump, depending upon the configuration of the site.

Importance, special requirements and benefits for solar sustainable housing

The design issues have specific requirements:

- Electric pumps are an important energy drain on the overall energy consumption of the house. These can use up to 1kWh per day and increase energy use and associated 'greenhouse' emissions if fossil fuels are used to generate energy.
- The tank systems are large (3.6m to 4m in diameter); hence, their location must be considered at the sketch design phase. Tanks can be metal, concrete or polypropylene.
- If site conditions permit, tanks can be buried.

Overview of variations/typology and selection criteria

Variations in this main concept depend upon the owners' need, the site and constraints and opportunities provided by the climate; they are wide ranging and require an inventive approach by designers. Several examples are provided.

Type 1: rainwater collection and grey water recycling only. In urban areas where site space is critical, black water can be disposed of to the sewers; black water is a minor fraction of the wastewater system (about 100 litres per day), so the cost of onsite treatment for such a small amount can often not be justified. Grey water from showers, baths and washing machines comprises a larger volume and poses a simpler problem for engineers due to low quantities of solids; hence, it is simpler and more cost effective to treat just this part of the waste.

Type 2: innovative diffusion systems. These can be used to promote and enhance the biodiversity of the site or to minimize impact. In arid sites the nutrient load from the diffusion system can disturb the balance of the ecosystem. In such cases, wastewater can be evaporated, using an exposed bed system. In other climates, where microclimate control is needed and where there is a pronounced dry and wet season, the use of diffusion to increase the biodiversity through irrigation improves the microclimate.

Type 3: innovative water storage systems other than tanks. Depending upon the substructure of the soil and rock formations, freshwater lenses can be formed in the site substructure to create a water storage aquifer. A further potential in large apartment complexes is *double functioning* of active fire protection and water storage systems; the large water tanks provided onsite for fire fighting and sprinkler systems can also be used for water storage.

Example of a cost-effective system

The following example illustrates the advantages and limitations of a low cost system, located on the Gold Coast in Queensland, Australia. The project is being monitored as part of the Healthy Home Project research programme. The strategies and elements used are:

1   Water production:
    *   collection, storage and treatment of rainwater for household use; and
    *   treatment and storage of grey water for future garden use.
2   Water consumption:
    *   water-efficient appliances, AAA shower heads and toilets with dual-flush cisterns;
    *   irrigation monitoring devices; and
    *   selection of plant species suited to the area.

Elements used

Elements used include the following:

*   *Rainwater tanks*: rainwater is collected from the roof and stored in tanks. The main tank is a 22,000 litre tank buried beneath the house. Water quality is protected by a 20 micron filter and first-flush devices, which divert the first 120 litres (equivalent to a fall of 8mm) of each downpour into the local storm-water system. The rainwater supply is supplemented by the town supply; the local council requires that there is no connection between the town supply and the rainwater tank, and this is achieved by allowing an air break between the tap from the town supply and the rainwater tank. Following water quality tests, an ultraviolet (UV) pasteurization system was installed to improve the quality of the rainwater for drinking.

- *Wastewater*: grey water (water from basins, laundry and showers) is collected and treated onsite in a 6000 litre tank, which settles out solids and sand filters the grey water before releasing it for reuse. The water would normally be used for the garden; but since this is a sandy site, the local council requires stringent testing before it will permit the grey water to be reused onsite.
- *Water for irrigation*: native plants were selected, locally found and, hence, suited to the local environment and requiring minimal watering. The garden was designed according to permaculture principles (Mollison, 1988) so that the garden would eventually be self-sustaining. Two devices were installed to help monitor and minimize water use in the garden. The first, a tensiometer, measures the dryness of the soil and indicates when watering is required. The second is a water sensor, developed by the Commonwealth Scientific and Industrial Research Organisation (CSIRO), which sends a signal to the household when water has reached the level of the roots of the plants.

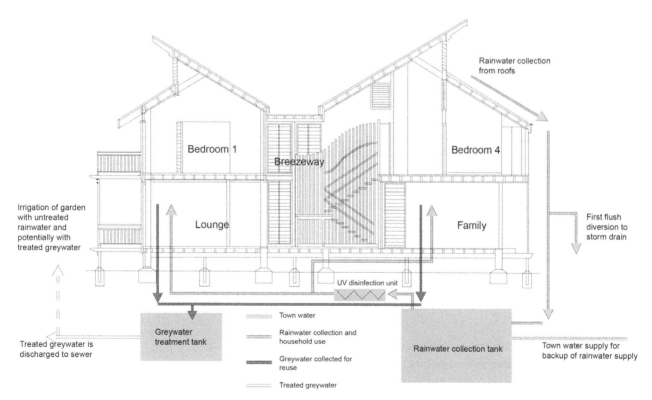

Source: Richard Hyde

10.7 Schematic diagram of the water system used in the Healthy Home Project, Gold Coast, Australia. Rainwater is harvested from the roof and stored in a 22,000 litre tank located under the house. In the dry season, town water is used to top up the storage system; in the wet season, rainwater is used and sterilized using a UV filter. Grey water can be harvested, treated and used for sub-surface garden irrigation to reduce health risks

Construction and details

The construction system involved a number of elements, designed to reduce the impact of construction on the environment. These include:

- *Tank installation*: the tanks were buried in the site; the sandy consistency of the soil facilitated this.
- *Raised ground floor*: a raised ground floor allows easy access to the tanks and the distribution system, allowing future flexibility and promoting easy maintenance.
- Polypropylene piping is used for the pipes to reduce the PVC content of the piping system and to avoid metal-based pipes, which have high extraction impacts.

Water-sensitive urban design (WSUD)

Development of decentralized systems for water is being advanced by initiatives in water-sensitive urban design (WSUD). Principles for WSUD can be developed for local areas, as in the following example by Melbourne Water, Australia (see www.melbournewater.com.au):

- Protect natural systems – protect and enhance natural water systems within urban developments.
- Integrate storm water treatment within the landscape – use storm water in the landscape by incorporating multiple-use corridors that maximize the visual and recreational amenity of developments.
- Protect water quality – improve the quality of water draining from urban developments into the receiving environment.
- Reduce runoff and peak flows – reduce peak flows from urban development by using local detention measures and minimizing impervious areas.
- Add value while minimizing development costs – minimize the drainage infrastructure cost of the development.

MATERIAL SYSTEMS

Richard Hyde

The uses of green materials and technologies are seen as a significant part of the approach to solar sustainable houses from a number of perspectives. First, green materials and technologies reduce a development's impacts on the planet's ecosystems. Second, there are numerous benefits for homeowners in using these materials and technologies. Current perceptions are that although the economic benefits of using these systems are long term, there is still interest in the holistic notion of value added by employing these green systems since they are perceived as creating a sense of harmony and well-being in the home. This comes from the use of materials and systems that do not create any of the following:

- *'Sick' home*: the materials and systems used do not contain harmful chemicals and substances potentially dangerous to health. There is

(a)

Conventional

Water sensitive

Source: Melbourne Water (2007)

10.8 (a) Water-sensitive urban design (WSUD) in comparison to a traditional subdivision

New footpath alignment allows for integrated storm water management and responds to natural features

Variation in width of the reserve facilitates integrated design of storm water management

Traditional setback creates unusable space which reduces the function and aesthetics of the street

Conventional

Water sensitive

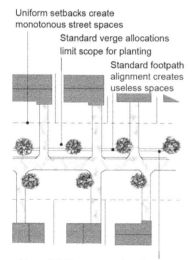

Uniform setbacks create monotonous street spaces

Standard verge allocations limit scope for planting

Standard footpath alignment creates useless spaces

Unpredictable crossover locations limit scope for retention of existing vegetation and new planting

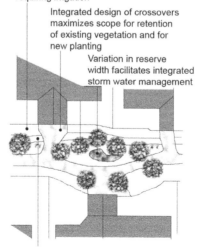

Narrow road reserve reduces area requiring irrigation

Integrated design of crossovers maximizes scope for retention of existing vegetation and for new planting

Variation in reserve width facilitates integrated storm water management

Footpath alignment response to natural feature and storm water management to create spaces that are easy to maintain and efficient to irrigate

(b)

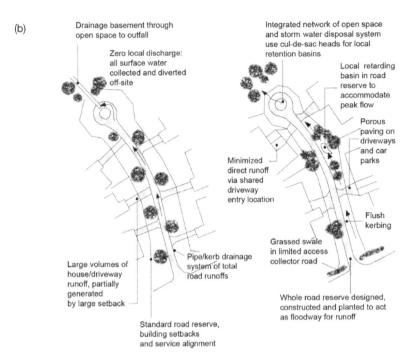

Drainage basement through open space to outfall

Zero local discharge: all surface water collected and diverted off-site

Integrated network of open space and storm water disposal system use cul-de-sac heads for local retention basins

Local retarding basin in road reserve to accommodate peak flow

Porous paving on driveways and car parks

Minimized direct runoff via shared driveway entry location

Flush kerbing

Grassed swale in limited access collector road

Large volumes of house/driveway runoff, partially generated by large setback

Pipe/kerb drainage system of total road runoffs

Whole road reserve designed, constructed and planted to act as floodway for runoff

Standard road reserve, building setbacks and service alignment

Source: Melbourne Water (2007)

10.8 (b) WSUD street layout compared to a traditional street layout. A water sensitive streetscape integrates the road layout, vehicular and pedestrian requirements with storm water management needs. Design measures include reduced frontages, zero lot lines, local detention of storm water in road reserves

increasing interest in returning to natural materials, where the impacts are more predictable, as opposed to new industrialized materials.

- *'Polluting' home*: materials and systems are favoured which reduce pollution and protect the air, water and earth.
- *'Wasteful' home*: the materials and systems used cannot only reduce the quantity of waste, but also retain and process it onsite. Control of air quality, processing of solid wastes and waste of energy and water can all be affected by the choice of materials and systems.

The careful selection of materials and systems in relation to their effects on the planet's ecology is of primary concern, although there is also increasing interest in the way in which the supply chain for house building can be more broadly managed to address construction impacts.

Material typologies

Material typologies can assist with making environmental design decisions on the use of materials. These typologies are fairly crude as yet, representing only the tip of the iceberg in terms of the complexity of performance impacts associated with materials. Three types of materials can be identified:

- natural;
- recycled and reused; and
- industrially manufactured.

Natural materials

Natural materials, such as timber, stone and earth, are the main construction materials used in pre-industrialized societies and they remain the cornerstone of sustainable building. The scale and demands of modern construction has seen them fall from favour as primary sources for the building of homes; natural materials have had particular structural and maintenance liabilities and have been relegated to specialized uses, such as for features and internal finishes.

Yet, during recent years, the limitations of many of these materials have been addressed. For example, the earth construction called *pisé* has been improved through better quality control of the base materials, such as clay, and through additions of small quantities of other materials, such as cement, not found in the traditional material. This has transformed the material for modern conditions, while still retaining its intrinsic qualities.

Recycled and reused materials

Recycled and reused materials come from many sources and provide benefits over first-use materials coming directly from a factory.

Significant benefits of recycled materials come from savings in the amount of processing energy and water used to create them. Recycled and reused materials only have a small component of processing energy and water other than that required to recycle or prepare them; for instance, an aluminium window, taken from one building and used in another, produces none of the impacts attached to its first life.

One approach to reducing environmental impacts is to examine the elements of the building that create the most impact. Studies have shown that the building structure is a significant element, second only to the façade of the building. A common strategy is to use recycled steel sections to form the main frame and the support structure for the cladding, meaning that the engineer works from what is available rather than using the optimum structural sections.

Industrialized materials

Research into the industrial processes by which building materials are made has identified a number of areas where environmental impacts are created. These are:

- raw material extraction impacts;
- raw material availability;
- environmental impacts during production;
- embodied energy and water efficiency;

- lifespan;
- maintenance impacts during use;
- ability to be recycled;
- potential for reuse; and
- effects on human health.

These impacts can be used to devise criteria to help rate materials and systems in buildings. There is increasing concern that materials should be assessed according to their environmental performance; this is a specialized area of research, called life-cycle assessment, not normally carried out by designers. Industry groups are involved in establishing life-cycle data for their products and systems; it is expensive and time consuming. The complexity of the process can be seen in the early work of Lawson that identified the key components in the process (Lawson, 1996). Despite a lack of field data, existing information has been used to create eco-profiles of construction materials.

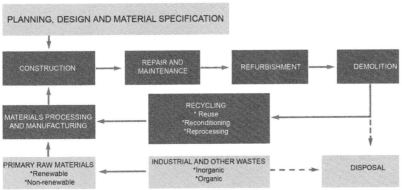

Source: Bill Lawson

10.9 Life-cycle model of materials showing the complexity of the manufacturing process: quantitative data from this model is used to make qualitative judgements of the relative performance of materials in terms of environmental impacts

Eco-profiling of materials

Early work on classifying materials centred on a single criterion for assessment; embodied energy, the energy used in production, transportation and installation, was seen as the major factor in the impact of materials because most industrialized materials are produced using fossil fuels and, hence, are linked to greenhouse emissions. If we classify materials solely on this basis then most commonly used materials in buildings are problematic – for example, aluminium windows with solar glass, where both the glass and the frames require a large amount of energy to produce, compared with timber windows and clear glass. However, professionals in the industry have pointed out that the aluminium and solar glass system has better maintenance performance and thermal efficiency.

Table 10.1 Materials ratings

Element	Specification	Raw material availability	Minimal environmental impact	Embodied-energy efficient	System lifespan	Freedom from maintenance	System recyclability	System potential for reuse	Effects on human health
Substructure	Concrete with recycled aggregates and fly ash; pad footings to support frame	3	3	4	5	5	5	4	5
Primary structure	Glue laminated timber from renewable sources	4	5	4	3	3	4	4	4
Roof	Timber roof purlins, insulation, bracing and metal roofing deck	4	5	4	3	3	4	4.5	5
Ground floor	Recycled timber suspended floor with natural sealer	3	3	4	3	3	4	4	5
Staircases	Recycled timber with natural sealer	5	5	5	3	3	4	4	5
External walls	75mm timber stud frame, insulated walls with fibre cement sheeting and plasterboard internal cladding with breathe-easy paint (zero VOC emissions)	4	4	4	5	5	3	4	5
Internal walls	75mm timber stud frame, insulated walls with fibre cement sheeting and plasterboard internal cladding with breathe-easy paint (zero VOC emissions)	4	5	4	3	3	3	4	5
Windows	Timber frame with single-glazed windowpanes, body tinted, 6mm thick	4	4	4	3	3	3	4	5
Doors	Timber solid core doors in timber frames with breathe-easy paint (zero VOC emissions)	3	3	5	3	3	3	4	5
Floor finishes	Recycled timber with natural sealer	4	4	3	4	3	3	4	5
Ceiling finish	Plaster board finish with breathe-easy paint (zero VOC emissions)	4	4	3	3	3	4	3	5
Total		3.7	4.0	3.9	3.5	3.4	3.6	4.0	4.9

Source: Watson and Hyde (2000)

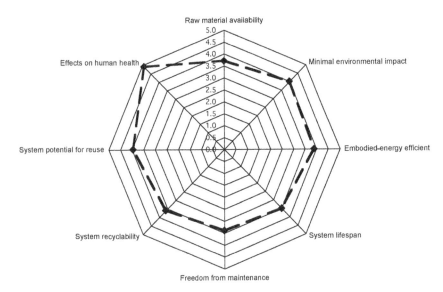

Source: Richard Hyde

10.10 Prosser House: Green materials rating (output from Table 10.1). The rating shows the aim of creating a healthy internal environmental and low embodied materials. For example, the concrete systems used in the pad footings use fly ash and recycled aggregates to incorporate recycled materials within the material to reduce embodied energy

As a result, there is a compelling argument for using a multi-criteria system for assessing materials that includes the main dimensions of both the physical and human systems. Four main criteria can be used:

1   environmental performance;
2   life-cycle cost;
3   maintenance and durability; and
4   effects on human health and well-being.

The integration of these dimensions within nearly all assessment systems is under way; the systems found to be most useful, to date, are those involving material rating systems and green specifications.

Material rating systems

A profile of the building can be created; if a rating of the materials and systems used in each element of the building is carried out, a fair representation of the strengths and weaknesses of the selections made can be seen (see Table 10.1). A spider graph is used to aggregate the results, as shown in Figure 10.10. This is a valuable tool for assessing the whole project.

Product assessment and green specification guides

Green materials rating systems are broad guides to the environmental performance of the materials and systems used. Attention is now turning to more product-specific assessment to form the basis of green specification systems for housing. One problem with the present approach is validating the environmental performance of the products and systems. Five aspects of quality data have been identified:

1   eco-labels (Australian and international);
2   life-cycle assessment;
3   independent verification;
4   manufacturer's declaration (questionnaire certified by a director of the company under trade practices legislation);
5   expert assessment (EcoSpecifier, 2003).

The cost of assessing products with regard to these data sources is one of the barriers to advancing green specification. Despite the lack of data in the materials area for making effective product assessments, building design professionals and builders are moving forward through a number of strategies; this involves rethinking both the design and procurement of buildings, particularly the supply chain.

Greening the supply chain

Greening the supply chain amounts to examining the way in which buildings are procured, and the materials systems supplied and deployed on the building site. This involves an investigation of the construction process, with a view to reducing the environmental impact of materials and systems, but also to observing the integration of building methods and construction practices onsite.

Case studies of the holistic performance of environmental materials

There is an increasing need to undertake design research into the holistic performance of materials used in buildings, involving both the upstream and downstream impacts. Examples of this can be seen in the Beddington Zero project in the UK, where each item of material is checked for its environmental performance, life-cycle cost, maintenance and durability, and its effects on human health and well-being (Lazarus, 2002).

Sustainable construction

There are five main criteria for sustainability:

1   minimal environmental impact;
2   resource efficiency;
3   personnel efficiency;
4   duration of activity; and
5   environmental protection.

These can be related to nine elements of construction:

1   site works;
2   excavation;
3   foundations;
4   wet trades;
5   dry trades;
6   external finishes;
7   internal finishes;
8   services; and
9   external works.

Involving the builder in design and construction decisions is an important prerequisite for sustainable construction. In some cases, an environmental management plan for the construction of the development can be implemented. An innovative way of managing this is to integrate green principles within the scope of the design so that green building issues are considered in the early design stage (Watson and Hyde, 2000).

INTEGRATION

Richard Hyde

The aim of this part of the chapter is to examine the issues and opportunities for integrating active systems provided by the elements of the house. As solar sustainable houses have become more technically complex, the issue of how the technologies in the homes become integrated is a major consideration. Integration can be thought of with respect to four levels:

1   *Technical*: simply applying active systems to serve a single service function with little thought of integration.
2   *Spatial*: simply providing space for active systems without fully understanding the functions of the systems or the opportunities within the building's spaces, such as attics and sub-floors to accommodate technical systems.
3   *Elemental*: the meshing of systems within elements to provide multiple functions.
4   *Conceptual*: issues of integration underlie all the thinking and concepts of the design. The designers are fully engaged with both a technical understanding of the systems and their design impact on the overall project so that synergies between active and passive systems can be realized.

There are a number of elements of a home that provide opportunities for integrating green technologies.

Examples of integration

Roof and attics

Conceptual: the roofs of houses in warm and hot climates are principally defensive elements against heat and rain intrusion; however, their functions have been adapted to include not only protection, but also collection of water and heat. For the roof to act as a primary element for harvesting the natural available solar radiation and water supply, its configuration and form must be adapted to better accommodate these new functions.

Current thinking on the roof's *technical integration* is that it provides an opportunity to use PV, solar hot water and rainwater collection systems. PV cells are very easy to install on a range of roof types and protect the roof area that they cover. The PV array can be mounted above and parallel to the roof surface with a standoff of several inches for cooling purposes. Sometimes, in the case of a flat roof, a separate structure can be mounted to give optimum tilt; but very often these systems look unsightly.

Four main design criteria apply:

1 The orientation, shading and angle of the roof for optimal yield: this depends upon the latitude, but is typically at 35°, facing the equator.
2 Ventilation is needed for PV cells since overheating can reduce performance.
3 The roof needs to be adequate structurally to take the additional load of the systems; with PV systems that are integrated with the roof, it is critical that strong materials such as steel are used to avoid shifting of the systems.
4 Self-cleansing of the surface is needed to maintain the yield of the systems; an angle of inclination of at least 5° is needed so that runoff will remove dirt from the face of the systems.

Technical integration has given way to *elemental integration* with the development of PV shingles. These are made from complementary roof tile materials such as concrete or steel. These are less visually intrusive and conform to conventional roofing materials, but impose a trade-off in output efficiency.

The use of *spatial integration* in the roof can facilitate the storage of rainwater. The attic space created by the roof is a large void that is often underutilized.

Atria

PV atrium designs are a progressive architectural concept, opening possibilities for harnessing translucent PV/glass technologies that are integrated within the roof and wall elements of atrium spaces so that a balanced passive and active solar outcome is achieved. Careful design of these spaces is needed to avoid excessive heat gain.

(b)

(c)

(d)

The use of solar energy is most effective when the household is fairly energy efficient. Passive design strategies, effective insulation, star-rated appliances and occupants' roles should also be taken into account when considering PV installations.

Walls

PV façades are an option for using the walls to generate energy. There are some disadvantages; energy yields are typically lower for wall and façade applications because of suboptimal orientation and shading influences from surrounding buildings. Nevertheless, PV applications on façades can take advantage of available surface areas and replace conventional façade cladding materials.

Opaque PV panels installed as window awnings, eaves, patio covers and louvres shield direct sunlight while allowing diffuse light to penetrate the interior spaces of the building. These use both the wall and window area of a building's façade.

Despite being suboptimal, west-oriented applications have been demonstrated to be very effective in reducing cooling needs, particularly during afternoon conditions in the summer months. They need to be combined with tracking systems that follow the sun for optimum output.

Source: Richard Hyde

10.11 Casa Solar project shows the integration of a solar system with other passive elements of the building: (a) a renovated farm building is earth integrated; (b) an external pergola is provided to give an external living area; (c) the pergola has integrated PV solar systems; (d) the roof is used for solar thermal hot-water heating

Case study 10.1 Prototype Solar House in Kuala Lumpur, Malaysia

Lim Chin Haw

Source: SIRM Bhd

10.12 Prototype Solar House is primarily a demonstration test bed for applications of photovoltaic (PV) technology in a domestic setting in Malaysia

PROFILE

Table 10.2 Profile of the Prototype Solar House in Kuala Lumpur, Malaysia

Country	Malaysia
City	Shah Alam, Selangor
Building type	Detached
Year of construction	2002
Project name	Prototype Solar House
Architect	SIRM Bhd Architects, Lim Chin Haw

Source: Richard Hyde

PORTRAIT

The Prototype Solar House is primarily a demonstration test bed for applications of photovoltaic (PV) technology in a domestic setting in Malaysia.

The photovoltaic installation consists of three arrays integrated within the principal roofs of the house. The PV arrays include one example each of the current types available: mono-crystalline, polycrystalline and amorphous technology. The electrical power generated was continuously monitored from mid 2003 to mid 2004; the installation is grid connected. The house is only under occasional occupancy for demonstration purposes. The design is of a conventional two-storey house, typical of the end lot in terrace or link housing, or one half of a semi-detached development. Since only one unit was built, it can be regarded as performing as a detached house.

The design intention was to produce a conventional, familiar house with some additional features, primarily to make it a low energy house consistent with ecological sustainable design (ESD) practice. It is not, however, seen as an autonomous or passive design; it needs some energy inputs to maintain environmental comfort, and the potential renewable energy gains will not entirely match the energy demands. The intention is to provide a demonstration of PV application, along with other sensible energy strategies consistent with modern living.

ORIENTATION AND LAYOUT

The long axis is east–west. The reason for this orientation is so that the largest surface of the house is facing north and south position minimizing heat gain. Locating the carport on the west provides shading to the lounge and protects from overheating through solar gain. An external deck is used at the first-floor level for similar reasons.

CONSTRUCTION

The solid construction uses lightweight aerated block work with a high thermal resistance, to reduce thermal transmittance, particularly when cooling is used. The glazing has external reefing louvre blinds to the exposed windows on the east and west elevations. The blades are light-coloured perforated aluminium sections, achieving a shading coefficient of over 50 per cent, while allowing sufficient daylight penetration to minimize the need for electric lighting in daytime.

The glazing to the south and north elevations has a sloping projection at the window head to achieve solar exclusion. The glazing is set back in the window wall to provide additional shading for intermediate sun angles.

ROOF

The roof of the main dwelling area is elevated to give a large roof-ceiling void and is ventilated on all sides. While this measure is primarily to assist in reducing the roof temperature and to enhance PV collector performance, it will reduce the solar gain. Secondary roofs over occupied spaces will have conventional insulation and radiation barriers.

(a)

Source: (a) Richard Hyde; (b) and (c) SIRM Bhd

10.13 (a) Three types of photovoltaic systems are tested in this building: Level 1 = System 3, polycrystalline modules integrated with the carport roof; Level 2 = System 2, amorphous thin film modules integrated with the roof over the first floor deck; and Level 3 = System 1 monocrystalline modules integrated with the house roof; (b) display panel showing performance of each of the systems; (c) electrical layout

(b)

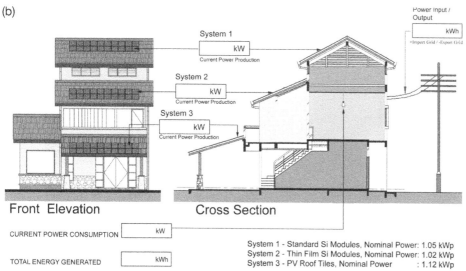

IGS Project : Prototype Solar House - Photovoltaic Systems

(c)

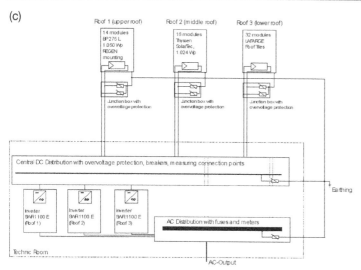

(a)

Source: Richard Hyde

10.14 Prototype Solar House: (a) the roof structure is designed with a steel truss system to provide a stable base to the PV installations and to provide sufficient ventilation to reduce overheating; (b) integration of the amorphous modules is easier since they are bonded to a steel roofing system; (c) integration of the polycrystalline system in plane with the roof tile provides more cost-effective solutions than placing above the tiles, as is the normal method of installation; (d) installation of monocrystalline modules; (e) installation of emorphous modules; (f) and (g) polycrystalline installation of modules

(b)

(c)

(d)

(e)

(f)

(g)

Case study 10.2 Prasad Residence, Sydney, Australia

Deo Prasad and Shailja Chandra

PROFILE

Table 10.3 Profile of the Prasad Residence, Sydney, Australia

Country	Australia
City	Sydney
Building type	Detached
Year of construction	2005
Project name	Prasad Residence
Architect	Professor Deo Prasad

Source: Deo Prasad

Source: Deo Prasad

10.15 Prasad Residence, Sydney, Australia: main entry to the house

CONTEXT

The deployment of novel design features, such as advanced glazing systems, sunspaces and photovoltaic arrays, is slowly becoming a trademark for green buildings. However, in contemporary green residential developments where there is an increasing emphasis on energy consumption and comfort, environmental performance cannot be judged merely by looking at isolated components. A residential building can function sustainably only when it is designed to take account of the collective effect of all aspects of living. Environment, architecture and lifestyle – in no particular order, a fusion of all of them – constitute the essence of 33 Romani Parade. The house was not meant to be a showcase for green strategies, but rather to demonstrate a balance between the three paradigms. The design employs an approach where each of these aspects is addressed in such a way that the whole is greater than the sum of the parts.

The decision to renovate the existing house, rather than to rebuild it, was the first step towards a sustainability outcome; but then improving the functionality of the house meant that major changes would be needed. Conflicts like this prevail, perhaps, in most decision-making; but what was notable in this case was a concerted effort of the occupants and architects, coupled with the owner's extensive experience in sustainability, which was instrumental in obtaining a balance that suited the needs of occupants (for functionality and comfort) and architects (for aesthetics). And while achieving this balance, the house is intrinsically in synch with the environment for its resources and loadings.

The following sections discuss some of the key features of the house.

SITE DESCRIPTION

The house is situated on a quiet street in one of the low-density suburbs of Sydney, with a site area of 596 square metres. The area is fully developed and in close proximity to all necessary amenities.

The site offers panoramic views of Botany Bay and the central business district (CBD) towards the west and north directions, respectively. Total exclusion of unwanted sun on the west façade is not feasible since in that direction are some most beautiful views and vistas, as well as views of the garden. This means that there is a special need for effective sun controls for the west side; the task of optimizing the sun for the northern façade was included in the agenda at the same time.

BUILDING STRUCTURE

The decision to refurbish rather than to rebuild occurred at the beginning of a design exercise that resulted in an award-winning sustainable house.

The existing house was a two-storey, four-bedroom dual occupancy building that has been converted into a four-bedroom single-occupancy house. The building orientation of the existing house is, of course, fixed; but it has been reconfigured by placing key living areas to the north, bedrooms to the east and kitchen and services to the south.

The overall building envelope of the existing building has largely been retained. The major changes have been carried out so that it fulfils all three needs – lifestyle, passive design and architectural character. Some of the highlights are as follows:

- Roof: the whole roof has been redesigned with a lower pitch to suit the aesthetics; it also houses the solar photovoltaic arrays in its northern slope.
- Dormer roof: to ensure good daylighting, natural ventilation and spatial character, a dormer roof, higher than the rest of the roof, has been added to the middle portion. This allows a small mezzanine floor to be added, overlooking the kitchen, dining and living area, and serving as a useful library/study area. While this allows a better visual connection throughout the house, it also provides additional height for a row of ventilators, essentially using buoyancy effects to provide enhanced natural ventilation in the summer; fresh air is drawn in through the windows and hot air is expelled through the ventilators.
- Part of the balcony has been converted into bay windows on the north and west façades. This provides some interesting seating areas, with city views out to the north. This adds architectural character, as well as

Source: Deo Prasad

10.16 Prasad residence: extended overhangs for effective sun shading were used to allow solar gain in winter but to exclude sun in summer

daylight and sunlight during winter, also bringing the previously outside balcony noise into the house, sheltering neighbours from potential noise. A similar bay window is also added to the master bedroom on the east façade, which effectively adds to the architectural character of the street and brings in the morning sun.

- The *stairs* on the north side are relocated to the east side for better integration within the overall design. It also masks the noise from neighbours and improves the architectural presentation of the façade.

BUILDING CONSTRUCTION

The decision to refurbish rather than to rebuild was taken in order to minimize construction and demolition waste by reusing all existing material in the envelope and interior:

Source: Deo Prasad

10.17 The western side of the house has used a number of strategies to address heat gain in summer: the first-floor balcony has retractable shading and a deck is created downstairs facing the pool and garden

- Materials from the demolition of structural components (mainly roof) or interior storage, etc. were salvaged with utmost care for reuse in this or other projects. The roof is the only major component that was replaced; but the tiles were reused in another project and the framing was also reused for the patio structure.
- All walls were retained. On the first floor walls, an additional layer of ecoply cladding was fixed, and painted for aesthetic reasons, for thermal performance.

- Simplicity of the design and form also resulted in optimum use of materials, particularly by omitting embellishments.
- The existing house had no ceiling insulation; a minimum R value of 2.5 was used in the new ceiling. The wall R value was also improved by additional cladding and the enclosed air gap and insulation that this provided.
- During construction, perimeter fencing was maintained so that all construction activities could be contained within the site.

VENTILATION AND HEATING SYSTEM

The house enjoys the key benefits of passive solar design all year round, including solar gains, daylighting, thermal comfort control and ventilation/ infiltration. This substantially reduces the heating and cooling requirements and therefore the overall energy consumption.

The building works entirely on natural ventilation. An optimum area of openable windows located on opposite walls allows for natural ventilation. Moreover, the spaces are also juxtaposed so that the layout promotes airflow throughout the house.

In summer, the double-height living area and a ventilator placed high up on the wall work together to syphon out the hot air, allowing fresh air to enter from a number of cross-windows.

In winter, all three sunlit façades – east, north and west – are used for heat gain. This is further assisted by a ceiling fan that mixes the air thoroughly

Source: Deo Prasad

10.18 Prasad Residence: a two-storey space connects the ground and first floor to promote cooling

Source: Deo Prasad

10.19 Prasad Residence: mezzanine study area overlooking the living room and kitchen

so that warm air is shared between the two levels of the house (which are visually connected). This is of benefit both in winter and summer.

In addition to all this, the green behaviour of the occupants also plays its part in keeping the bills down.

No active heating is utilized at this stage; the living area enjoys sun from the north-western façades all day long.

APPLIANCES

All appliances are energy smart, rated at a 4 to 5 star level. The energy efficiency of appliances was taken into account because the low energy requirements for ventilation and heating meant that the amount of energy consumed by household appliances became significant.

Source: Deo Prasad

10.20 Prasad Residence: solar panels have been used on the northern slope of the roof

The lighting design places high emphasis on daylighting and the high efficiency lamps; all rooms have compact fluorescent lights.

A solar air heater is to be installed that pumps warm air into the space in winter, with a cut-off in summer, redirecting the flow into water and/or pool heating.

ENERGY SUPPLY SYSTEM

A mix of utility grid and solar energy has been used; a photovoltaic system of 1.5kW output has been installed on the northern slope of the roof.

Instant gas heaters are used for all hot water requirement (with respect to greenhouse gas emissions, these are superior to solar heaters with electric boosters).

SOLAR ENERGY UTILIZATION

The sun is used in the winter for direct heat gain and throughout the year for daylighting. Patios have been added to ensure summer sun control, and the eave overhangs have also been optimally designed to manage summer sun and winter gains. On the western balcony, drop-down shading has been installed for summer sun control. The glass areas have also been designed so as to optimize them for both better sun control and for views.

BUILDING HEALTH AND WELL-BEING

Low volatile organic compound (VOC) and low static carpet was chosen in order to reduce dust mites and VOCs. Carpet is used only in bedrooms and hallways; living room floors are of reused plantation timber. In addition to these measures, open living with natural ventilation also eliminates a number of pollutants, particularly in summer.

Almost all enclosed spaces have been provided with openable windows and daylighting, including all bathrooms and the laundry and kitchen, significantly enhancing the occupants' indoor experience and their general well-being. In addition, wherever possible, distant views and a leafy green outlook are provided for visual comfort and amenity.

WATER RECYCLING AND CONSERVATION

All water fixtures are AAA rated.

A rainwater tank with a 2500 litre capacity is used to harvest rainwater from the roof. This water is used for garden irrigation.

Diverted grey water from the laundry is also used to irrigate the garden. This was implemented as part of the garden design, with pipes laid beneath the lawn.

A significant amount of water loss is prevented by using a cover on the swimming pool that not only saves evaporation loss, but also traps heat and prevents impurities from entering the pool.

SUSTAINABLE GARDEN

One of the highlights of the house is its garden. The sustainable garden was implemented by the *Better Homes and Gardens* team in collaboration with the Cool Community team from the Ecoliving Centre at the University of New South Wales.

Source: Deo Prasad

10.21 Prasad Residence: ample daylight is a central aspect of the interior design concept

Source: Deo Prasad

10.22 Prasad Residence: the garden is designed around a sustainable garden concept

The essence of the garden is its simple and relevant design, with features such as:

- Native vegetation: local plant and tree species require low levels of irrigation and suit local soil conditions.
- Selected plants and trees meet the household's requirement for various herbs, vegetables and fruits, such as lemon, passion fruit and banana. Citrus and banana trees were used for shade and privacy, while mango and avocado trees were planted in the chicken yard. Sandy coastal soils and full sunshine called for a combination of native species, such as lomandra, kangaroo paw and shrubby westringia; santolina was planted as a low border. All of these survive on minimal watering (*Better Homes and Gardens*, 2004).
- The garden beautifully integrates the existing trees into its design and also provides interesting outdoor seating for the occupants.
- The irrigation for the garden is done with the collected rainwater and diverted grey water from the laundry – salinity wasn't a critical issue, the site being in the eastern suburbs of Sydney. The tank is positioned so that it doesn't require an energy-consuming pump.
- The garden boasts two green waste recycling systems: a compost heap and a chook shed. An old cubby house was converted into a home for four chickens: all fresh fruit and vegetable scraps are fed to the chickens, to be recycled as manure for the garden (and eggs for the house). The rest of the green waste is composted for spreading on the garden (*Better Homes and Gardens*, 2004).

PLANNING TOOLS APPLIED

Three-dimensional (3D) modelling was deployed during the design of the refurbishment to help predict the indoor quality that would be achieved. This aided in the design process, given that the existing building imposed structural constraints.

It was anticipated that a reduction of 60 per cent of energy use would be possible; the design aimed at achieving a better than 5 star performance using a rating system such as NatHERS. The NatHERS assessment that was carried out achieved 4 stars; however, the consultants, as experts in NatHERS management, vouched for a much higher level of performance than that indicated, due primarily to the inability of the NatHERS engine to present a holistic picture of performance.

The green economics

Since the renovations, the annual energy costs dropped from about Aus$1000 to approximately Aus$400. The energy generated by the rooftop solar module is sold back to the New South Wales grid system. The total cost of the environmental revamp is put at Aus$200,000, including several years' lead-up work. Although this seems to be a significant capital cost, much of the cost had been for the routine job of replacing old fittings and would be absorbed over the next three to five years by energy savings.

Ecological features have also added to the commercial value of the home. The property is valued at more than twice the cost of purchase in 1998. While part of that is due to the general increase in house prices in the area, its sustainable edge over other houses in the neighbourhood accounts for a 5 to 15 per cent impact on the value of the home. The house succeeds in disposing of the common belief that making a house sustainable means that it will be architecturally less appealing. The house has lent a noticeable architectural character to the street, together with a subtle demonstration of its sustainability features.

AWARD

The Prasad Residence was awarded the Inaugural Randwick City Council Urban Design Award in 2005.

INFORMATION

Further information can be found in *Better Homes and Gardens* (2004).

Source: Deo Prasad

10.23 Prasad Residence: plant selection criteria were established based on drought tolerance and soil conditions

Case study 10.3 Mobbs House, Sydney, Australia

Deo Prasad

PROFILE

Table 10.4 Profile of Mobbs House, Sydney, Australia

Country	Australia
City	Sydney
Building type	Row house
Year of construction	1996
Project name	Sustainable House
Designer	Michael Mobbs

Source: Deo Prasad

Source: Michael Mobbs

10.24 Mobbs House, Sydney, Australia: front elevation

PORTRAIT

Michael Mobbs's house is an excellent illustration of the practicalities of retrofitting an inner urban terrace house for energy, water and materials efficiency. The initial idea of the sustainable house seemed to be too optimistic due to the odd proportions of the house block: 35m long and 5m wide. This very ordinary house provides an opportunity to illustrate the efforts by which a waste-generating house was turned into a self-contained one – without major changes in lifestyle for the household. The sustainability objective behind the renovation of this 100-year-old house in the inner suburb of Chippendale, Sydney, was to achieve a low impact design incorporating water and energy saving devices by integrating the environmental techniques used by the project team. This example of energy, water and waste conservation equipment fitted in a normal house can be considered to present a major marketing opportunity for fostering sustainable practices in mainstream housing in Australia.

CONTEXT

The renovation of the house, originally purchased in 1979, started with simple plans for refurbishing the kitchen and bathroom. However, based on the strong conviction of the owners and a four-year research programme, the renovated house now features passive solar modifications, solar water heating, rainwater harnessing, a composting toilet and a grey water system. The chief intentions of these refurbishment plans were:

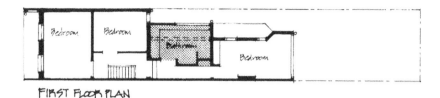

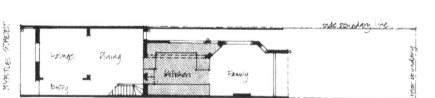

FIRST FLOOR PLAN

GROUND FLOOR PLAN

New Work shown shaded.

0 1 2   4 Metres

North

Source: Michael Mobbs

10.25 Mobbs House: plans

- no storm water to leave the site;
- no sewage to leave the site;
- to collect sufficient potable water onsite;
- to be a net exporter of electricity over 12 months;
- to use recycled or sustainable materials wherever possible; and
- to minimize waste.

The project followed an integrated approach between the client/architect and builders. Project management techniques facilitated a universal mission and helped to achieve conformity to the aims and objectives. The following techniques are worth mentioning:

- Environmental goals and benchmarks were included in claims of progress.
- The builders had to demonstrate that a reasonable attempt was being made to achieve set targets and goals, such as no waste leaving the site, before the progress claims were paid.
- Design meetings between the client, architect and tradespersons assisted with problem solving on a small site where many features were to be integrated.

ECONOMIC CONTEXT

The project cost Aus$165,000 including some other refurbishment costs. Cost of energy, water and waste systems were an additional Aus$48,000. The clients commissioned accounting personnel to undertake a cost-benefit analysis of the project. Using unaudited data, it was found out that, considering the constraints posed by the site, technologies and current pricing structure, the exercise was not economically attractive. However, if the house's excess capacity was used to service another neighbourhood building, the net present value (NPV) of the equipment rises.

Furthermore, the recurrent projected annual savings of more than Aus$1100 are substantial, and when improvements in commercially available renewable technologies are made the capital costs are expected to drop. A wide-scale use of simple systems provides opportunities to make savings on the costs of public infrastructure. A small degree of financial assistance was given by the New South Wales government and by suppliers in the form of discounts.

SITE DESCRIPTION

The house is situated in the inner Sydney suburb of Chippendale, which is located 2km south-west of the Sydney central business district (CBD). The house is situated in a 5m by 35m block with the back of the house facing directly north, which is considered to be an advantage for good passive solar performance.

BUILDING STRUCTURE

The building is two storeys and has a linear configuration mainly due to the site shape. There are three bedrooms on the first floor, with the living and family areas on the ground floor. The kitchen and bathroom are newly constructed and incorporate some passive solar techniques.

BUILDING CONSTRUCTION

The new western wall of the bathroom and kitchen are heavily insulated with R value 3.5 natural wool insulation batts; the roof of the bathroom has the same wool insulation. Plantation timber from regrowth forests was used; PVC plastics were prohibited and no toxic materials were used.

VENTILATION SYSTEM

The glass louvres of the new kitchen window facilitate greater air circulation through the ground floor of this narrow terraced house.

APPLIANCES

High efficiency appliances are used: the oven and stove are gas fired. A solar fan ventilates the sub-floor to reduce the rising damp exacerbated by the sandstone footings. The solar hot water system is gas boosted during overcast days. High efficiency refrigerators, compact fluorescent lighting and halogen lighting are used.

(a)

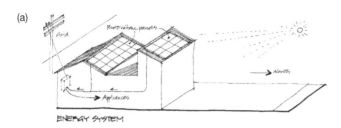

(b)

Source: Michale Mobbs

10.26 Mobbs House – electric supply system: (a) schematic diagram; (b) layout of photovoltaic cells

ENERGY SUPPLY SYSTEM

Eighteen 120W solar panels sit in two banks on the roof. The pitch and orientation of the roof allows energy from the panels to be fed into the electricity grid. A two-way solar/grid interactive inverter, provided by Sustainable Energy Development Authority (SEDA), allows energy from the panels to be fed into the electricity grid. At night, electricity is drawn from the grid in the conventional manner. The energy is used to run the normal range of household appliances.

The solar hot water system is 99 per cent efficient and is gas boosted for two hours a day to provide hot water for showers. The water is heated to boiling point and is stored in a hot water tank. The 300 litre, two-panel solar hot water system used in this house is calculated to save 7923kWh per year of power compared to an electric hot water system (Mobbs, 1998).

The total energy use fell from 24 to 6kWh per day after the refurbishment and changeover to energy efficient appliances.

SOLAR ENERGY UTILIZATION

The pitch and north face of the roof allows a high degree of solar energy generation. Horizontal shading devices were not permitted to extend the full width of the rear glazing in order to prevent their extremities from shading neighbouring properties. Shading of the central section of glass is being considered to reduce the heat build-up in the north bedroom.

PLANNING TOOLS APPLIED

Cost-benefit analysis was commissioned to estimate the capital costs and paybacks.

INCLUSION OF ENVIRONMENTAL GOALS AND BENCHMARKS WITHIN CLAIMS OF PROGRESS

The builders had to demonstrate that a reasonable attempt was being made to achieve set targets and goals, such as no waste leaving the site, before the progress claims were paid.

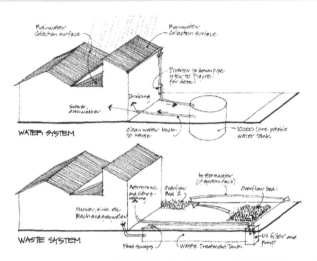

Source: Michael Mobbs

10.27 Mobbs House: waste and water systems

WASTE AND WATER SYSTEMS

Water is collected from the roof and diverted via SmartFlo gutters into a 10,000 litre water tank, with measures taken to exclude or remove small dust particles from the tank. An on-demand pump, located at the back of the garden, pumps potable water up to the house for drinking, showers and dishwashing. All black and grey water and compostible waste from the house are transferred to a Dowmus composting wastewater treatment steel tank that measures 1.2 x 0.6 x 8m. A combination of microbes, bacteria and compost worms break down the waste materials in the tank.

The products are water, carbon dioxide (CO_2) and compost. A lamp emits a particular wavelength of ultraviolet (UV) light to kill any bacteria in the processed water; its germicidal action is by destruction of the bacterial cell wall and DNA. The water is then pumped back to the house for flushing the toilet and washing clothes. The CO_2 is released into the atmosphere.

INDICATORS OF SUSTAINABILITY

Indicators of sustainability for the project include the following:

- More than 80,000 litres per annum of storm water are retained onsite.
- More than 60,000 litres per annum of sewage are kept out of the Pacific Ocean.
- Daily water use has dropped from 310 litres to 230 litres.

- A saving of 102,000 litres of water per annum has been estimated.
- The house produces Aus$1119.30 of clean energy per year, or Aus$3.06 a day, and saves the burning of 4.3 tonnes of coal per year to produce mains electricity.

INFORMATION

Case study for the national seminar series Making Energy Pay, www.sustainablehouse.com.au

Case study on Michael Mobbs's house by Tracy Loveridge, architecture student, University of New South Wales, Australia

Mobbs, M. (1998) *Sustainable House: Living for the Future*, Choice Books, New South Wales, Australia

Prasad, D. and Veale, J. (1998) *Armstrong-Mobbs Sustainable House: Environment Design Guide*, Royal Australian Institute of Architects, Canberra Australia

---

Case study 10.4 Sunny Eco-house, Nara, Japan

Yoshinori Saeki

Source: Yoshinori Saeki

10.28 Sunny Eco-house, Nara, Japan: front elevation

PROFILE

Table 10.5 Profile of Sunny Eco-house, Nara, Japan

Country	Japan
City	Nara
Building type	Detached
Year of construction	2000
Project name	Kankyo Kobo, Sunny Eco-house Project
Architect	Daiwa House Industry Co Ltd

Source: Yoshinori Saeki

PORTRAIT

Protecting the environment is now vital for every human being on Earth. Daiwa House, as a housing company, proposes a solution to the matter.

Improving energy efficiency, by utilizing natural energy and making the most use of natural resources, was our policy when we developed our Eco-house: Kankyo Kobo.

Although ecology-conscious housing is usually expensive, the prefabricated house that we developed has realized an affordably priced eco-house. In addition, a drastic reduction of volatile organic compounds (VOCs) has been achieved. With its barrier-free design, Kankyo Kobo supports a healthy and comfortable life.

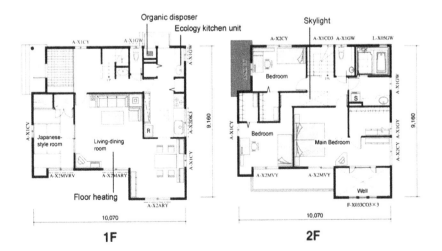

Source: Yoshinori Saeki

10.29 Sunny Eco-house: plans

ENERGY SAVING

We are constantly consuming huge amounts of energy in our daily life; it is not only possible, but also advisable, to reduce such loads on the environment by means of a properly designed house. Proper design aspects include positioning the structure suitably on the building site according to local conditions and applying architectural inventiveness to such elements as building methods, materials and elements, adequate insulation and daylight controls.

MORE EFFECTIVE USE OF NATURAL RESOURCES

Natural resources are consumed by housing during its entire life cycle, from construction through useful life and until final disposal. Consequently, longevity and flexibility in conforming to changes in residents' lifestyles and family structures are essential requirements for housing. Efforts in these directions, together with a reduction in the use of resources and recycling of subsequent waste, will help to create a productive and enduring society.

COMPATIBILITY AND HARMONY WITH THE LOCAL ENVIRONMENT

Regardless of whether an environment is urban or rural, it reflects natural aspects and the heritage and culture of its people. Understanding these features of the site, and taking them into consideration when designing a house, can create comfortable dwellings that are compatible and in harmony with the local environment.

HEALTH AND AMENITY: BE SAFE, FEEL SAFE

The home is a base for its occupants' daily lives and a place of healing. So that residents can lead a healthy life, safety measures against indoor accidents and air pollution should be applied to the housing design to suit the site and planning conditions.

BUILDING STRUCTURE

This house has two storeys and six rooms as follows:

- ground floor: Japanese-style room, living room, dining room and kitchen;
- first floor: three bedrooms.

BUILDING CONSTRUCTION

Industrialized house

Kankyo Kobo is a prefabricated house, a high-quality stable structure, built from steel frames and proof-stress panels. All panel frames, exterior wall materials, heat insulating materials and window sash frames are prefabricated in the factory. Painting of the exterior walls is also carried out in

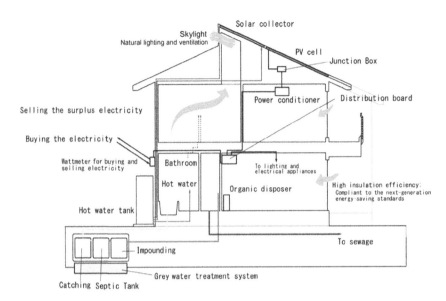

Solar collector

Skylight
Natural lighting and ventilation

PV cell

Junction Box

Power conditioner

Distribution board

Selling the surplus electricity

Buying the electricity

Wattmeter for buying and
selling electricity

Bathroom

To lighting and
electrical appliances

Hot water

Organic disposer

High insulation efficiency:
Compliant to the next-generation
energy-saving standards

Hot water tank

To sewage

Impounding

Grey water treatment system

Catching Septic Tank

Source: Yoshinori Saeki

10.30 Sunny Eco-house: section

the factory to avoid possible air pollution to the surroundings. Air tightness is improved by connecting panels with high-precision bolts and by patching sheets and taping them scrupulously. Light gauge steel is used for the frames, ceramics for the exterior walls and high efficiency glass wool for heat-insulating materials.

In cold districts, greater amounts of insulating materials are used to improve insulation efficiency. Longer eaves exclude the fierce sunlight of summer and reduce the energy needed for air conditioning. In winter, the rays of sunlight can shine into the rooms and heighten the effectiveness of heating. Kankyo Kobo satisfies the 'Next Generation Standards' in every district of Japan.

ACTIVE SYSTEMS

Variable air control (VAC) system: Standard model

The variable air control (VAC) system is an ecology-conscious ventilation system with convergence control and an inlet grille that opens and shuts, sensing the air temperature and controlling the intake airflow.

Photocatalytic air cleaning (PAC) system: Cold district model

In cold districts, a photocatalytic air cleaning (PAC) system is used to heat the outside air up to the indoor temperature before taking it in, thus minimizing heat loss.

Ventilation

A motor-operated air-cleaning louvre is installed in the monitor roof.

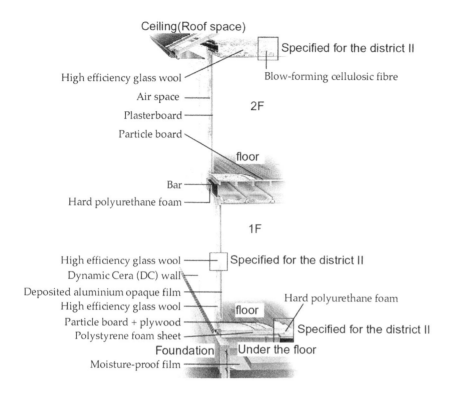

Ceiling(Roof space)

Specified for the district II

High efficiency glass wool

Blow-forming cellulosic fibre

Air space

Plasterboard

Particle board

2F

floor

Bar

Hard polyurethane foam

1F

High efficiency glass wool

Specified for the district II

Dynamic Cera (DC) wall

Deposited aluminium opaque film

High efficiency glass wool

Hard polyurethane foam

Particle board + plywood

floor

Polystyrene foam sheet

Specified for the district II

Foundation   Under the floor

Moisture-proof film

Source: Yoshinori Saeki

10.31 Sunny Eco-house: construction section. Note the Dynamic Cera siding, made of fibre reinforced cement

ENERGY SAVING DEVICES

Window glass

There are three types of glass: heat-sealed high adiathermic double-glazing glass; high adiathermic double-glazing glass; and double-glazing glass. The one best suited to the climate of the location and the orientation of the windows should be chosen.

Window sash

Resin-framed sashes are appropriate for cold districts, while high adiathermic sashes are suitable for other districts.

DOORS: ADIABATIC DOORS

Highly insulated doors are used, which have a U value of 2.33.

Lighting apparatus

Inverter lighting is used in corridors to reduce electricity. An automatic lighting system is also installed on the porch; the light is switched on by a sensor that detects when someone approaches.

ENERGY PERFORMANCES

Solar energy generation

The photovoltaic (PV) cells are a hybrid type, mono-crystal and amorphous. Conversion efficiency is favourable at 17.3 per cent and decrease of power generation by overheating is suppressed. Loading capacity is 3kW. The exterior appearance is designed to complement the plain roof tiles.

At night, when solar power cannot be obtained, electricity is bought from the power company. Surplus energy can also be sold to the company. With a 3kW system, annual production of electricity is estimated at 3382kWh. Currently, the average electricity consumption of a normal household is 6336kW per year; thus, 53 per cent of consumption will be generated by the household (these figures were derived from a simulation model of an average family of four with a total floor space of 150 square metres in Osaka). An indoor monitor displays the power generation in order to raise the residents' awareness.

Solar collector

Hot water from the solar collector with a controlled circulation system is potable and can be supplied to three or four places in the house. Solar utilization reduces the annual consumption of gas by 54 per cent.

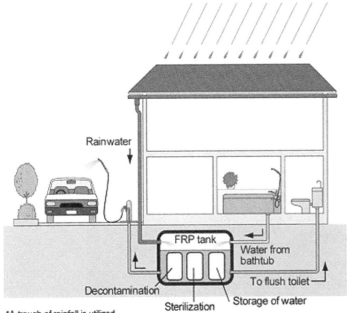

Source: Yoshinori Saeki

10.32 Sunny Eco-house: water system

*A trough of rainfall is utilized

*FRP(plastic) tank is installed in outer space of the house

RECYCLING FACILITIES

Recycling of grey water

Rainfall and discharged water from the bathtub are decontaminated and used for toilet flushing, watering plants in the garden or car washing. This recycling of water can reduce the use of clean water by 200 litres per day.

COSTS

The costs are as follows:

* with the solar system: 25,220,000 Japanese yen for 150.35 square metres, or 167,700 yen per square metre;
* without the solar system: 21,268,000 for 150.35 square metres, or 141,400 per square metre.

MARKETING STRATEGY

Kankyo Kobo is not an idealized prototype of a solar house, but an industrialized house with an affordable price. It supports the residents' healthy and comfortable eco-life with solar energy utilization, grey water recycling and waste recycling, and with devices to make life easier.

Housing with energy-saving efficiency measures, or a solar system, can attract an extra loan from the Housing Loan Corporation. In addition, the New Energy Foundation supplies a subsidy for the energy generated by solar power: 100,000 Japanese yen per kW with a limit of 10kW (April 2002 to March 2003). These advantages are part of our marketing strategy.

SUMMARY

Green technologies for housing have made advances in two main directions. First, life-support systems that use natural resources available onsite are now available – in particular, solar energy and water. From the case studies presented in this chapter, it can be seen that water use can be reduced to 60 to 80 per cent of normal consumption. In houses where occupants' behaviour reduces demand for energy, renewable systems can provide sufficient power through PV and solar hot water systems alone.

Second, there are advances in material systems that reduce environmental impact. The dominant factor is the energy embodied in these materials; but the selection of materials is complex, and single-criterion systems have been replaced by a holistic approach. A framework for assessing these materials was provided in this chapter.

Lessons from the case studies demonstrate a shift in the balance between passive and active systems. The current practice of using a low performance building envelope and air conditioning for climate control has been replaced by a high performance envelope and the use of microclimate control with high-efficiency active systems. Mixed-mode systems (providing both passive cooling and air conditioning that can be selected in appropriate conditions) are also effective.

High performance envelopes are focused principally on reducing heat gain. This involve the use of light-coloured external surfaces, shading, lower U values for transparent elements such as windows, appropriate glazing ratios (15 per cent) and higher levels of insulation (R values of 2.7 for walls and 3.5 for roofs). The use of insulation, including radiant barriers, is critical for rejecting heat.

The capital cost of improving the performance is offset against reduced life-cycle costs, including energy costs and maintenance.

REFERENCES

Adsten, M. (2002) *Solar Thermal Collectors at High Latitudes: Design and Performance of Non-Tracking Concentrators*, Department of Materials Science, Angstrom Laboratory, Uppsala University, Uppsala

AGO (Australian Greenhouse Office) (1999a) *Australian Residential Building Sector Greenhouse Gas Emissions, 1990–2010*, www.greenhouse.gov.au/buildings/publications/residential.html, p36

AGO (1999b) *Scoping Study for Minimum Energy Performance Requirements for Incorporating in the Building Code of Australia*, www.greenhouse.gov.au/buildings/publications/residential.html, pp62–63

Better Homes and Gardens (2004) 'Going Green', March, pp132–133

EcoSpecifier (2003) Materials in Context, Seminar, Brisbane, Australia

Edwards, B, (1999) *Green Architecture,* Architectural Press, Oxford, UK

Groenhout, N. K. (2004) *Design and Optimisation of Advanced Solar Water Heaters*, PhD thesis, University of New South Wales, New South Wales, Australia

Hyde, R. A. (1999) 'Prototype environmental/work home infrastructure building for warm climates', *Proceedings of PLEA 1999 Conference*, Brisbane, 22–24 September, pp289–293

Journal of the American Water Works Association (1998) 'Report on the use of evaporative coolers', vol 90, no 4 (April), available at www.ci.phoenix.az.us/WATER/evapcool.html

Lawson, B. (1996) *Building Materials Energy and the Environment*, RAIA, Red Hill, ACT, Australia

Lazarus, N. (2002) *Beddington Zero (Fossil) Energy Development: Construction Materials Report*, Part 1 and Part 2, Bioregional Development Group, UK

Melbourne Water (2007) *Sustainable Urban Design: Urban Layout,* available at www.wsud.melbournewater.com.au

Mollison, B. C. (1988) *Permaculture: A Designer's Manual,* Tagari Publications, Tyalgum, New South Wales

Newman, P. and Kenworthy, J. R. (1999) *Sustainability and Cities: Overcoming Automobile Dependence,* Island Press, Washington DC.

Photovoltaics Bulletin (2003) 'European PV market rises to world no 2 in BP energy review', no 7, July, pp6–6(1)

Sustainable Victoria (2007) 'Choosing a home cooling system', www.sustainable-energy.vic.gov.au/, accessed 6 April 2007

USDOE (US Department of Energy) (2003) *A Consumer's Guide: Get your Power from Sun*, National Renewable Energy Laboratory, Washington, DC

Watson, S. and Hyde, R. A. (2000) 'An environmental prototype house: A case study of holistic environmental assessment', *Proceedings of the PLEA 2000 Conference*, Cambridge, UK, July, pp170–175

Appendix

International Energy Agency Solar Heating and Cooling Programme Task 28 on Solar Sustainable Housing: Annex 38

TASK DESCRIPTION

Duration: April 2000–April 2005

Objectives: the goal of this task is to help participating countries achieve significant market penetration of sustainable solar housing by the year 2010 by providing home builders and institutional real estate investors with the following:

- a *task website* that illustrates built projects which are exemplary in design, living quality, low energy demand and environmental impact;
- documentation sets of *Exemplary Sustainable Solar Housing* as a basis for local language publications to communicate the experience from built projects and to motivate planners to develop marketable designs;
- a handbook, *Marketable Sustainable Solar Housing*, with guidelines, graphs and tables derived from building monitoring, lab testing and computer modelling;
- *demonstration buildings* with press kits for articles and brochures in local languages to increase the multiplication effect beyond the local region;
- *workshops* after the task has concluded presenting overall results.

Sharing the work of the task over its five-year duration are experts from 16 countries:

- Australia;
- Austria;
- Belgium;
- Brazil;

- Canada;
- Czech Republic;
- Finland;
- Germany;
- Italy;
- Japan;
- The Netherlands;
- New Zealand;
- Norway;
- Sweden;
- Switzerland;
- UK.

ENERGY CONSERVATION IN BUILDINGS AND COMMUNITY SYSTEMS

Approximately one third of primary energy is consumed in non-industrial buildings, such as dwellings, offices, hospitals and schools, where it is utilised for space heating and cooling, lighting and the operation of appliances. In terms of the total energy end use, this consumption is comparable to that used in the entire transport sector. Hence, energy use in buildings represents a major contributor to fossil-fuel use and carbon dioxide production. Following uncertainties in energy supply and concern over the risk of global warming, many countries have now introduced target values for reducing energy consumption in buildings. Overall, these are aimed at reducing energy consumption by between 15 to 30 per cent. To achieve such a target, international cooperation, in which research activities and knowledge can be shared, is seen as an essential activity.

In recognition of the significance of such energy use, the International Energy Agency (IEA) has established an Implementing Agreement on Energy Conservation in Buildings and Community Systems (ECBCS). The function of ECBCS is to undertake research and provide an international focus for building energy efficiency. Tasks are undertaken through a series of annexes that are directed at energy saving technologies and activities that support their application in practice. Results are also used in the formulation of international and national energy conservation policies and standards.

ECBCS undertakes a diverse range of activities both through its individual annexes and through centrally organized development and information exchange. ECBCS countries are free to choose which annexes to take part in. Activities usually take the form of a 'task shared' annex in which each country commits an agreed level of effort. Occasionally, an annex may be either jointly or part jointly funded. More informal activities take place through working groups.

ECBCS participating countries include the following:

- Australia;
- Belgium;
- Canada;
- Czech Republic;
- Denmark;
- Finland;
- France;
- Germany;
- Greece;
- Israel;
- Italy;
- Japan;
- The Netherlands;
- New Zealand;
- Norway;
- Poland;
- Portugal;
- Sweden;
- Switzerland;
- Turkey;
- UK;
- US.

The IEA Secretariat is located in both France and the US and may be contacted at the following addresses.

Alan Meier
Office of Energy Efficiency,
Technology, R&D
9, rue de la Federation
75739 Paris Cedex 15
France
Tel: +33 1 40 57 66 85
Email: alan.meier@iea.org

Pamela Murphy
Morse Associates, Inc
1808 Corcoran Street, NW
Washington, DC, 20009 US
Tel: +1-202-483-2393
Fax: +1-202-265-2248
Email:
pmurphy@MorseAssociatesInc.com

Index

Printed and bound by CPI Group (UK) Ltd, Croydon, CR0 4YY

23/10/2024

01777692-0006